GUT MICROBIOTA IN NEUROLOGIC AND VISCERAL DISEASES

"Death sits in the Bowels" and **"Bad Digestion is the Root of all Evil"**

—Hippocrates (400 BC)

GUT MICROBIOTA IN NEUROLOGIC AND VISCERAL DISEASES

Edited by

TAHIRA FAROOQUI

Research Scientist, Department of Entomology/Center for Molecular Neurobiology, College of Biological Sciences, The Ohio State University, Columbus, OH, United States

AKHLAQ A. FAROOQUI

Research Scientist, Department of Molecular and Cellular Biochemistry, The Ohio State University, Columbus, OH, United States

ELSEVIER

ACADEMIC PRESS

An imprint of Elsevier

Academic Press is an imprint of Elsevier
125 London Wall, London EC2Y 5AS, United Kingdom
525 B Street, Suite 1650, San Diego, CA 92101, United States
50 Hampshire Street, 5th Floor, Cambridge, MA 02139, United States
The Boulevard, Langford Lane, Kidlington, Oxford OX5 1GB, United Kingdom

Library of Congress Cataloging-in-Publication Data
A catalog record for this book is available from the Library of Congress

British Library Cataloguing-in-Publication Data
A catalogue record for this book is available from the British Library

ISBN 978-0-12-821039-0

For information on all Academic Press publications
visit our website at https://www.elsevier.com/books-and-journals

Publisher: Nikki Levy
Acquisitions Editor: Melanie Tucker
Editorial Project Manager: Kristi Anderson
Production Project Manager: Swapna Srinivasan
Cover Designer: Alan Studholme

Typeset by SPi Global, India

Dedication

Dedicated to our teachers

"Teachers open the door, but you must enter by yourself"

Contents

Contributors

Shaadi J. Abughazaleh
Department of Internal Medicine, Houston Methodist Hospital and Weill Cornell Medical College, Houston, TX, United States

Undurti N. Das
UND Life Sciences, Battle Ground, WA, United States; BioScience Research Centre and Department of Medicine, GVP Medical College and Hospital, Visakhapatnam, India

Lindsay A. Euers
Department of Internal Medicine, Houston Methodist Hospital and Weill Cornell Medical College, Houston, TX, United States

Akhlaq A. Farooqui
Department of Molecular and Cellular Biochemistry, The Ohio State University, Columbus, OH, United States

Tahira Farooqui
Department of Entomology, The Ohio State University, Columbus, OH, United States

Evelina Maria Gosav
Department of Internal Medicine, "Grigore T. Popa" University of Medicine and Pharmacy, Internal Medicine Clinic, "St. Spiridon" County Clinical Emergency Hospital, Iasi, Romania

Gokul Krishna
Department of Child Health, University of Arizona College of Medicine—Phoenix, Phoenix, AZ, United States

Alina Mihaela Leustean
Department of Gastroenterology, "Grigore T. Popa" University of Medicine and Pharmacy, Institute of Gastroenterology and Hepatology "St. Spiridon" County Clinical Emergency Hospital, Iasi, Romania

M. Muralidhara
Department of Biochemistry and Nutrition, CSIR—Central Food Technological Research Institute (CFTRI), Mysuru, Karnataka, India

Eamonn M.M. Quigley
Department of Internal Medicine; Lynda K and David M Underwood Center of Digestive Disorders, Houston Methodist Hospital and Weill Cornell Medical College, Houston, TX, United States

P.S. Rajini
Food Protectants and Infestation Control, CSIR-CFTRI, Mysore, India

Meghashri Sridhar
Wiley APAC Services LLP, New Delhi, India

M.M. Srinivas Bharath
Department of Clinical Psychopharmacology and Neurotoxicology, National Institute of Mental Health and Neuro Sciences (NIMHANS), Bangalore, India

Daniela Maria Tanase
Department of Internal Medicine, "Grigore T. Popa" University of Medicine and Pharmacy, Internal Medicine Clinic, "St. Spiridon" County Clinical Emergency Hospital, Iasi, Romania

Preface

The human gastrointestinal tract harbors a complex and dynamic population of microorganisms (more than 1000 species) called gut microbiota, which have evolved with humans through a symbiotic relationship. Gut microbiota provide many benefits to humans, including performing numerous physiological activities such as producing vitamins, regulating gene expression, fighting against pathogenic bacteria, absorbing nutrients, regulating metabolic disorders and host immunity, and harvesting energy. Four dominant phyla (Firmicutes, Bacteroidetes, Actinobacteria, and Proteobacteria) occur in the gut. Firmicutes and Bacteroidetes account for > 90% of the bacterial population in the colon, whereas Actinobacteria and Proteobacteria (which includes Enterobacteriaceae) are regularly present but are scarce (< 1%–5%). The gut microbiota population is affected by environment, diet, lifestyle, medications, age, geographic factors, surgical interventions, and host genetics, particularly genes related to the immune system and metabolism. Among these factors, diet plays a predominant role in modulating the composition of gut microbiota. Alterations in microbiota, including a reduction in health-promoting bacteria such as *Lactobacilli* and *Bifidobacteria*, may lead to dysbiosis, a process caused by Gram-negative bacteria. Dysbiosis is categorized into three different types: (1) loss of beneficial microorganisms, (2) unwarranted growth of potentially harmful microorganisms, and (3) reduced overall microbial diversity.

There is a continuous bidirectional communication between the brain and the gut through the microbiota gut-brain axis (GBA). The bidirectional communication between the gut microbiota and the brain occurs not only through the microbiota GBA and vagus nerve but also through the immune system, neuroendocrine pathways, and microbiota-derived metabolites. This bidirectional communication modulates gut functions (nausea, satiety, and pain) as well as neurotransmission. Alterations in neurotransmission are related to the pathogenesis of neurological and neuropsychiatric disorders. Recent studies have also linked microbiota dysbiosis to visceral diseases (insulin resistance, type 2 diabetes, metabolic syndrome, and heart disease). Mechanistic studies in rodent models have revealed an ever-increasing multitude of molecular mechanisms whereby the gut microbiota interacts with various host-sensing and signaling pathways, leading to modulation of the endocrine system, immune responses, nervous system activity, and hence the predisposition to metabolic diseases. Remarkably, the microbiota-driven immune responses in metabolic tissues and the host nutrient-sensing mechanisms of gut microbial metabolites, in particular short-chain fatty acids, have been significantly associated with proneness to diabetes and related disorders. Similarly, microbiota dysbiosis is linked to neurological disorders such as Parkinson's disease (PD), Alzheimer's disease (AD), multiple sclerosis (MS), stroke, and autism. The central nervous system (CNS) connects with the gut via sympathetic and parasympathetic nerves. It is becoming increasingly evident that gut microbiota modulates the development and maintenance of obesity and insulin resistance. The molecular mechanisms involved in the induction of obesity and insulin resistance are regulated by host food intake behavior, energy harvest, energy expenditure, and fat storage. Several mechanisms are associated with obesity-mediated inflammation in the peripheral tissues and hypothalamic region of the brain. These mechanisms include the activation of TLR4, induction of endoplasmic reticulum stress, and activation of serine/threonine kinases such as IKKβ. Although the relative contribution of these mechanisms remains unknown, early onset of inflammation in the hypothalamus relative to that in peripheral tissues suggests that different processes may cause inflammation in these areas. Moreover, the nature of the hypothalamic inflammation occurring during the first days of consuming a high-fat diet may differ fundamentally from that involved with chronic consumption of a high-fat diet. Several studies propose that the rapid onset of mediobasal hypothalamus (MBH) inflammation is a manifestation of neuronal injury and associated neuroprotective responses. This hypothesis is consistent with previous evidence of apoptosis and glial ensheathment of hypothalamic arcuate nucleus (ARC) neurons in animals exposed to high fat diet.

In light of the preceding information, this book provides readers with a comprehensive and cutting-edge description of molecular aspects of the contribution of microbiota in the pathogenesis of visceral and neurological disorders. Chapter 1 discusses the classification of gut microbiota and factors responsible for changing its composition and roles, as well as ways by which human gut microbiota can be manipulated to treat and/or prevent human diseases. Chapter 2 examines the effect of different types of diet patterns on gut microbiota composition. Chapters 3 and 4 address the importance of fiber (present in human diet) and fermented food on human health. Chapter 5 discusses the contribution of gut microbiota in the pathogenesis of Amyotrophic lateral sclerosis and Huntington's disease. Chapter 6 reviews the role of the gut microbiome in neurodegenerative disorders. Chapter 7 presents cutting-edge information on the association of gut microbiota with the pathogenesis of PD, and Chapter 8 focuses on the contribution of gut microbiota, cerebral blood flow, and sleep deprivation in the pathogenesis of AD. Chapter 9 examines the implications of gut microbiota in the pathogenesis of diabetes mellitus and cardiovascular diseases, and Chapter 10 discusses the link between the gut microbiome and cardiometabolic diseases. Chapter 11 reviews the contribution of gut microbiota in obesity and obesity-related chronic diseases, and Chapter 12 comments on the contribution of microbiota in the pathogenesis of ischemic stroke. Chapter 13 reviews the interaction of bioactive lipids and gut microbiota to regulate health and disease. Chapter 14 is a brief review of the contribution of gut microbiota and multiple organ failure in the pathogenesis of Covid-19 infection. Chapter 15 deals with the contribution of gut microbiota in the pathogenesis of migraine headache. Chapter 16 presents insights on the modulatory role of Ayurveda-based herbal preparations on the gut microbiome and neuroprotection. Chapter 17 deals with the emerging roles of the gut microbiome on neurotoxic outcomes and their implications for neurological disorders. Lastly, Chapter 18 provides readers with an in-depth perspective and directions for future research.

We hope that *Gut Microbiota in Neurologic and Visceral Diseases* will spur further research on this important topic.

Tahira Farooqui
Akhlaq A. Farooqui
Columbus, Ohio

Acknowledgments

We thank all the authors of this book who shared their expertise by contributing high-quality chapters and thus making our editorial task much easier. We are also grateful to Melanie Tucker, Senior Acquisition Editor; Kristi Anderson, Senior Editorial Project Manager; and Swapna Srinivasan, Production Project Manager at Elsevier/Academic Press, for their full cooperation and professional handling of this manuscript.

Tahira Farooqui
Akhlaq A. Farooqui

CHAPTER 1

Gut microbiota: Implications on human health and diseases

Tahira Farooqui
Department of Entomology, The Ohio State University, Columbus, OH, United States

1 Introduction

The human gastrointestinal (GI) tract harbors a complex and dynamic population of microorganisms (archaea and eukarya) called "gut microbiota" that has been estimated to exceed 10^{14}. This number encompasses ~10 times more bacterial cells than human cells.[1] Microbiota coexists in close association with humans and most of them are not harmful but rather important for the host. The microbiome is the catalog of these microbes (bacteria, fungi, archea, and viruses) living in the human lower GI tract and their genes. The number of genes in the microbiome is more than 100 times greater than the number of genes in the human genome.[1]

Intestinal colonization begins after birth.[2–4] At birth, the gut microbiota is aerobic, with small numbers and low diversity, with the most common bacteria being facultative anaerobes and members of the Enterobacteriaceae family. Within a few days, the gut environment becomes anaerobic, which favors the growth of the dominant bacterium genus, such as *Bifidobacterium*, in the infant gut in the first months of life. The adult-like microbiome starts developing at 6 months, dominated by Firmicutes and Bacteroidetes.[5] The gut microbiota changes dramatically during pregnancy and intrinsic factors (such as stress), in addition to extrinsic factors (such as diet and drugs) as well as genetics, age, and hygiene factors (such as status, job security, salary, vacations, and work conditions), markedly influence the composition and activity of the gut microbiome throughout life.[5]

Under physiological conditions, a perfect equilibrium among commensals (bacteria that colonize the gut and protect the host from intruding pathogens by imposing a colonization barrier) and invasive pathogens (bacteria that bypass the epithelial barrier imposed by commensals disrupt tolerance to the resident microbiota causing stereotypic intestinal inflammation) maintains intestinal homeostasis.[6] Any imbalance in gut microbiota

Gut Microbiota in Neurologic and Visceral Diseases
https://doi.org/10.1016/B978-0-12-821039-0.00009-5

composition is thought to precipitate a pathological state known as gut dysbiosis,[7] which may be associated with the pathogenesis of many inflammatory diseases, infections, and visceral and neurological diseases. However, it is still unclear whether gut dysbiosis is a cause or consequence of these diseases.

Gut microbiota has been reported to influence human physiology, metabolism, nutrition, and immune function, thereby modulating the health of its host by providing crucial benefits in the form of immune system development, prevention of infections, nutrient acquisition, and perhaps even brain and nervous system functionality.[8] The microbiome has been shown to play a causal role in the development of pathologies in animal models of human disease, such as obesity (and associated pathologies), autoimmune diseases, and neurological disorders.[9–15] However, the importance of the microbiome has recently been significantly recognized in clinical and physiological research even for the diagnosis of diseases.[16–19]

This chapter discusses gut microbiota, its classification, factors responsible for changing its composition, microbial metabolites, and the effects of prebiotic and probiotic supplementation on host health. It also attempts to shed a light on the molecular mechanisms involved in an altered microbiome for causing or treating visceral and neurological diseases.

2 Classification

Gut microbiota is composed of several species of microorganisms such as bacteria, yeast, and viruses. Taxonomically, bacteria are classified according to phyla, classes, orders, families, genera, and species.[20] It is now known that more than 1000 different species colonize the human gut, all of which belong to a small number of phyla.[21] The most dominant gut microbial phyla are Firmicutes, Bacteroidetes, and Actinobacteria, which represent 98% of human gut microbiota communities (Fig. 1). Proteobacteria, Fusobacteria, Cyanobacteria, and Verrucomicrobia are usually less well represented phyla that account for only ~2% of human gut microbiota communities (Fig. 1).[21–23] The Firmicutes (a phylum of Gram-positive bacteria) is composed of more than 200 different genera such as *Enterococcus*, *Lactobacillus*, *Streptococcus*, *Bacillus*, *Clostridium*, and *Staphylococcus*. *Clostridium* genera represent 95% of the Firmicutes phyla. Bacteroidetes (a phylum of Gram-negative bacteria) consists of predominant genera such as *Bacteroides* and *Prevotella*, which do not exist in equal proportion in the gut. The Actinobacteria phylum (a phylum of Gram-positive

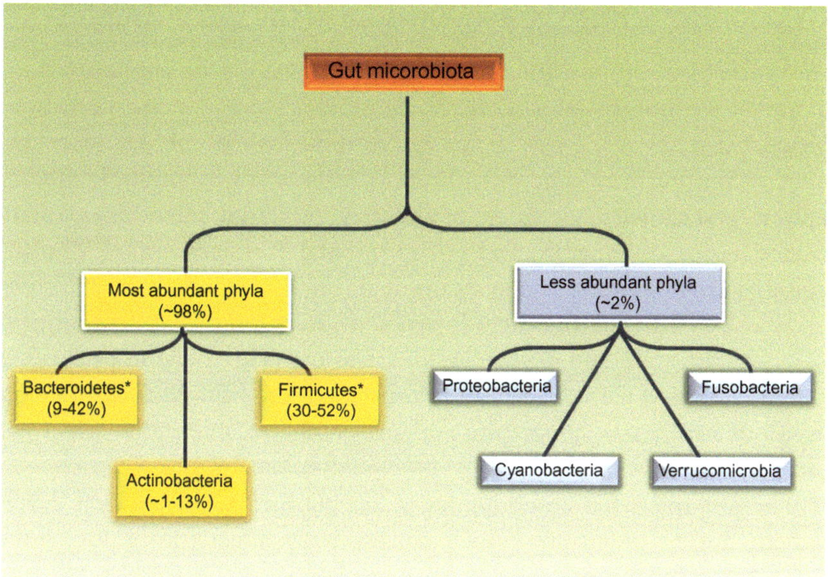

Fig. 1 Classification of gut microbiota.

bacteria) is proportionally less abundant, and is mainly represented by the *Bifidobacterium* genus.[22] In the less well-represented group of gut microbial phyla, Proteobacteria is a major phylum of Gram-negative bacteria, which includes a wide variety of pathogenic genera, such as *Escherichia*, *Salmonella*, *Vibrio*, *Helicobacter*, *Yersinia*, *Legionellales*, and many others. Fusobacteria is phylum of Gram-negative nonsporing bacteria similar to Bacteroides. The four species most commonly implicated as significant human pathogens are *Fusobacterium necrophorum*, *Fusobacterium nucleatum*, *Fusobacterium mortiferum*, and *Fusobacterium varium*. Cyanobacteria is a bacterial phylum (also referred to as blue-green algae and blue-green bacteria) that obtains energy through a process known as photosynthesis, but lacks a nucleus or membrane-bound organelles, like chloroplasts, but has cell walls containing peptidoglycan. There are an estimated 150 genera of cyanobacteria containing ~2000 species. Cyanobacteria often contain a variety of cyanotoxins that can exist as both cell-associated and free forms in the surrounding water, which can be highly neurotoxic and act through a variety of mechanisms.[24] Lastly, the Verrucomicrobia phylum of bacteria is abundant in aquatic and soil environments, but occasionally found in the human gut. It contains only a few described species. However, high-level colonization of the human gut by Verrucomicrobia has been reported following broad-spectrum antibiotic treatment.[25]

3 Factors affecting the composition of the gut microbiota

The microbial composition is host specific, evolving throughout an individual's lifetime in response to various intrinsic and extrinsic factors. Actually, it starts at the gestational age in mothers's womb. The development of the perinatal gut microbiota is influenced by multiple factors, such as gestational age, mode of delivery, maternal microbiota, infant feeding method, genetics, and environmental factors/stressors. The intrinsic factors include pregnancy, mode of birth, age, and genetic factors, whereas extrinsic factors include types of diet, drugs (e.g., antibiotics), infections, lifestyle, and environmental stressors (Fig. 2). Moreover, other stimuli including many other characteristics of the host, such as ethnicity, region of habitation, geography, and/or socioeconomic status, affect gut microbial composition (Fig. 2).

The gut microbial communities of the mother influence fetal development during pregnancy, which consequently affect the health of the

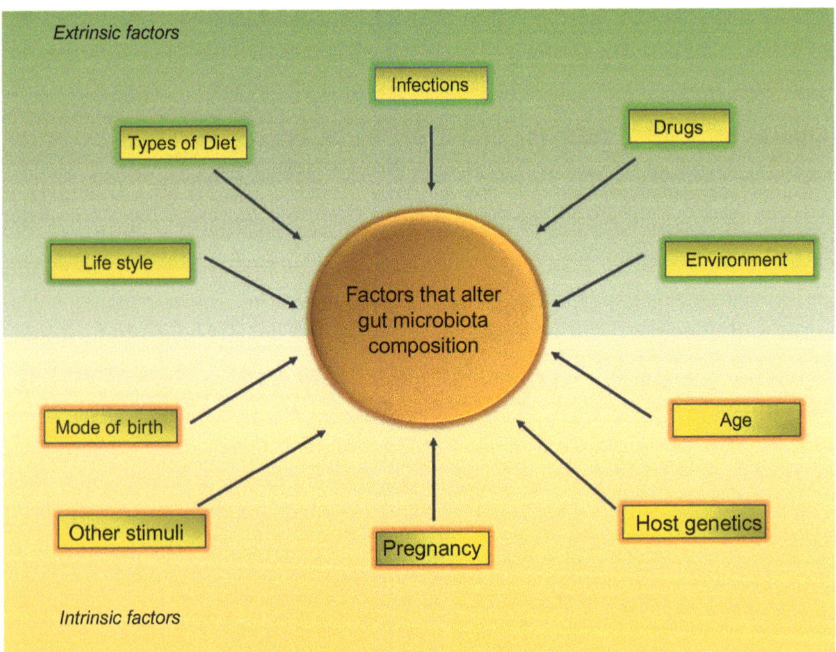

Fig. 2 Various intrinsic and extrinsic factors that are responsible for influencing gut microbiota composition.

offspring.[26] At birth, the initial microbiota composition is affected by the mode of delivery (vaginal or caesarean).[27–29] Babies delivered by natural birth show greater gut bacterial counts at 1 month of age than those delivered by caesarean section, implicating that gut colonization by microbes begins during pregnancy and is enhanced by natural birth, having a dramatic effect on the health of the infant.[27–29] Moreover, vaginally delivered infants acquire bacterial communities resembling their own mother's vaginal microbiota, dominated by *Lactobacillus*, *Prevotella*, or *Sneathia* spp., whereas infants delivered by C-section harbor bacterial communities similar to those found on the skin surface, dominated by *Staphylococcus*, *Corynebacterium*, and *Propionibacterium* spp.[30]

Whether a baby is breast-fed or formula-fed plays an important role in the infant's gut microbiota.[31–33] Breast-fed infants show a predominance of *Bifidobacteria* and *Lactobacilli*, whereas bottle-fed infants develop a mixed flora with a smaller number of *Bifidobacteria*.[33] The growth and development of a robust gut microbiota is important for the development of the immune system, which continues during breast-feeding (an important stage in which the transfer of immunity from mother to child that begins in the uterus is continued), and provides a nurturing environment required for the long-term health of the baby.[34–36] Subsequently, another significant shift in the composition of the gut microbe community occurs when the infant switches to solid food and varied diet.

It has been increasingly evident based on molecular analyses, including The Human Microbiome Project, that there are differences found in the composition of gut microbiota among each stage of human life (infants, toddlers, adults, and the elderly). This suggests that each individual has a distinct combination of gut microbial species. In fact, gut microbiota is strictly linked to the chronological age of each individual (Fig. 2). Bacteria of the Bacteroidetes phylum tend to dominate numerically during youth, but numbers decline significantly by old age. The reverse trend occurs for bacteria of the Firmicutes phylum.[37]

Diet has a dominant role over other possible variables, such as ethnicity, sanitation, hygiene, geography, and climate, on the microbial diversity in individual adults.[38] Interindividual differences in populations of gut microbiota may lead to different capacities to utilize dietary components and to different levels of disease risk. For example, the consumption of modern Western diets (containing high fat and low fiber due to low consumption of vegetables) in one group of adults tends to result in the loss of some important microbial species. A high-fat diet is considered a risk factor for disorders including obesity, diabetes, and metabolic syndrome, which are associated with profound modifications of gut microbiota composition.[39] In contrast, a different group of adults consuming a more healthy diet (containing high fiber and low fat) tends to have a smaller amount of pathogenic bacteria but a large amount of beneficial microbes, implicating that the gut microbiota and its composition can be altered by type of diet and its components.[40–42] Nearly 90% of the bacteria in the human gut can be mapped to just two phyla (Bacteroidetes and Firmicutes). However, the relative proportions of these two dominant phyla vary and can be influenced by a range of factors.[43] Both short-term and long-term dietary variations influence gut microbiota composition.[44]

Older adults have substantially different GI tract communities than younger adults, which indicates that gut microbiota changes throughout life. Gut microbiota composition is also affected by various genetic factors (Fig. 2), transfering from mother to child.[45, 46] Major concerns are the long-term use of drugs (e.g., antibiotics), which could result in long-term alteration of normal healthy gut microbiota (Fig. 2), and horizontal transfer of resistance genes, which could result in a reservoir of organisms with a multidrug-resistant gene pool.[47] Long-term use of antibiotics can bring changes in commensal bacteria associated with various conditions such as development of asthma, eczema, atopic dermatitis, other allergic senitization, and autoimmune encephalitis.[48] Therefore, the individual microbiota pattern is highly variable due to varying extrinsic and intrinsic factors, indicating a strong influence on the balance of the bacterial community.

4 Functional aspects of the gut microbiota

The GI tract is the main site where environmental microorganisms and antigens interact with the host through intensive cross talks. The gut microbiota benefits the host in many ways, including protecting against pathogenic colonization, which is essential for the maintenance of intestinal epithelial integrity (exerting trophic effects on the intestinal epithelium by favoring the development of intestinal microvilli), modulating host physiology and metabolism by harvesting of energy from ingested food but not digested by the host, playing a fundamental role in the maturation of the host's innate and adaptive immune responses, and regulating energy homeostasis and neurobehavioral development (Table 1).[21, 49–53]

Table 1 Functions of the gut microbiota.

Roles	The gut microbiota products and metabolites
Colonization resistance[54–56]	–Secretion of AMPs, bacteriophage, nutrients competition, and gut barrier integrity
Protection against epithelial integrity[53–60]	–Diet affects intestinal integrity. A balanced diet of SFAs, MUFA, MACs, protein, phytochemicals, vitamins, and minerals with limited sources of *n*-6 PUFAs, simple carbohydrates, BAPs, and iron restores intestinal homeostasis
Metabolism	
• Synthesis of essential and nonessential amino acids[50, 64]	–Glutamate family: glutamate, glutamine, proline, arginine –Serine family: serine, glycine, cysteine –Aspartate family: aspartate, asparagine, lysine, threonine, methionine –Pyruvate family: isoleucine, valine, leucine, alanine –Aromatic family: phenylalanine, tyrosine, tryptophan –Histidine
• *de novo* synthesis of essential vitamins[61–64]	–Biotin, folate, and riboflavin
• Participation in drugs and xenobiotics metabolism[50]	–Gut microbiota contributes to the metabolism of numerous widely used drugs (e.g. Benzylpenicillin, Chloramphenicol, Clonazepam, Digoxin, Fluorouracil, and Insulin) –Xenobiotic (e.g. arsenic exposure) significantly decreases the abundance of Firmicutes (producers of butyrate) and alters the composition of indole and glucuronide metabolites
• Metabolism of nondigestible carbohydrates and dietary fiber[49–53]	–Bacterial fermentation of dietary fiber and carbohydrate polymers results in increased SCFAs, which is sensed by SCFARs (FFA2, FFA3, GPR109a, and Olfr78), regulating a variety of physiological and hormonal processes, contributing to whole body energy sensing
• Regulation of BAs metabolism[55, 66, 67]	–GI microbiota participates in the biotransformation of primary BAs into secondary BAs by promoting deconjugation, dehydrogenation, and dehydroxylation reactions –BAs synthesis is under negative feedback control through activation of the FXR in the ileum and liver –BAs-activated FXR plays role in BA synthesis and metabolism, glucose, and lipid metabolism –BAs also exert antimicrobial activity

Continued

Table 1 Functions of the gut microbiota—cont'd

Roles	The gut microbiota products and metabolites
• Digestion of complex plant-derived polysaccharides[49]	–Gut microbiota has a larger repertoire of carbohydrate degradative enzymes
• Host physiology and metabolism[43, 50, 51]	–Gut microbiota provides the vital biochemical pathways for the metabolism of nondigestible carbohydrates, including large polysaccharides (such as resistant starches, cellulose, hemicellulose, pectins, and gums); some oligosaccharides that escape digestion; unabsorbed sugars and alcohols; and host-derived mucins
Influence on energy gain and expenditure[68–75]	–SCFAs activate GPCR (GPR41 is activated by propionate and butyrate; GPR43 is responsive to propionate > acetate > butyrate) –Butyrate inhibits HDACs, inducing histone hypermethylation that changes gene transcription, affecting genes and pathways involved in FA oxidation, epithelial integrity, and apoptosis –Each SCFA reacts differently on gut peptides (incretins and hormones) and energy homeostasis –Butyrate (orally): ↑ GIP, GLP-1, PYY, insulin, and amylin in plasma, which slows down digestion and nutrient intestinal transit, promotes satiety, and increases plasma insulin –Modest effect of propionate on GIP, insulin, and amylin –Acetate has no effect, increases leptin release by fat cells –Butyrate and propionate increase PYY and GLP secretion
Regulation of the innate & adaptive immune system[21, 49–54]	–Pathogenic invasion activates two innate immune sensors (TLRs and NLRs) inducing the recruitment of innate immune cells (such as macrophages and neutrophils), initiating tissue repair processes, resulting in adaptive immune activation
Communication between the microbiota and HPA axis[86-90]	–CRF is released in response to stress, induces release of ACTH in systemic circulation, thus enhancing the release of cortisol from adrenal cortex –Gut microbiota may influence host HPA axis

Table 1 Functions of the gut microbiota—cont'd

Roles	The gut microbiota products and metabolites
Modulation of the GBA[93–96]	–GBA is influenced by diet, lifestyle, genes, and anatomy; also influenced by activation of vagus nerve; production of microbial antigens and microbial SCFAs; and enteroendocrine signaling from gut epithelial cells (I-cells release CCK, and L-cells release GLP-1, PYY and other peptides) –Gut-brain-endocrine metabolic interaction
Dysbiosis of the gut microbiota[105-110]	–Alterations in the composition and diversity of the gut microbiota may result from various environmental and genetic factors, resulting in the intestinal, metabolic, and brain disorders

A compound that is foreign to a living organism (Xenobiotic); Nuclear receptor farnesoid X receptor (FXR); antimicrobial products (AMPs); hypothalamic-pituitary-adrenal axis (HPA), corticotropin-releasing factor (CRF); toll-like receptors (TLRs); microbe-associated molecular patterns (MAMPs); Nod-like receptors (NLRs); gastric inhibitory peptide (GIP); G protein-coupled receptors (GPCRs, such as GPR41 and GPR43); a class of antidiabetic agents that mimic the actions of the glucagon-like peptide (GLP-1); paraventricular nucleus of the hypothalamus (VN); adrenocorticotropic hormone (ACTH); fatty acid (FA); short-chain fatty acids (SFAs), SCFA receptors (SCFARs), monounsaturated fatty acid (MUFA), microbiota-accessible carbohydrates (MACs), polyunsaturated fatty acids (*n*-6 PUFAs), biologically active proteins (BAPs) like trypsin and chymotrypsin inhibitors.

4.1 Protection against pathogenic colonization (colonization resistance)

On one hand, gut microbiota protects the host from infection by pathogenic microorganisms and contributes to both nutrient metabolism and the development and function of the GI immune system. On the other hand, the host provides a nutritious and hospitable environment to ensure the survival of resident microbial communities, thereby implicating a symbiotic relationship between gut microbiota and host. In addition to metabolic benefits, symbiotic commensals provide the host with several functions that promote immune homeostasis, immune responses, and protection against pathogen colonization.[53,54] The ability of symbiotic commensals bacteria to inhibit pathogen colonization is called colonization resistance (CR), which is mediated via two mechanisms: (1) direct mechanisms (such as direct killing, competition for limited nutrients, as well as space) and (2) indirect mechanisms (including stimulation of innate or adaptive immune system, as well as by other nonimmune defenses).[53] In direct mechanisms, symbiotic bacteria scavenge nutrients that would otherwise be available to pathogens. Bacteriophages, type VI secretion systems (T6SS), and bacteriocins may target and kill pathogens. Microbiota-dependent digestion supports the growth of the bacterial community and generates byproducts that are utilized by the host, such as short-chain fatty acids (SCFAs), which can inhibit pathogen growth. Symbiotic bacteria produce enzymes that convert conjugated, primary bile acids (BAs) to secondary BAs, which can kill some pathogens. However, in indirect mechanisms, microbes can compete with one another indirectly by acting on the host.[53,54]

CR is carried out by secretion of antimicrobial products (AMPs), nutrient competition, bacteriophage deployment, and support of gut barrier integrity (Table 1).[54–56] Moreover, changes in microbiota composition and potential subsequent disruption of CR can also be caused by various drugs, such as antibiotics, proton pump inhibitors, antidiabetics, and antipsychotics, thereby providing opportunities for exogenous pathogens to colonize the gut and ultimately causing infection.[57] In addition, the most prevalent bacterial enteropathogens, including *Clostridioides difficile*, *Salmonella enterica* serovar Typhimurium, enterohemorrhagic *Escherichia coli*, *Shigella flexneri*, *Campylobacter jejuni*, *Vibrio cholerae*, *Yersinia enterocolitica*, and *Listeria monocytogenes*, can employ a wide array of mechanisms to overcome colonization resistance.[57]

4.2 Maintenance of the intestinal epithelium integrity

The gut intestinal epithelial layer forms the major barrier that separates the host body from the external environment. Gut microbiota are known to influence epithelial homeostasis by preventing pathogenic bacterial invasion by maintaining intestinal epithelium integrity (Table 1).[57–60] The first layer of defense in the epithelium of the gut is formed by a mucus layer, which is critical for limiting the exposure of epithelial cells to the microbiome. The absence of a highly glycosylated polymeric protein (mucin) in the mucus layer makes the host vulnerable to intestinal inflammation. The intestinal epithelial system is also home to immune cells, including dendritic cells, T cells, B cells, and macrophages, which function in close relation with the intestinal epithelial cells (IECs) to maintain intestinal homeostasis.[58] Gut microbiota maintains a symbiotic relationship with immune cells. The maintenance of the intestinal epithelial barrier is the essential function of the IECs, which integrate positive and negative interactions from the microbiota living in the gut and signal the immune cells to accommodate the microbiota, thereby perpetuating the normal function of the body.[59]

The intestinal epithelial monolayer is composed of different types of specialized epithelial cells (such as enterocytes, Paneth, goblet, endocytes, and microfold cells), each with a distinct function. The most abundant of these are IECs or enterocytes, which plays a major role in maintaining epithelial barrier integrity. Paneth cells are only found in the small intestine, particularly in the ileum. Paneth cells can directly sense enteric bacteria via cell-autonomous MyD88-dependent toll-like receptor (TLR) activation, triggering expression of multiple antimicrobial factors. They can synthesize and secrete AMPs, such as α-defensin, to impede microbial entry to the intestinal lumen. Goblet cells secrete mucus, trefoil peptides, and resistin-like molecule-β, which are central to both the defense and repair of the epithelial layer, and play significant roles in epithelial homeostasis. Endocytes regulate incoming antigens, whereas microfold cells secrete IgA, which, in addition to goblet cells, helps present bacterial antigens to dendritic cells. IECs are capable of phagocytosis and can sequester and neutralize bacterial toxins. Moreover, they recognize bacterial-derived molecules known as prokaryotic-associated molecular patterns (PAMP) with the help of the Toll-like receptors (TLRs) on the cell surface and the nucleotide-binding oligomerization domain-like receptors (NODRs) in the cytoplasm, which activate defense mechanisms by the secretion of AMPS.

IECs also maintain two-way communication with the underlying immune cells to regulate the inflammatory response against bacterial toxins. In conjunction with the mucosal layer (chemical barrier) and specialized cells, the epithelial layer forms a well-equipped, intricately regulated, and stringent barrier with continuous scrutiny by immune cells to create an immune-silent environment. IECs form a layer that functions as a physical barrier facilitated by tight connections between each cell. The tight junction proteins (such as occludin, claudins, and zonula occludens) are crucial for maintaining epithelial barrier integrity. To allow the transport of essential molecules and restrict harmful substances, the intracellular signaling transduction system involves a number of extracellular stimuli (such as cytokines, small GTPases, and posttranslational modifications), which markedly modulate the tight junction protein complexes. Any imbalance in these regulations, or dysfunction of IECs, may result in compromised barrier integrity causing several diseases. Thus, IECs maintain barrier integrity by allowing the permeability of essential ions, nutrients, and water, but restricting the entry of bacterial toxins and pathogens.[60] Moreover, genetic susceptibility, diet, and a number of environmental conditions may also affect the maintenance of the intestinal epithelium integrity directly or indirectly through changes in microbiota.

4.3 Modulation of host metabolism

As the gut microbiota encodes a substantively larger number of genes than its human host, it follows that they undertake a variety of metabolic functions that humans are unable to do or are only able to do in a limited capacity. The GI microbiota is crucial to the de novo synthesis of essential vitamins. Lactic acid bacteria are key organisms in the production of vitamin B12, which cannot be synthesized by either animals, plants, or fungi.[61, 62] *Bifidobacteria* are main producers of folate, a vitamin involved in vital host metabolic processes.[63] Gut microbiota also synthesize vitamins such as vitamin K, riboflavin, biotin, nicotinic acid, pantothenic acid, pyridoxine, and thiamine (Table 1).[64] Moreover, GI microbiota also synthesizes all essential and nonessential amino acids, and carries out biotransformation of BAs (Table 1).[65]

BAs are hydroxylated steroids, synthesized in the liver, from cholesterol. BAs are ligands for the nuclear farnesoid X receptor (FXR) as well as a G protein-coupled receptor (TGR5). BAs are synthesized under negative feedback control through activation of the FXR in the ileum and liver (Table 1). BAs appear to inhibit hepatic phosphoenolpyruvate carboxykinase (PEPCK) activity. PEPCK allows hepatic parenchymal cells to produce glucose from pyruvate derived from amino acid metabolism. Primary BAs, such as cholic and chenodeoxycholic acids, are synthesized from cholesterol in the liver, conjugated to glycine, taurine, or sulfate. Glycine from glycocholic acid serves as an energy, carbon, and/or nitrogen source, whereas taurine from taurodeoxycholate serves as a sulfur source. Deconjugation reaction (hydrolysis of the C-24 *N*-acyl amide bond) is catalyzed by bile salt hydrolase (BSH) enzymes, detected in several bacterial genera, including *Clostridium*, *Bifidobacterium*, *Lactobacillus*, *Bacteroides*, and *Enterococcus*. Deconjugation of bile salts makes them less soluble and less efficiently reabsorbed.[66] While most BAs are efficiently absorbed and recycled back to the liver, only about 5% of the total pool of BAs serve as a substrate for bacterial metabolism in the GI tract, then constitute the major route for cholesterol excretion from the body.[66, 67] The composition and size of the pool of BAs can be altered by intestinal microbiota via the biotransformation of primary bile acids to secondary bile acids, subsequently killing some pathogens. BA-activated FXR plays important roles in synthesis of BAs and metabolism, glucose and lipid metabolism, and even hepatic autophagy. The distribution of BAs in the small and large intestine can also affect the bacterial community dynamics in the gut. Primary BAs, such as taurocholate, can provide homing signals to gut bacteria and promote germination of spores, and may facilitate recovery of microbiota after dysbiosis induced by antibiotics or toxins. On the other hand, reduced levels of BAs in the gut may play an important role in allowing pro-inflammatory microbial taxa to expand, implicating the role of BAs in shaping the GI microbiota. Gut microbial enzymes contribute to BAs metabolism, generating unconjugated and secondary BAs that act as signaling molecules and metabolic regulators to influence important host pathways.

4.4 Influence on energy gain and expenditure

The gut microbiota produces an array of metabolites through the breakdown of nondigestible carbohydrates, which include large polysaccharides such as resistant starches, cellulose, hemicellulose, pectins, and gums; some oligosaccharides that escape digestion; unabsorbed sugars and alcohols from the diet[68]; and host-derived mucins.[69] The most abundant of these metabolites are the short chain fatty acids (SCFAs) acetate, propionate, and butyrate, existing at a ratio of approximately 3:1:1 in the human intestinal lumen, respectively. Acetate and propionate are mainly produced by the Bacteroidetes phylum, whereas butyrate is produced by the Firmicutes phylum. SCFAs can inhibit pathogen growth, however, they exert multiple beneficial effects on various aspects of host energy metabolism. SCFAs are sensed by specific membrane-bound receptors, FFA2, FFA3, GPR109a, and Olfr78, which are expressed uniquely throughout the gut and play significant role in the regulation of intestinal motility, hormone secretion, maintenance of the epithelial barrier, and immune cell function.[70] Acetate, the most abundant SCFA and an essential metabolite for the growth of other bacteria, is used in cholesterol biosynthesis and lipogenesis, and may play a role in central appetite regulation.[71, 72] Propionate is transferred to the liver where it regulates gluconeogenesis and satiety signaling through interaction with the gut fatty acid receptors.[73] After absorption, SCFAs (particularly butyrate) are partially used as energy sources by the colonocytes.[74] Greater production of SCFAs correlates with less diet-induced obesity and reduced insulin resistance.[75]

The production of trimethylamine (TMA) from dietary phosphatidylcholine and carnitine (from meat and dairy) depends on the gut microbiota and thus its amount in blood varies between people. Intestinal microbes metabolize choline into TMA, a gas at body temperature, which is further metabolized by the hepatic flavin mono-oxygenases (FMOs), especially by FMO3, into trimethylamine oxide (TMAO),[76, 77] which is positively associated with an increased risk of atherosclerosis and major adverse cardiovascular events.[76,77] Indolepropionic acid (IPA), a microbial metabolite of tryptophan, has been suggested to ameliorate inflammation and oxidative stress, thereby resulting in better insulin secretion due to preservation of β-cells, and consequently lowering the risk of type 2 diabetes. The putative beneficial effects of IPA on lessening type 2 diabetes risk might be mediated by the interplay between dietary fiber intake, reduction in inflammation, or by direct effect of IPA on β-cell function.[78]

4.5 Host protection and immune-system development by gut microbiota

The intestinal epithelium is the main interface between the immune system and the external environment. The development of a host's immune system is affected by continuous and dynamic interactions with the GI microbiota and its metabolites. Bacteria are integral to the early development of the gut mucosal immune system, both in terms of its physical components and its function, and continue to play a role later in life in its operation.[79] The innate immune system provides critical mechanisms for the rapid sensing and elimination of pathogens. Commensal microorganisms are required for the maturation of the immune system, which "learns" to differentiate between commensal bacteria and pathogens. Toll-like receptors (TLRs) from the membrane of the epithelial and lymphoid cells of the small intestine participate in the differential recognition process, being responsible for the normal development of the intestinal mucosal immune system. TLRs suppress the occurrence of an inflammatory response and promote immunological tolerance to normal microbiota components. The role of TLRs is to recognize different general microbe-associated molecular patterns (MAMPs) and to trigger the innate intestinal immunity (Table 1).[80,81] MAMPs stimulate host innate immunity via TLRs and MyD88 (an innate immune signal transduction adaptor) on epithelial cells or dendritic cells. MyD88 transfers signals from certain TLRs and interleukin-1 (IL-1) receptors, which are important for an early immune response to foreign invaders such as bacteria. In response to signals from these receptors, the MyD88 adapter protein stimulates signaling molecules, a complex cascade of signals is initiated, leading to the release of nuclear factor kappa-light-chain-enhancer of activated B cells (NF-κB), which activates a variety of genes coding for chemokines, cytokines, acute phase proteins, and other effectors of the humoral immune response (Table 1).[82,83]

TLR activity decreases during the first weeks of life, potentially allowing the development of a stable gut bacterial community. Furthermore,

TLR activation by antigens belonging to the normal intestinal microbiota signals the inhibition of inflammatory reactions, thus being essential to maintain intestinal homeostasis.[84] Complementarily, NOD-like receptors (NLRs) recognize various microbial specific molecules and trigger the assembly of inflammasomes, which can act as sensors of damage-associated patterns (Table 1). NLPRP6 deficiency has been associated with an altered immune response (e.g., decreased IL-18 levels), dysbiosis, and intestinal hyperplasia.[84,85]

4.6 Gut microbiota-mediated modulation of host hypothalamic-pituitary-adrenal (HPA) axis

It is rapidly becoming apparent that the gut microbiota plays a major role in the development and regulation of neuroendocrine systems, such as the hypothalamic-pituitary-adrenal axis (HPA), which is the principal neuroendocrine component of stress response. Activation of the HPA axis represents a primary hormonal response to homeostatic challenge. It causes secretion of glucocorticoids, primarily cortisol, that act on multiple organ systems to redirect energy resources to meet real or anticipated demand.[86] The HPA stress response is driven primarily by neural mechanisms, invoking corticotrophin releasing hormone (CRH) release from hypothalamic paraventricular nucleus (PVN) neurons of the hypothalamus in response to stress, which then induces the release of adrenocorticotropic hormone (ACTH) from the anterior pituitary. Pathways activating CRH release are stressor dependent: reactive responses to homeostatic disruption frequently involve direct noradrenergic or peptidergic drive of PVN neurons by sensory relays, whereas anticipatory responses use oligosynaptic pathways originating in upstream limbic structures.[86] Anticipatory responses are driven largely by disinhibition, mediated by trans-synaptic silencing of tonic PVN inhibition via GABAergic neurons in the amygdala. Stress responses are inhibited by negative feedback mechanisms, whereby glucocorticoids act to diminish drive (brainstem) and promote trans-synaptic inhibition by limbic structures (e.g., hippocampus).[86] A bidirectional communication between the gut microbiota and the HPA axis is closely interrelated with other systems, including the immune system, intestinal barrier and blood-brain barrier, microbial metabolites, and gut hormones, as well as the sensory and autonomic nervous systems.[87]

Both the HPA axis and the intestinal microbiota show rapid and profound developmental changes during the first years of life. Early environmental disturbances can affect the development of both systems.[88] Early adversity, for example, is known to lead to later imbalances in both as well as to psychopathological behavior and emotions.[88] Thus gut microbiota plays an important role in regulating glucocorticoid production by activating the HPA axis.[89] In contrast, host-derived hormones increase the bacterial proliferative capacity and pathogenicity.[90] A cross-talk between microorganisms and HPA axis, and the role of gut luminal catecholamines and γ-aminobutyric acid (GABA) that were strongly correlated with anxiety, depression, and functional GI disorders,[91] contradict early suspicions that the high prevalence of anxiety and depression in the clinic was mainly due to neurotic health-seeking behavior. The communication between microorganisms of the gut lumen and the host is presumed to be performed continually through various kinds of luminal molecules, as numerous types of bacteria and host cells are in close proximity in the GI tract of mammals.

4.7 Gut microbiota-mediated alteration in the gut-brain axis (GBA)

The gut-brain axis (GBA) is defined as a two-way communication system between the central nervous system (CNS) and the enteric nervous system (ENS), linking emotional and cognitive centers of the brain with peripheral intestinal functions.[92, 93] The gut microbiota plays a very important role in the GBA through signaling from gut microbiota to brain and from brain to gut microbiota by means of neural, endocrine, immune, and humoral links (Table 1).[93, 94] Based on epidemiological evidence and animal studies, it was suggested that brain-gut pathways include the autonomic nervous system and HPA axis, including CRH directly acting on the gut. According to this group, gut-brain pathways include ascending pain pathways, cytokines, including tumor necrosis factor alpha (TNF-α), in response to bacterial translocation and inflammation, 5-hydroxytryptamine (serotonin) secretion by entero-endocrine cells, and psychoactive chemicals of bacterial origin, which may enter the blood stream, such as gamma-aminobutyric acid (GABA), fatty acids, and serotonin precursors.[95] Obrenovich et al. suggested the concept of co-metabolism or "leaky gut" playing a role in the movement of metabolites.[94] In that regard, this group has explored the gut-brain metabolic interaction, which is one segment of the "gut-brain-endocrine axis."[96] The concept of the microbiota-gut-brain axis is simply not the endocrine system and the stress axis, nor is it the HPA axis or even the immune system, but they all play a part in its modern and current definition, defined as a "gut-brain-endocrine metabolic interaction" emphasizing how bacteria can affect the brain and the hormonal axes in the process of co-metabolism.[96]

Microbiota-gut-brain-endocrine axis communication occurs through small molecules, such as serotonin, leptin, ghrelin, and cortisol among others. Furthermore, choline, a trimethylamine-containing compound and part of the head group of phosphatidylcholine, is metabolized by gut microbiota to produce an intermediate compound known as TMA, which is further oxidized by hepatic flavin monooxygenases to form TMAO. Recently, TMAO has been identified as a risk factor for promoting atherosclerosis through the induction of vascular inflammation.[97, 98] TMAO levels in cerebrospinal fluid (CSF) are found to be greater in individuals with mild cognitive impairment (MCI) and Alzheimer's disease (AD) dementia compared to cognitively unimpaired individuals, and elevated CSF TMAO is considered biomarkers of AD pathology (phosphorylated tau and phosphorylated tau/$A\beta_{42}$) and neuronal degeneration (total tau and neurofilament light chain protein), suggesting the involvement of gut microbiota in AD.[99] TMAO has been implicated in human disease pathogenesis, including known risk factors for AD, metabolic, cardiovascular, and cerebrovascular disease.[97–99] Moreover, the cyanobacterial neurotoxin β-*N*-methylamino-L-alanine (BMAA, the nonprotein amino acid) is an excitotoxin, which has been shown to be a neurotoxin in mouse and other laboratory animals, reinforcing the notion that L-BMAA can cause serious neurodegenerative diseases in humans. BMAA can pass through the blood-brain barrier, and can be transferred from mother to offspring through the mother's milk.[100, 101] Protein-bound BMAA occurs in the brains of Guamanians dying of amyotrophic lateral sclerosis/Parkinsonism dementia complex (ALS/PDC) in North American AD patients, but not in control brains, implicating a possible role for BMAA in neurodegenerative diseases.[102]

Furthermore, the substances produced by the gut microbiota can be effective at relatively low concentrations, but can impact distant organs (e.g., brain). In addition to SCFAs, tryptophan metabolites, and BAs, the gut microbiota produces a number of neurotransmitters, such as serotonin, dopamine, and noradrenaline, as well as tryptophan that is converted to 5-hydroxytryptamine. Therefore, the gut microbiota can be considered an endocrine organ and "Second Brain" as it is capable of generating a number of chemical substances that can directly interact and activate specific receptors, and can impact host physiological functions.[103] The gut microbiota also generates metabolites, such as 4-ethylphenyl sulfate (aryl sulfate), SCFA, and ammonia, which may exert adverse neurobehavioral effects. All these findings suggest that the gut microbiota and its metabolic byproducts could play a role in the susceptibility and pathogenesis of brain diseases.

4.8 Dysbiosis of the gut microbiota

The gut microbiota plays a crucial role in maintaining homeostasis. As such, alterations in the gut microbiota composition and diversity (dysbiosis) are associated with the pathogenesis of dysbiosis-associated diseases (Table 1), which can be intestinal as well as extra-intestinal disorders. Intestinal disorders include inflammatory bowel disease, irritable bowel syndrome, and celiac disease, while extra-intestinal disorders include allergy, asthma, metabolic syndrome, cardiovascular disease, and obesity, as well as neuropsychiatric and neurodegenerative diseases (Fig. 3). The mechanisms leading to disease development involve the pivotal mutualistic relationship between the colonic microbiota, their metabolic products, and the host immune system.[104] However, dysbiotic patterns are context- and disease-specific and often not consistently detected among different studies. In addition, for most diseases, it is unclear if dysbiosis is the cause or the consequence of disease. Moreover, the molecular mechanisms by which the altered microbiome causes disease are lacking. Due to these complications, it is extremely difficult to define what constitutes a healthy microbiome. However, several bacteria

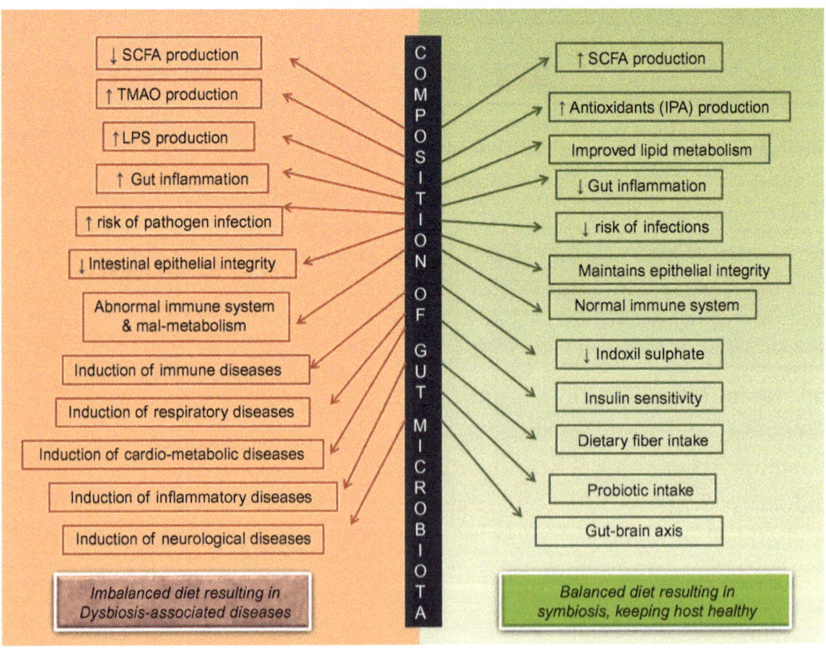

Fig. 3 Schematic representations of the roles of gut microbiota in health and dysbiosis-associated diseases. Short-chain fatty acids (SCFAs); trimethylamine *N*-oxide (TMAO); lipopolysaccharide (LPS); indolepropionic acid (IPA); indole sulfate (kidney toxin).

taxa found in AD patients are different from those in controls at taxonomic levels, such as Bacteroides, *Actinobacteria*, *Ruminococcus*, *Lachnospiraceae*, and *Selenomonadales*, suggesting that the gut microbiota is altered in AD patients and may have a crucial role in the pathogenesis of AD.[105] In a transgenic mouse model of ALS, the animals showed an abnormal intestinal microbiome, damaged tight junctions, and superoxide dismutase 1 (SOD1) deficiency before the onset of disease. After taking butyrate in their drinking water, transgenic mice recovered intestinal microbial homeostasis, improved gut integrity, and prolonged life span compared to control mice, suggesting that butyrate treatment is not a cure for genetic defect in mice but is associated with decreased aggregation of the SOD1 (mutated protein) in transgenic mice.[106] Recent preclinical studies have compared ALS subjects and healthy controls, revealing a variation in the intestinal microbial composition with a greater abundance of *E. coli* and *enterobacteria* and a low abundance of total yeast in patients, confirming a difference in the microbiota profile in ALS patients.[107] Exposure to stressors has long been known to increase susceptibility to and severity of GI diseases.[108] Another group used a different strategy by comparing the fecal microbiomes of PD patients with healthy controls, and reported that PD patients displayed a reduced abundance of *Prevotellaceae* and elevated levels of *Enterobacteriaceae* compared to healthy control groups.[109] The relative abundance of *Enterobacteriaceae* is positively associated with the severity of postural instability and gait difficulty, suggesting that the intestinal alteration of the microbiome in PD is related to motor phenotype. Synucleinopathies are characterized by aggregation of the protein α-synuclein (α-Syn), often resulting in motor dysfunction as exemplified by PD. Sampson et al. demonstrated that gut microbiota are required for motor and GI deficits, microglia activation, and α-Syn pathology in a mouse model of PD.[110] All these findings strongly suggest that the gut microbiota has important roles in maintaining human health, and any alteration in their composition could play a role in the pathogenesis of neurodegenerative diseases. Thus regardless of whether dysbiosis arises first and causes disease, or whether it is an outcome of disease, once it arises, it almost certainly contributes to the perpetuation of disease.[111]

5 Manipulation of the gut microbiota

The gut microbiome plays a crucial role in metabolic homeostasis and may contribute to health and disease states of the host. Therefore, one should treat/prevent these diseases through the modulation of the composition and diversity of gut microbiota through dietary manipulations. Several strategies are available, including antibiotics, administration of probiotics, prebiotics as food supplements, changes in diet, and fecal microbiota transplantation. Of these strategies, diet is perhaps the most pervasive, but it is an often-ignored modulator of the microbiota. Failure to recognize its impact complicates the interpretation of many microbiota studies. The impacts of antibiotics on the microbiota are more complex than originally thought and though antibiotics can be life-saving, their effects on commensal bacterial populations can be clinically significant.[112] The prebiotics concept was introduced for the first time in 1995 by Glenn Gibson and Marcel Roberfroid.[113] The prebiotics are nondigestible oligosaccharides in general, and fructooligosaccharides in particular. They are nondigestible food ingredients that beneficially affect the host by selectively stimulating the growth and/or activity of one or a limited number of bacterial species already resident in the colon, and thus attempt to improve host health. The intake of prebiotics significantly modulates the colonic microbiota by increasing the number of specific bacteria, thereby changing the composition of the microbiota. The prebiotics stimulate the growth of endogenous bifidobacteria, which, after a short feeding period, become predominant in human feces. Prebiotics modulate lipid metabolism, most likely via fermentation products.[113]

Probiotics do not always colonize the intestinal tract to exert their effects. Some probiotics like *Bifidobacterium longum* become part of the human intestinal microflora, whereas others like *Lactobacillus casei* indirectly exert their effects in a transient manner as they pass through by remodeling or influencing the existing microbial community.[114] Probiotics are defined as microbial food supplements that beneficially affect the host by improving its intestinal microbial balance; therefore, they have been used to change the composition of colonic microbiota. However, such changes may be transient, and the implantation of exogenous bacteria therefore becomes limited. In addition to improving cognitive ability, probiotics modulate the response to stress, thus they are also called psychobiotics (live organisms when ingested in adequate amounts produce a health benefit in patients suffering from psychiatric illness).[115] For example, preclinical evaluation in rodents suggests that certain psychobiotics possess antidepressant or anxiolytic activity. The risks associated with probiotic microbes include probiotic sepsis, contamination of the probiotic product, and lack of efficacy related to decrease in viability. Probiotic sepsis is quite rare with a variety of commercial products. Administration of probiotics and prebiotics has been widely used in order to manipulate gut microbiota. Most prebiotics and probiotics can be safely consumed without side effects. However, only in few cases, side effects include abdominal discomfort, bloating, and gas. The combination of pro- and prebiotics, the concept of synbiotics, is proposed to characterize some colonic foods with interesting nutritional properties that make these compounds candidates for classification as health-enhancing functional food ingredients.[113]

Several studies have shown the importance of fecal microbiota transplantation (FMT) as an effective treatment for recurrent *Clostridium difficile* infection when antibiotic treatment is failing. FMT involves transplanting the stool of a healthy donor into the GI tract of a sick patient to restore good bacteria in the gut microbiome. It is often performed via colonoscopy, but can also be done through retention enema or by consuming an oral capsule, among other methods. Constipation is a common complaint in PD patients. Huang et al. reported the first case of FMT to treat PD patient with tremors and constipation. The neurological and GI symptoms (like tremors and constipation) in the PD patient were effectively reduced by FMT, suggesting that gut microbiota reconstruction may have a therapeutic effect.[116] Furthermore, little is known about the role of FMT in AD. Cognitive deficits, brain deposits of amyloid-β (Aβ) and phosphorylation of tau, synaptic plasticity as well as neuroinflammation were assessed in a transgenic mouse model of AD.[117] This group observed that FMT treatment improved cognitive deficits and reduced the brain deposition of Aβ in transgenic mice. These improvements were accompanied by decreased phosphorylation of tau protein and the levels of Aβ40 and Aβ42, and reversed the changes of gut microbiota and SCFAs after FMT treatment, indicating that FMT may be a potential therapeutic strategy for AD.[117]

FMT and pre- and probiotics are considered as potential options for treating neurodegenerative diseases. However, clinical data are scarce and more research is needed to support their use.

6 Conclusion

Gut microbiota plays major roles in human health and diseases. Over the years, many studies have found correlations between the gut microbiota and HPA axis as well as GBA. However, we still do not fully understand the mechanisms by which GI microbiota is modified or how these modifications are associated with visceral and neurological diseases. These questions should be fully addressed in future studies. Most findings suggest beneficial clinical effects of prebiotics due to safe ingredients, but the evidence on prebiotics having significant health benefits through alteration of gut microbiota is still limited. Probiotics are known to modulate the response to stress, a condition that has been increasingly associated with diseases. Probiotics are referred to as psychobiotics, however, the microbial composition and molecular mechanism associating gut microbiota and the brain need to be understood in much greater depth before using probiotics as targeted therapeutics. Moreover, the mechanism by which the transplanted microbiota out-compete the *C. difficile* infection in FMT also requires further investigation.

Some methodological issues concerning the evaluation of the microbiota composition and function including metabolomic profiling techniques need also to be considered. Moreover, the involvement of other microbiotas, apart from the microbiota in the gut and the oral and nasal cavities, in the pathophysiology of neurodegenerative disorders has not yet been explored.

References

1. Thursby E, Juge N. Introduction to the human gut microbiota. *Biochem J.* 2017;474(11):1823–1836.
2. Koenig JE, Spor A, Scalfone N, et al. Succession of microbial consortia in the developing infant gut microbiome. *Proc Natl Acad Sci.* 2011;108:4578–4585.
3. Matamoros S, Gras-Leguen C, Le Vacon F, Potel G, de La Cochetiere M-F. Development of intestinal microbiota in infants and its impact on health. *Trends Microbiol.* 2013;21:167–173.
4. Flaherman VJ, Narayan NR, Hartigan-O'Connor D, Cabana MD, McCulloch CE, Paul IM. The effect of early limited formula on breastfeeding, readmission, and intestinal microbiota: a randomized clinical trial. *J Pediatr.* 2018;196:84–90.
5. Mohajeri MH, Brummer RJM, Rastall RA, et al. The role of the microbiome for human health: from basic science to clinical applications. *Eur J Nutr.* 2018;57(suppl 1):1–14.
6. Pédron T, Sansonetti P. Commensals, bacterial pathogens and intestinal inflammation: an intriguing ménage à trois. *Cell Host Microbe.* 2008;3(6):344–347.
7. Zeng MY, Inohara N, Nunez G. Mechanisms of inflammation-driven bacterial dysbiosis in the gut. *Mucosal Immunol.* 2017;10(1):18–26.
8. McBurney M, Davis C, Fraser CM, et al. Establishing what constitutes a healthy human gut microbiome: state of the science, regulatory considerations, and future directions. *J Nutr.* 2019;149(11):1882–1895.
9. Holmes MV, Newcombe P, Hubacek JA, et al. Effect modification by population dietary folate on the association between MTHFR genotype, homocysteine, and stroke risk: a metaanalysis of genetic studies and randomised trials. *Lancet North Am Ed.* 2011;378:584–594.
10. Holmes E, Li JV, Marchesi JR, Nicholson JK. Gut microbiota composition and activity in relation to host metabolic phenotype and disease risk. *Cell Metab.* 2012;16:559–564.
11. Fei N, Zhao L. An opportunistic pathogen isolated from the gut of an obese human causes obesity in germfree mice. *ISME J.* 2013;7:880–884.
12. Surana NK, Kasper DL. Moving beyond microbiome-wide associations to causal microbe identification. *Nature.* 2017;552:244–247.

13. Round JL, Palm NW. Causal effects of the microbiota on immune mediated diseases. *Sci Immunol.* 2018;3. eaao1603.
14. Tremlett H, Bauer KC, Appel-Cresswell S, Finlay BB, Waubant E. The gut microbiome in human neurological disease: a review. *Ann Neurol.* 2017;81:369–382.
15. Fischbach MA. Microbiome: focus on causation and mechanism. *Cell.* 2018;174:785–790.
16. Levy M, Blacher E, Elinav E. Microbiome, metabolites and host immunity. *Curr Opin Microbiol.* 2016;35:8–15.
17. Tilg H, Cani PD, Mayer EA. Gut microbiome and liver diseases. *Gut.* 2016;65:2035–2044.
18. Wehkamp J, Frick JS. Microbiome and chronic inflammatory bowel diseases. *J Mol Med.* 2016;95:21–28.
19. Klymiuk I, Bilgilier C, Stadlmann A, et al. The human gastric microbiome is predicated upon infection with *Helicobacter pylori*. *Front Microbiol.* 2017;8:2508.
20. Rinninella E, Raoul P, Cintoni M, et al. What is the healthy gut microbiota composition? A changing ecosystem across age, environment, diet, and diseases. *Microorganisms.* 2019;7(1):E14.
21. Qin J, Li R, Raes J, et al. A human gut microbial gene catalogue established by metagenomic sequencing. *Nature.* 2010;464:59–65.
22. Arumugam M, Raes J, Pelletier E, et al. Enterotypes of the human gut microbiome. *Nature.* 2011;473:174–180.
23. D'Argenio V, Salvatore F. The role of the gut microbiome in the healthy adult status. *Clin Chim Acta.* 2015;451(2015):97–102.
24. Metcalf JS, Codd GA. Cyanobacteria, neurotoxins and water resources: are there implications for human neurodegenerative disease? *Amyotroph Lateral Scler.* 2009;10(suppl 2):74–78.
25. Dubourg G, Lagier JC, Armougom F, et al. High-level colonisation of the human gut by Verrucomicrobia following broad-spectrum antibiotic treatment. *Int J Antmicrob Agents.* 2013;41(2):149–155.
26. Sanz Y. Gut microbiota and probiotics in maternal and infant health. *Am J Clin Nutr.* 2011;94(6 suppl):2000S–2005S.
27. Huurre A, Kalliomaki M, Rautava S, Rinne M, Salminen S, Isolauri E. Mode of delivery—effects on gut microbiota and humoral immunity. *Neonatology.* 2008;93:236–240.
28. Grölund MM, Lehtonen OP, Eerola E, Kero P. Fecal microflora in healthy infants born by different methods of delivery: permanent changes in intestinal flora after cesarean delivery. *J Pediatr Gastroenterol Nutr.* 1999;28:19–25.
29. Grölund MM, Grześkowiak Ł, Isolauri E, Salminen S. Influence of mother's intestinal microbiota on gut colonization in the infant. *Gut Microbes.* 2011;2(4):227–233.
30. Dominguez-Bello MG, Costello EK, Contreras M, et al. Delivery mode shapes the acquisition and structure of the initial microbiota across multiple body habitats in newborns. *Proc Natl Acad Sci U S A.* 2010;107:11971–11975.
31. Harmsen HJ, Wildeboer-Veloo ACM, Raangs GC, et al. Analysis of intestinal flora development in breast-fed and formula-fed infants by using molecular identification and detection methods. *J Pediatr Gastroenterol Nutr.* 2000;30:61–67.
32. Kelly D, King T, Aminov R. Importance of microbial colonization of the gut in early life to the development of immunity. *Mutat Res.* 2007;622:58–69.
33. Coppa GV, Zampini L, Galeazzi T, Gabrielli O. Prebiotics in human milk: a review. *Dig Liver Dis.* 2006;38:S291–S294.
34. Walker WA, Iyengar RS. Breast milk, microbiota, and intestinal immune homeostasis. *Pediatr Res.* 2015;77(1–2):220–228.
35. Le Doare K, Holder B, Bassett A, Pannaraj PS. Mother's milk: a purposeful contribution to the development of the infant microbiota and immunity. *Front Immunol.* 2018;9:361.

36. Hanson LA, Korotkova M, Lundin SB, et al. The transfer of immunity from mother to child. *Ann NY Acad Sci.* 2003;987(1):199–206.
37. Nicholson JK, Holmes E, Kinross J, et al. Host-gut microbiota metabolic interactions Jeremy. *Science.* 2012;108:1262e8.
38. De Filippo C, Cavalieri D, Di Paola M, et al. Impact of diet in shaping gut microbiota revealed by a comparative study in children from Europe and rural Africa. *Proc Natl Acad Sci U S A.* 2010;107:14691–14696.
39. Statovci D, Aguilera M, MacSharry J, Melgar S. The impact of Western diet and nutrients on the microbiota and immune response at mucosal interfaces. *Front Immunol.* 2017;8:838.
40. Rajoka R, Shahid M, Junling S, et al. Interaction between diet composition and gut microbiota and its impact on gastrointestinal tract health. *Food Sci Human Wellness.* 2017;6(13):121–130.
41. Schnorr SL, Candela M, Rampelli S, et al. Gut microbiome of the Hadza hunter-gatherers. *Nat Commun.* 2014;5:3654.
42. Ou J, Carbonero F, Zoetendal EG, et al. Diet microbiota, and microbial metabolites in colon cancer risk in rural Africans and African Americans. *Am J Clin Nutr.* 2013;98(1):111–120.
43. Jeffery IB, Claesson MJ, O'Toole PW, Shanahan F. Categorization of the gut microbiota: enterotypes or gradients? *Nat Rev Microbiol.* 2012;10:591–592.
44. Wu GD, Chen J, Hoffmann C, et al. Linking long-term dietary patterns with gut microbial enterotypes. *Science.* 2011;334:105–108.
45. Dore J, Corthier G. The human intestinal microbiota. *Gastroenterol Clin Biol.* 2010;34(suppl 1):S7–15.
46. Goodrich JK, Waters JL, Poole AC, et al. Human genetics shape the gut microbiome. *Cell.* 2014;159(4):789–799.
47. Jandhyala SM, Talukdar R, Subramanyam C, Vuyyuru H, Sasikala M, Reddy DN. Role of the normal gut microbiota. *World J Gastroenterol.* 2015;21(29):8787–8803.
48. Willing BP, Vacharaksal A, Croxen M, Thanachayanont T, Finlay BB. Altering host resistance to infections through microbial transplantation. *PLoS ONE.* 2011;6(10):e26988.
49. Flint HJ, Scott KP, Duncan SH, Louis P, Forano E. Microbial degradation of complex carbohydrates in the gut. *Gut Microbes.* 2012;3:289–306.
50. Quigley EMM. Gut bacteria in health and disease. *Gasteroenterol Hepatol (NY).* 2013;9(9):560–569.
51. Sekirov I, Russell SL, Antunes LCM, Finlay BB. Gut microbiota in health and disease. *Physiol Rev.* 2010;90:859–904.
52. Cryan JF, O'Mahony SM. The microbiome-gut-brain axis: from bowel to behavior. *Neurogastroenterol Motil.* 2011;23:187–192.
53. Brestoff JR, Artis D. Commensal bacteria at the interface of host metabolism and the immune system. *Nat Immunol.* 2013;14:676–684.
54. Pickard JM, Zeng MY, Caruso R, Núñez G. Gut microbiota: role in pathogen colonization, immune responses, and inflammatory disease. *Immunol Rev.* 2017;279(1):70–89.
55. Lawley TD, Walker AW. Intestinal colonization resistance. *Immunology.* 2013;138:1–11.
56. Buffie CG, Pamer EG. Microbiota-mediated colonization resistance against intestinal pathogens. *Nat Rev Immunol.* 2013;13:790–801.
57. Ducarmon QR, Zwittink RD, Hornung BVH, Schaik WV, Young VB, Kuijper EJ. Gut microbiota and colonization resistance against bacterial enteric infection. *Microbiol Mol Biol Rev.* 2019;83(3).
58. Khosravi A, Mazmanian SK. Disruption of the gut microbiome as a risk factor for microbial infections. *Curr Opin Microbiol.* 2013;16(2):221–227.
59. Chelakkot C, Ghim J, Ryu SH. Mechanisms regulating intestinal barrier integrity and its pathological implications. *Exp Mol Med.* 2018;50:103.

60. Rescigno M. The intestinal epithelial barrier in the control of homeostasis and immunity. *Trends Immunol.* 2011;32(6):256–264.
61. Martens JH, Barg H, Warren M, Jahn D. Microbial production of vitamin B-12. *Appl Microbiol Biotechnol.* 2002;58:275–285.
62. LeBlanc JG, Milani C, de Giori GS, Sesma F, van Sinderen D, Ventura M. Bacteria as vitamin suppliers to their host: a gut microbiota perspective. *Curr Opin Biotechnol.* 2013;24:160–168.
63. Pompei A, Cordisco L, Amaretti A, Zanoni S, Matteuzzi D, Rossi M. Folate production by bifidobacteria as a potential probiotic property. *Appl Environ Microbiol.* 2007;73:179–185.
64. Hill MJ. Intestinal flora and endogenous vitamin synthesis. *Eur J Cancer Prev.* 1997;6:S43–S45.
65. Vyas U, Ranganathan N. Probiotics, prebiotics, and synbiotics: gut and beyond. *Gastroenterol Res Pract.* 2012;2012, 872716.
66. Kriaa A, Bourgin M, Potiron A, et al. Microbial impact on cholesterol and bile acid metabolism: current status and future prospects. *J Lipid Res.* 2019;60:323–332.
67. Gérard P. Metabolism of cholesterol and bile acids by the gut microbiota. *Pathogens.* 2013;3:14–24.
68. Cummings JH, Pomare EW, Branch WJ, Naylor CP, Macfarlane GT. Short chain fatty acids in human large intestine, portal, hepatic and venous blood. *Gut.* 1987;28(10):1221–1227.
69. Koropatkin NM, Cameron EA, Martens EC. How glycan metabolism shapes the human gut microbiota. *Nat Rev Microbiol.* 2012;10(5):323–335.
70. Priyadarshini M, Kotlo KU, Dudeja PK, Layden BT. Role of short chain fatty acid receptors in intestinal physiology and pathophysiology. *Compr Physiol.* 2018;8(3):1091–1115.
71. Frost G, Sleeth ML, Sahuri-Arisoylu M, et al. The short-chain fatty acid acetate reduces appetite via a central homeostatic mechanism. *Nat Commun.* 2014;5:3611.
72. van der Beek CM, Canfora EE, Lenaerts K, et al. Distal, not proximal, colonic acetate infusions promote fat oxidation and improve metabolic markers in overweight/obese men. *Clin Sci (Lond).* 2016;130(22):2073–2082.
73. Byrne CS, Chambers ES, Morrison D, Frost G. The role of short chain fatty acids in appetite regulation and energy homeostasis. *Int J Obes.* 2005;39(9):1331–1338.
74. Roediger WE. Role of anaerobic bacteria in the metabolic welfare of the colonic mucosa in man. *Gut.* 1980;21:793–798.
75. Schönfeld P, Wojtczak L. Short- and medium-chain fatty acids in energy metabolism: The cellular perspective. *J Lipid Res.* 2016;57:943–954.
76. Lang DH, Yeung CK, Peter RM, et al. Isoform specificity of trimethylamine N-oxygenation by human flavin-containing monooxygenase (FMO) and P450 enzymes: selective catalysis by FMO3. *Biochem Pharmacol.* 1998;56:1005–1012.
77. Zhang AQ, Mitchell SC, Smith RL. Dietary precursors of trimethylamine in man: a pilot study. *Food Chem Toxicol.* 1999;37:515–552.
78. Tuomainen M, Lindström J, Lehtonen M, et al. Associations of serum indolepropionic acid, a gut microbiota metabolite, with type 2 diabetes and low-grade inflammation in high-risk individuals. *Nutr Diabetes.* 2018;8(1):35.
79. Guarner F, Malagelada J-R. Gut flora in health and disease. *Lancet.* 2003;361(9356):512–519.
80. Francino MP. Early development of the gut microbiota and immune health. *Pathogens.* 2014;4:769–790.
81. Shanahan F. Physiological basis for novel drug therapies to treat the inflammatory bowel diseases: I. Pathophysiological basis and prospects for probiotic therapy in inflammatory bowel disease. *Am J Physiol Gastrointest Liver Physiol.* 2005;288(3):G417–G421.

82. Thomas CM, Versalovic J. Probiotics-host communication modulation of signaling pathways in the intestine. *Gut Microbes.* 2010;1(3):1–16.
83. Belkaid Y. Role of the microbiota in immunity and inflammation. *Cell.* 2014;157(1):121–141.
84. Rakoff-Nahoum S, Paglino J, Eslami-Varzaneh F, Edberg S, Medzhitov R. Recognition of commensal microflora by toll-like receptors is required for intestinal homeostasis. *Cell.* 2004;118:229–241.
85. Clemente JC, Ursell LK, Parfrey LW, Knight R. The impact of the gut microbiota on human health: an integrative view. *Cell.* 2012;148(6):1258–1270.
86. Herman JP, McKlveen JM, Ghosal S, et al. Regulation of the hypothalamic-pituitary-adrenocortical stress response. *Compr Physiol.* 2016;6(2):603–621.
87. Farzi A, Fröhlich EE, Holzer P. Gut microbiota and the neuroendocrine system. *Neurotherapeutics.* 2018;15:5–22.
88. Tsigos C, Chrousos GP. Hypothalamic-pituitary-adrenal axis, neuroendocrine factors and stress. *J Psychosom Res.* 2002;53:865–871.
89. Weerth CD. Do bacteria shape our development? Crosstalk between intestinal microbiota and HPA axis. *Neurosci Biobehav Rev.* 2017;83:458–471.
90. Sudo N. Microbiome, HPA axis and production of endocrine hormones in the gut. *Adv Exp Med Biol.* 2014;817:177–194.
91. O'Malley D. Immunomodulation of enteric neural function in irritable bowel syndrome. *World J Gastroenterol.* 2015;21(24):7362.
92. Mayer EA. Gut feelings: the emerging biology of gut-brain communication. *Nat Rev Neurosci.* 2011;12(8):453–466.
93. Carabottia M, Sciroccoa A, Antonietta Masellib M, Severia C. The gut-brain axis: interactions between enteric microbiota, central and enteric nervous systems. *Ann Gastroenterol.* 2015;28:203–209.
94. Collins SM, Surette M, Bercik P. The interplay between the intestinal microbiota and the brain. *Nat Rev Microbiol.* 2012;10(11):735–742.
95. Keightley PC, Koloski NA, Talley NJ. Pathways in gut-brain communication: evidence for distinct gut-to-brain and brain-to-gut syndromes. *Aust N Z J Psychiatry.* 2015;49(3):207–214.
96. Obrenovich M, Mana TSC, Rai H, et al. Recent findings within the microbiota–gut–brain–endocrine metabolic interactome. *Pathol Lab Med Int.* 2017;9:21–30.
97. Chen ML, Zhu XH, Ran L, Lang HD, Yi L, Mi MT. Trimethylamine-N-oxide induces vascular inflammation by activating the NLRP3 inflammasome through the SIRT3-SOD2-mtROS signaling pathway. *J Am Heart Assoc.* 2017;6, e006347.
98. Chen K, Zheng X, Feng M, Li D, Zhang H. Gut microbiota-dependent metabolite trimethylamine N-oxide contributes to cardiac dysfunction in western diet-induced obese mice. *Front Physiol.* 2017;8:139.
99. Vogt NM, Romano KA, Darst BF, et al. The gut microbiota-derived metabolite trimethylamine N-oxide is elevated in Alzheimer's disease. *Alzheimers Res Ther.* 2018;10(1):124.
100. Duncan MW, Villacreses NE, Pearson PG, et al. 2-Amino-3-(methylamino)-propanoic acid (BMAA) pharmacokinetics and blood-brain barrier permeability in the rat. *J Pharmacol Exp Ther.* 1991;258(1):27–35.
101. Andersson M, Karlsson O, Bergström U, Brittebo EB, Brandt I. Maternal transfer of the cyanobacterial neurotoxin β-N-methylamino-L-alanine (BMAA) via milk to suckling offspring. *PLoS ONE.* 2013;8(10), e78133.
102. Bradley WG, Mash DC. Beyond Guam: the cyanobacteria/BMAA hypothesis of the cause of ALS and other neurodegenerative diseases. *Amyotroph Lateral Scler.* 2009;10(suppl 2):7–20.
103. Ridaura V, Belkaid Y. Gut microbiota: the link to your second brain. *Cell.* 2015;161(2):193–194.

104. Carding S, Verbeke K, Vipond DT, Corfe BM, Owen LJ. Dysbiosis of the gut microbiota in disease. *Microb Ecol Health Dis.* 2015;26:26191.
105. Zhuang ZQ, Shen LL, Li WW, et al. Gut microbiota is altered in patients with Alzheimer's disease. *J Alzheimers Dis.* 2018;63:1337–1346.
106. Zhang YG, Wu S, Yi J, Xia Y, Jin D, et al. Target intestinal microbiota to alleviate disease progression in amyotrophic lateral sclerosis. *Clin Ther.* 2017;39:322–336.
107. Mazzini L, Mogna L, Marchi FD, Amoruso A, Pane M, Aloisio I. Potential role of gut microbiota in ALS pathogenesis and possible novel therapeutic strategies. *J Clin Gasteroenterol.* 2018;52(suppl 1):S68–S70. Proceedings from the 9th Probiotics, Prebiotics and New Foods, Nutraceuticals and Botanicals for Nutrition & Human and Microbiota Health Meeting, held in Rome, Italy from September 10 to 12, 2017.
108. De Palma G, Collins S, Bercik P, Verdu E. The microbiota–gut–brain axis in gastrointestinal disorders: stressed bugs, stressed brain or both? *J Physiol.* 2014;592(14):2989–2997.
109. Scheperjans F, Aho V, Pereira P, et al. Gut microbiota are related to Parkinson's disease and clinical phenotype. *Mov Disord.* 2014;30(3):350–358.
110. Sampson TR, Debelius JW, Thron T, et al. Gut microbiota regulate motor deficits and neuroinflammation in a model of Parkinson's disease. *Cell.* 2016;167:1469–1480.
111. Messer JS, Cheng EB. Microbial physiology of the digestive tract and its role in inflammatory bowel diseases. In: *Physiology of the Gastrointestinal Tract.* 6th ed. Academic Press; 2018:795–810. [chapter 36].
112. Giron F, Quigley EMM. Pharmabiotic manipulation of the microbiota in gastrointestinal disorders: a clinical perspective. *J Neurogastroenterol Motil.* 2018;24(3):355–366.
113. Gibson GR, Roberfroid MB. Dietary modulation of the human colonic microbiota: introducing the concept of prebiotics. *J Nutr.* 1995;125:1401–1412.
114. Ohland CL, Macnaughton WK. Probiotic bacteria and intestinal epithelial barrier function. *Am J Physiol Gastrointest Liver Physiol.* 2010;298(6):G807–G819.
115. Dinan TG, Stanton C, Cryan JF. Psychobiotics: a novel class of psychotropic. *Biol Psychiatry.* 2013;74(10):720–726.
116. Huang H, Xu H, Luo Q, et al. Fecal microbiota transplantation to treat Parkinson's disease with constipation A case report. *Medicine.* 2019;98(26):e16163.
117. Sun J, Xu J, Ling Y, et al. Fecal microbiota transplantation alleviated Alzheimer's disease-like pathogenesis in APP/PS1 transgenic mice. *Transl Psychiatry.* 2019;9:189.

CHAPTER 2

Effect of different types of diet patterns on the gut microbiota composition

Akhlaq A. Farooqui
Department of Molecular and Cellular Biochemistry, The Ohio State University, Columbus, OH, United States

1 Introduction

The human gastrointestinal tract harbors a complex and dynamic population of microorganisms called the gut microbiota.[1] Gut microbiota have evolved with humans through a symbiotic relationship that provides many benefits for humans. These benefits include numerous physiological activities such as vitamin production, regulation of gene expression, protection against pathogenic bacteria, absorption of nutrients, regulation of metabolic disorders, regulation of host immunity, and harvesting energy (Fig. 1).[2–6] The human gastrointestinal tract microbiota becomes established in stages during early life, beginning antenatally.[7] Interestingly, initial bacterial colonizers of the gut are largely determined by the mode of delivery; infants born naturally are initially inoculated by bacteria typically present in the vaginal and fecal microbiota, such as *Lactobacillus* and *Prevotella* spp., while those born by caesarean section are colonized by bacteria from the skin and environment.[8] Indeed, the most significant change in microbiota composition occurs during weaning with introduction to solid foods resulting in a shift within the first 2–3 years of life towards an adult microbiota.[9] The mammalian gastrointestinal tract microbiota is highly variable at lower taxonomic levels. There are four dominant phyla: Firmicutes, Bacteroidetes, Actinobacteria, and Proteobacteria.[10] Firmicutes and Bacteroidetes account for more than 90% of the bacterial population in the colon,[11] while Actinobacteria and Proteobacteria (which includes Enterobacteriaceae) are regularly present but are scarce (< 1%–5%).[10] The Bacteroidetes/Firmicutes phylum ratio remains remarkably stable over time.[12] The gut microbiota population is affected by the environment, diet, medications, age, geographic factors, surgical interventions, and host genetics, particularly genes related

Gut Microbiota in Neurologic and Visceral Diseases
https://doi.org/10.1016/B978-0-12-821039-0.00002-2

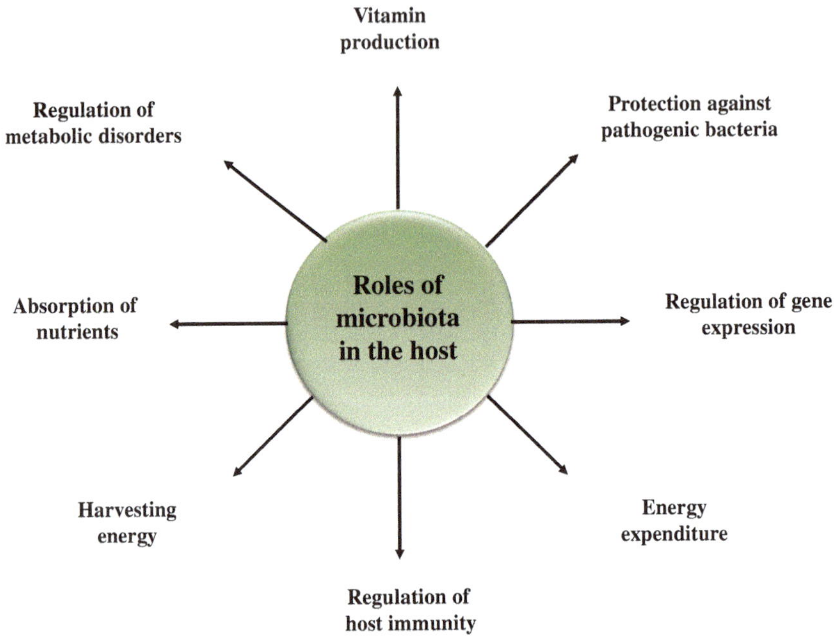

Fig. 1 Roles of microbiota in host tissues.

to the immune system and metabolism. Alterations in the relative abundance of each microbiota species in the gut microbial community not only influences immune performance and physiological homeostasis but also healthy aging.[13, 14] Converging evidence suggests that gut microbiota composition is modulated by several factors, including genetics, lifestyle, diet, and antibiotherapy.[10, 15] Among these factors diet plays a predominant role in modulating the composition of gut microbiota. Alterations in microbiota, including a reduction in health-promoting bacteria such as Lactobacilli and Bifidobacteria, may lead to dysbiosis, a process caused by Gram-negative bacteria. Dysbiosis can be categorized into three different types: (1) loss of beneficial microorganisms, (2) unwarranted growth of potentially harmful microorganisms, and (3) reduced overall microbial diversity. In most cases, these three types of dysbiosis are not mutually exclusive and can occur concurrently.[16] In most cases, dysbiosis produces immunogenic endotoxins such as lipopolysaccharide (LPS). Levels of circulating LPS are increased in cardiovascular disease (CVD) patients.[17] Gut dysbiosis suppresses the expression of tight junction proteins, leading to an increase in intestinal permeability and subsequently the translocation of LPS into the blood.[18, 19] Gut dysbiosis-derived LPS may play important roles in the activation and modulation of Toll-like receptors (TLRs) and their downstream targets.[20,21] As part of the pattern-recognition receptors family, TLRs can recognize bacterial products and modulate the host immune system.[22] Circulating LPS can bind to cell-surface receptor complexes composed of TLR4 and its co-receptor's cluster of differentiation 14 (CD14).[23–25] Using TLR4 and LDL receptor double knockout mice, Ding et al. reported that a TLR4 deficiency reduces atherosclerosis without affecting inflammation.[26] Moreover, clinical investigations have revealed that upregulation of TLRs not only contributes to inflammation but also development of atherosclerosis.[27] Collectively these studies indicate that dietary composition appears to have long-term and acute effects on the gut microbiota ecosystem,[28] and unsolicited perturbations in gut microbiota diversity and composition are key elements underlying chronic diseases, including several low-grade inflammatory disorders of the human gastrointestinal tract.

There is a continuous bidirectional communication between the brain and the gut through the microbiota gut-brain axis. The bidirectional communication between the gut microbiota and the brain occurs not only through the microbiota gut-brain axis and vagus nerve, but also through the immune system, neuroendocrine pathways, and microbiota-derived metabolites.[29,30] This bidirectional communication not only modulates gut functions (nausea, satiety, and pain) but also neurotransmission. Alterations in neurotransmission are related to the pathogenesis of neurological and neuropsychiatric disorders.[29,30] Recent studies have also linked microbiota dysbiosis to visceral diseases (insulin resistance, type 2 diabetes, metabolic syndrome, and heart disease). Mechanistic studies in rodent models have revealed an ever-increasing multitude of molecular mechanisms whereby the gut microbiota interacts with various host sensing and signaling pathways, leading to modulation of the endocrine system, immune responses, nervous system activity, and hence the predisposition to metabolic diseases. Remarkably, the microbiota-driven immune responses in metabolic tissues and the host nutrient-sensing mechanisms of gut microbial metabolites, in particular short-chain fatty acids (SCFAs), have been significantly associated with the proneness to diabetes and related disorders.[31] Similarly, microbial dysbiosis is linked to neurological disorders (Parkinson's disease (PD), and Alzheimer's disease (AD), multiple sclerosis (MS), stroke, and autism). The central nervous system (CNS) connects with the gut via sympathetic and parasympathetic nerves. It is increasingly evident that gut microbiota modulates the development and maintenance of obesity and insulin resistance.[32] The molecular mechanisms involving induction of obesity and insulin resistance are regulated by host food intake behavior, energy harvest, energy expenditure, and fat storage. Several mechanisms are associated with obesity-mediated inflammation in peripheral tissues and the hypothalamic region of the brain. These mechanisms include the activation of TLR4, induction of endoplasmic reticulum stress, and activation of serine/threonine kinases such as IKKβ. Although the relative contribution of these mechanisms remains unknown, early onset of inflammation in the hypothalamus relative to that in peripheral tissues suggests that different processes may cause inflammation in these areas.[32] Moreover, the nature of hypothalamic inflammation occurring during the first days of a high-fat diet may differ fundamentally from that due to a chronic high-fat diet.

2 Microbiota-derived metabolites in host tissues

Gut microbiota modulate host metabolism through the production of bioactive metabolites, which regulate many biological pathways involved in immunity and energy production. The population of microbiota in the rectum digests carbohydrates, proteins, and lipids left undigested by the small intestine. Undigested substances, named "microbiota-accessible carbohydrates" (MACs), are components of the plant cell wall, which is enriched in cellulose, hemicelluloses, pectin, and resistant starch; these polymers undergo microbial degradation and subsequent fermentation.[33] The genome of gut bacteria, which is different from the human genome, encoded several highly specified enzymes able to digest and ferment complex biomacromolecules by hydrolyzing glycosidic bonds.[34]

Alterations in gut microbiota composition (dysbiosis) have important repercussions on the regulation of host biochemical pathways and metabolic networks. Multiple studies have indicated that gut microbiota produce and secrete biologically active molecules (tryptophan, indole, hydrogen peroxide, choline, phosphatidylcholine, L-carnitine, betaine, trimethylamine, γ-butyrobetaine, crotonobetaine, and SCFAs including butyrate, propionate, and acetate) that enter the circulation and affect circulatory system homeostasis.[35,36] To enter the circulation, gut microbiota-derived metabolites pass through the intestinal barrier. The integrity and permeability of the intestinal barrier depend on numerous factors, including intestinal blood flow.[37] These microbiota-derived metabolites not only play crucial roles in host metabolic homeostasis, immunological processes, and neurobiology, but also contribute to human health and disease.[38,39] The gut microbiome associated with healthy lean individuals has greater microbial biodiversity and a greater Bacteroidete-to-Firmicute ratio compared to the microbiome of obese individuals. These gut microbial associations are similar to those seen in individuals with high and low dietary fiber intakes, respectively. Metabolites generated by Bacteroidetes and Firmicutes include the three main SCFAs related to obesity, namely, butyrate, acetate, and propionate. However, neither Bacteroidetes nor Firmicutes is purely causative or purely preventative of obesity. It is suggested that sustained consumption of a high-fiber diet may help in preventing obesity by promoting a microbiome associated with a lean phenotype. SCFAs are the byproducts of fiber fermentation and have both obesogenic and antiobesogenic properties. The production of specific forms of SCFAs depends on the microbes available in the gut and the type of fiber ingested. SCFAs play distinct physiological roles in host metabolism. They not only shape the gut environment and influence the physiology of the colon, but they can also be used as energy sources by host cells.[40] In addition to these functions, SCFAs play regulatory roles in insulin sensitivity and glucose and lipid metabolism (Fig. 2).[41] Butyrate and acetate protect against diet-induced obesity without causing hypophagia, while propionate reduces food intake. The underlying mechanisms for these effects are unclear. However, it is suggested that SCFAs produce their effects on signal transduction processes not only by interacting with

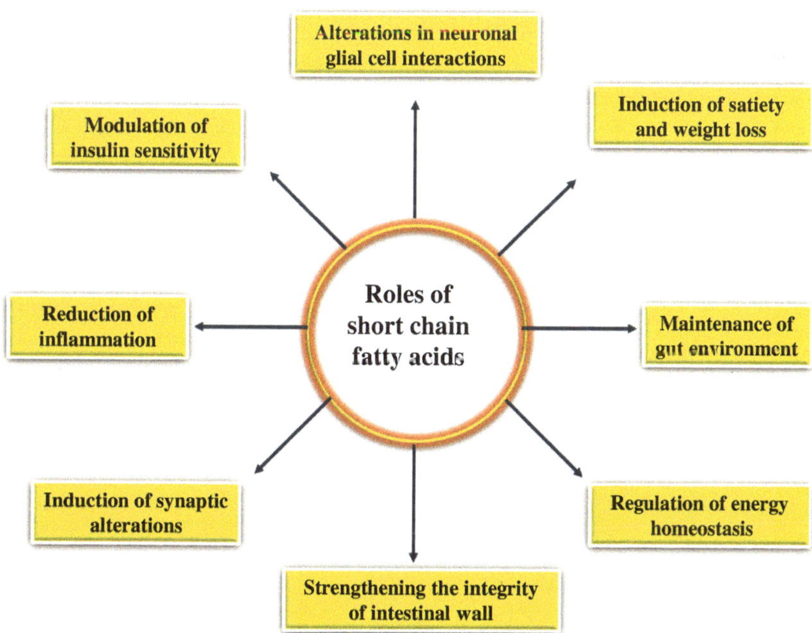

Fig. 2 Roles of short-chain fatty acids in the host tissues.

G protein-coupled receptors (FFAR2, FFAR3, OLFR78, GPR109A), but also by acting as epigenetic regulators of gene expression. This process involves the inhibition of histone deacetylase (HDAC).[42] Butyrate and propionate not only inhibit HDAC but also alter the expression of specific genes via conformational changes in the active site of HDAC leading to its inactivation.[43, 44] Gut microbiome-derived butyrate reduces glucose intolerance and insulin resistance. It also resists the increase in the relative abundance of microbiota belonging to the Lachnospiraceae, Rikenellaceae, and Paraprevotellaceae families induced by a high-fat diet.[45] Butyrate in fermented food increases the levels of α-linolenate, all-trans-retinal, resolvin E1, and leukotriene in plasma, and the differential pathways show enrichment in resolvin E_1 biosynthesis, histidine degradation, lipoxin biosynthesis, and leukotriene biosynthesis. In addition, butyrate also increases the levels of phosphorylated-adenosine 5′-monophosphate-activated protein kinase (p-AMPK) and facilitates glucose transporter member 4 (GLUT4) in the adipose tissue.Based on this information, it is proposed that oral supplements of the natural fermentation product (butyrate) can potentially be used to prevent cardiovascular disease.[45] Collective evidence suggests that acetate, propionate, and butyrate are important metabolites that are not only associated with production of energy by the gut mucosal cells but also contribute to intestinal and liver gluconeogenesis (propionate), de novo lipid synthesis, and protein synthesis through distinct mechanisms.[46, 47] In contrast, acetate is a precursor of cholesterol synthesis.

Choline, L-carnitine, γ-butyrobetaine, phosphatidylcholine, betaine, glycerophosphine, and crotonobetaine are metabolized by the intestinal microbiota to produce trimethylamine and trimethylamine oxide (TMAO). This microbial metabolite is known to increase the risk of atherosclerosis and cardiovascular diseases such as stroke.[48, 49] It is also proposed that metabolism of TMAO is linked to insulin sensitivity and glucose metabolism.[50] High levels of TMAO are known to induce vascular inflammation and prothrombosis by exaggerating platelet hyperresponsiveness to multiple agonists, and can be implicated in the pathogenesis of obesity and type 2 diabetes.[51]

3 Effects of long-term consumption of various types of diets on the host

Diet modulates gut microbiota composition. Western diet is typically high in fat (saturated and n-6 fatty acids), cholesterol, refined sugar, and salt, but low in fiber (Fig. 3). This diet is also high in calories. Long-term consumption of a Western diet is associated with a decrease in richness of microbial

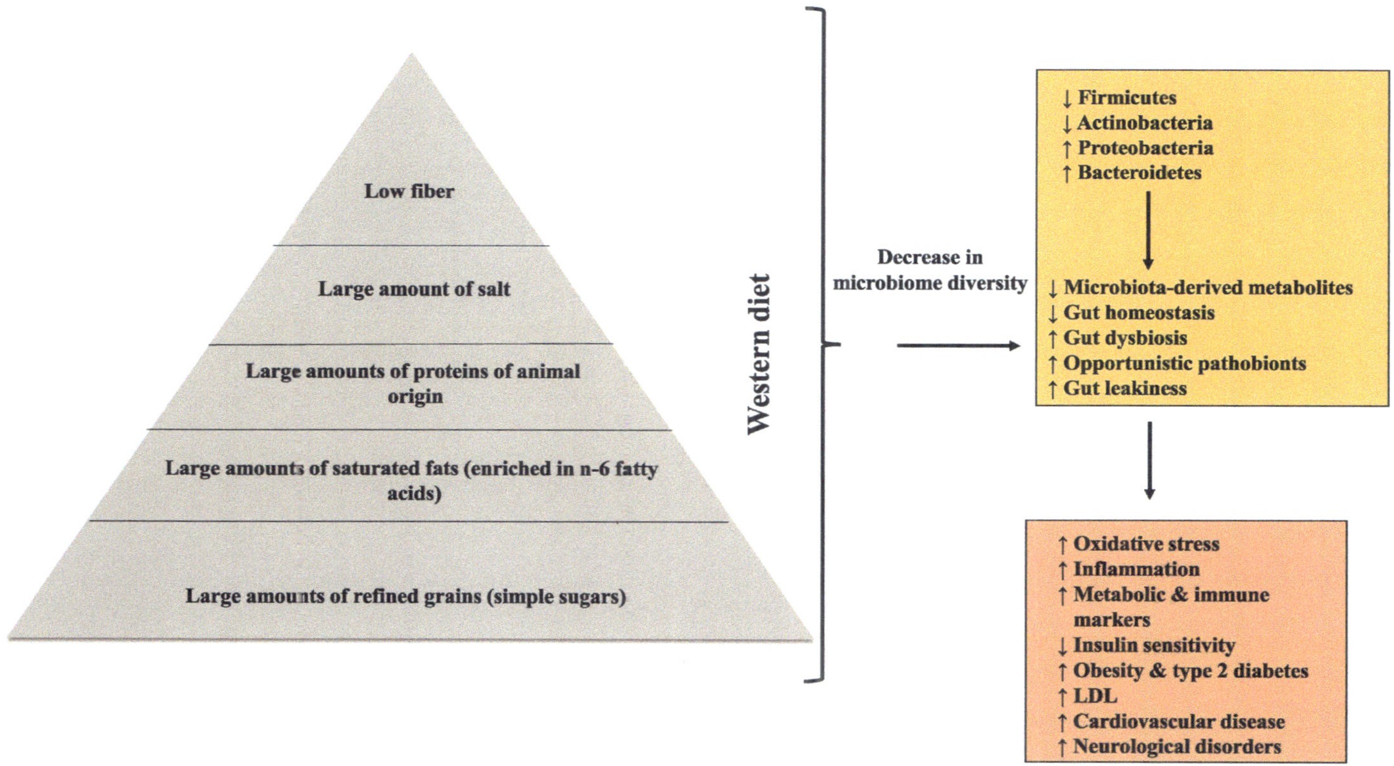

Fig. 3 Composition of Western diet and its effects on host microbiota.

communities at the individual level (alpha diversity) and is detrimental to human health. In contrast, the consumption of Mediterranean and vegetarian diets by humans and rodents contributes to greater variation in microbiota (beta diversity). These diets are high in fiber and relatively low in calories (Table 1).[52] As such, they promote and maintain human health.

4 Effects of long-term consumption of the Western diet

An animal fed with Western diet displays a significant shift in both bacterial and metagenomic profiles as compared to an animal on a normal, chow diet. Western diet-associated cecal microbiota is characterized by a reduction in the relative abundance of Bacteroidetes and an increase in the relative abundance of Firmicutes. In particular, Mollicutes, a class of Firmicutes, has been found to be significantly more prevalent in conventionally raised mice (CONV) fed Western-type diet.[53] An increase of genes involved in the import and processing of sugars in the gut metagenome was also found in Western diet-fed mice.[53] Examination of the role of Western diet-associated cecal microbiota in facilitating weight gain has revealed that Ex-germfree (ex-GF) mice receiving Western diet-associated gut microbiota gained significantly more body fat than mice receiving the chow diet-associated gut microbiota.[53] Murphy et al. reported that C57BL/6J mice responded to high-fat feeding with a progressive increase in abundance of Firmicutes over time.[54] Ravussin et al. also reported that high-fat feeding in weight-matched animals is associated with an increase in the family of Lachnospiraceae and genera *Bacteroides* and *Mucispirillum*, whereas low-fat feeding is associated with an increase in the genus *Allobaculum*.[55] These changes in microbiota composition alter metabolic and immune biomarkers, resulting in higher body mass index and greater levels of inflammatory markers than those who follow a high-fiber, low-calorie diet.[56] Furthermore, the consumption of Western diet also increases the abundance of bile-tolerant microorganisms (*Alistipes, Bilophila, and Bacteroides*). Converging evidence suggests that pro-obesity microbiota promote energy harvest and storage via influencing bile acid transformation, fermenting, and providing vitamins and amino acids for the host.[57] The molecular mechanisms associated with Western diet-mediated detrimental effects on human health are not fully understood. However, it is becoming increasingly evident that a Western diet, which is high in fat and simple sugars and low in fiber, reduces the production of SCFAs shifting the gastrointestinal microbiota metabolism to the production of detrimental metabolites, favoring the

Table 1 Effects of various diets patterns on gut microbiota.

Diet	Componerts	Total bacteria	Bifidobacteria	Lactobacilli	Prevotella	Eubacteria	Rosebu
Western	High animal fat and protein	↓	–	↓	↓	↓	↓
Mediterranean	High fiber, antioxidants	↑	↑	↑	↑	↑	↑
Vegetarian	High fiber	↑	↑	↑	↑	↑	↑
Vegan	High fiber	↑	↑	↑	↑	↑	↑
Gluten-free	No gluten	↓	–	↓	↓	↓	↓

expansion of bacteria associated with chronic inflammation.[58] In addition, long-term consumption of a Western diet results in induction of oxidative stress and development of inflammation not only in the hypothalamus and hippocampus but also in the peripheral tissues including the liver, adipose tissue, skeletal muscle, and intestine.[58, 59] These conditions contribute to obesity, type 2 diabetes, metabolic syndrome (MetS), heart disease, osteoporosis, rheumatoid arthritis, and cancer.[32, 60] In the development of heart disease, the consumption of a Western diet results in conditions that exacerbate dyslipidemia. This condition is characterized by increased triglyceride (TG) levels and decreased high-density lipoprotein cholesterol (HDL-C) levels.[61] In addition, diets rich in foods with high glycemic index decrease fatty acid beta-oxidation and increase adipogenesis and fat accumulation by downregulating carnitine palmitoyltransferase 1 (CPT-1) mRNA levels.[62] The consumption of Western diet also results in type 2 diabetes and MetS. These conditions are risk factors for neurological disorders (stroke, AD, and depression).[63] Furthermore, long-term consumption of Western diet elevates basal levels of plasma corticosterone and increases glucocorticoid receptor (GR) immunoreactivity.[64] These parameters also play an important role in the pathogenesis of neurological disorders.

5 Effects of long-term consumption of the Mediterranean diet

Components of the Mediterranean diet include fresh vegetables and fruits, whole grain pasta, legumes, fish, olive oil, fresh garlic, low levels of dairy products (cheese and yogurt), nuts and seeds, garlic, and red wine (Fig. 4).[65–67] The beneficial effects of the Mediterranean diet are due to the intake of fresh colored fruits (flavonols, anthocyanins, and procyanindins), vegetables (fiber), fish (Omega-3 fatty acids), wine (resveratrol), extra-virgin olive oil (tyrosol, hydroxytyrosol, oleuropein, and oleocanthal), and garlic (sulfur compounds). These active components not only modulate signal transduction and neuronal communication but also delay the onset of dementia.[68] Chronic intake of the Mediterranean diet or a low-fat diet over a year is associated with an improved insulin profile and modified lipid metabolism in obese patients with coronary heart disease, an effect linked to an increased abundance of *Roseburia* and *Faecalibacterium prausnitzii* in the gut.[69] Consistently, other studies have shown that obese patients with metabolic dysfunction have reversed microbiota dysbiosis upon long-term consumption of the Mediterranean diet.[70] It is also reported that long-term

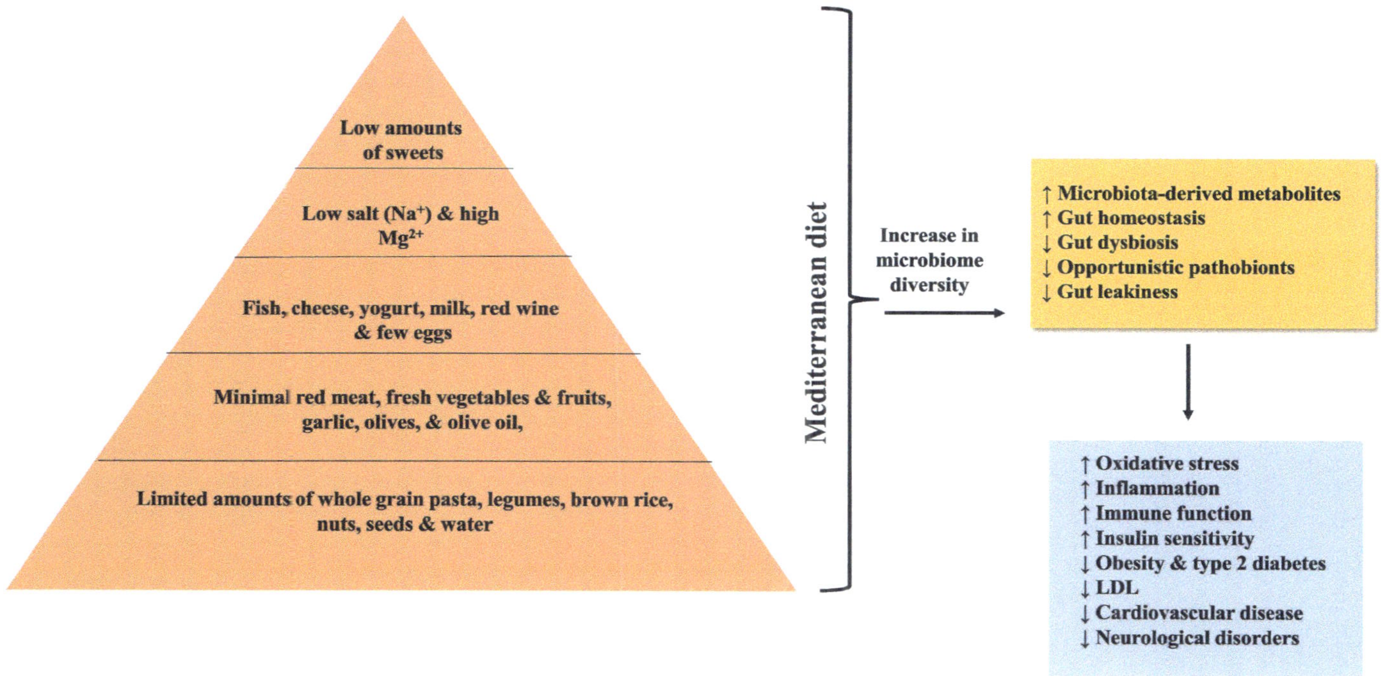

Fig. 4 Composition of Mediterranean diet and its effects on microbiota.

consumption of the Mediterranean diet increases levels of *Prevotella*.[28, 71–74] Indeed, regular ingestion of plant polysaccharides is integral for maintaining a healthy balance of microbes in the lower gastrointestinal tract.[75, 76] Large cohort studies have demonstrated that good adherence to the Mediterranean diet is related to reduced risk of developing several disorders, such as cardiovascular disease and type 2 diabetes.[77] It is also shown that consumption of the Mediterranean diet promotes reduction of inflammatory markers.[78] The Mediterranean diet may also reduce obesity.[79] In this study, a greater presence of Bacteroidetes and a smaller Firmicutes-to-Bacteroidetes ratio was found in individuals who consumed less animal protein and who have a higher Mediterranean diet score. A large Firmicutes-to-Bacteroidetes ratio is related to type 2 diabetes[80] and obesity.[81] A recent study on consumption of the Mediterranean diet indicated that subjects have greater proportions of Bacteroidetes in their gut.[82] In another study comparing fecal microbiota of European children with children from a rural African village found that those whose diets was enriched with fiber had significantly greater levels of Bacteroidetes and a unique abundance of bacteria from *Prevotella* and *Xylanibacter*.[75] It is concluded that the Mediterranean diet is highly regarded as a healthy, balanced diet. It is distinguished by a beneficial fatty acid profile that is rich in both monounsaturated (olive oil) and n-3 polyunsaturated fatty acids (eicosapentaenoic and docosahexaenoic acids), high levels of polyphenols and other antioxidants, high intake of fiber, other low glycemic carbohydrates, relatively greater vegetable than animal protein intake, and red wine. Based on many studies, it is proposed that long-term consumption of components of the Mediterranean diet may reduce the detrimental effects of oxidative stress and low-grade neuroinflammation not only by normalizing mitochondrial dysfunction and improving endothelial function, but also by increasing cognitive function on signal transduction processes. This process may delay the onset of visceral and neurological disorders (AD, PD, and stroke).[83]

6 Effects of long-term consumption of the vegetarian diet

A wholesome vegetarian diet consists of fresh vegetables, fruits, chickpeas, pulses, nuts, curcumin, garlic, and soy products (soy milk, soy yogurt, and tofu). The benefits of a vegetarian diet arise from lower intakes of saturated fat, cholesterol, and animal protein as well as higher intakes of complex carbohydrates, dietary fiber, and vitamins and minerals (magnesium, folic acid, vitamins C and E, carotenoids, and other phytochemicals including garlic and curcumin). Vegetables and fruits in vegetarian diet provide adequate levels of magnesium, potassium, nitrate, and fiber. Studies on the effects of a vegetarian diet on human health have indicated that this diet promotes development of more diverse and stable microbial systems. Additionally, vegetarian diet consumers have significantly greater counts of certain *Bacteroidetes*-related operational taxonomic units compared to omnivores. Presence of fibers in the diet increases lactic acid bacteria, such as *Ruminococcus, Eubacterium rectale*, and *Roseburia*, and reduces *Clostridium* and *Enterococcus* species.[84] Presence of polyphenols in a vegetarian diet increases levels of Bacteroidetes and *Akkermansia*, *Bifidobacteria*, *Lactobacillus*, *Bacteroides*, and *Prevotella* while reducing Firmicutes and the Firmicutes-to-Bacteroidetes ratio in the gut (Fig. 5). This composition of microbiota in the gut provides antipathogenic and antiinflammatory effects and cardiovascular protection.[85] High fiber intake in a vegetarian diet also promotes the growth of species that ferment fiber into SCFAs (acetate, propionate, and butyrate), bile acids, and lipopolysaccharides (LPS), which are now correlated with metabolic diseases such as type 2 diabetes. Converging evidence suggests that consumption of a vegetarian diet not only promotes a diverse ecosystem of beneficial bacteria to support both the human gut microbiome and overall health, but also improves immunity against pathogens, blood-brain barrier integrity, provision of energy substrates, and regulation of critical functions of the intestine.[84,85] It is realized that some important nutrients such as protein, iron, and Vitamin B12 are lacking in vegetarian diets.[86] Vitamin B12 is a crucial component involved in early brain development and maintaining normal CNS function,[87] and is suggested to be neuroprotective, particularly for memory performance and hippocampal microstructure.[88] One hypothesis is that high levels of homocysteine, which is associated with vitamin B12 deficiency, may be harmful to the body. Vitamin B12 is the essential cofactor required for the conversion of homocysteine into nonharmful components and serves as a cofactor in different enzymatic reactions. A person suffering from vitamin B12 insufficiency accumulates homocysteine, which promotes the formation of plaques in arteries and thereby increases atherothrombotic risk,[89] possibly facilitating symptoms in AD patients.[90] A metaanalysis found that vitamin B12 deficiency is associated with stroke, AD, vascular dementia, PD, and cognitive impairment (even in lower concentrations),[91] supporting the claim of its high potential for disease prevention when avoided or treated.[92] In fact, the vegetarian diet received the highest diet quality score from the Healthy Eating Index 2010 and the Mediterranean Diet

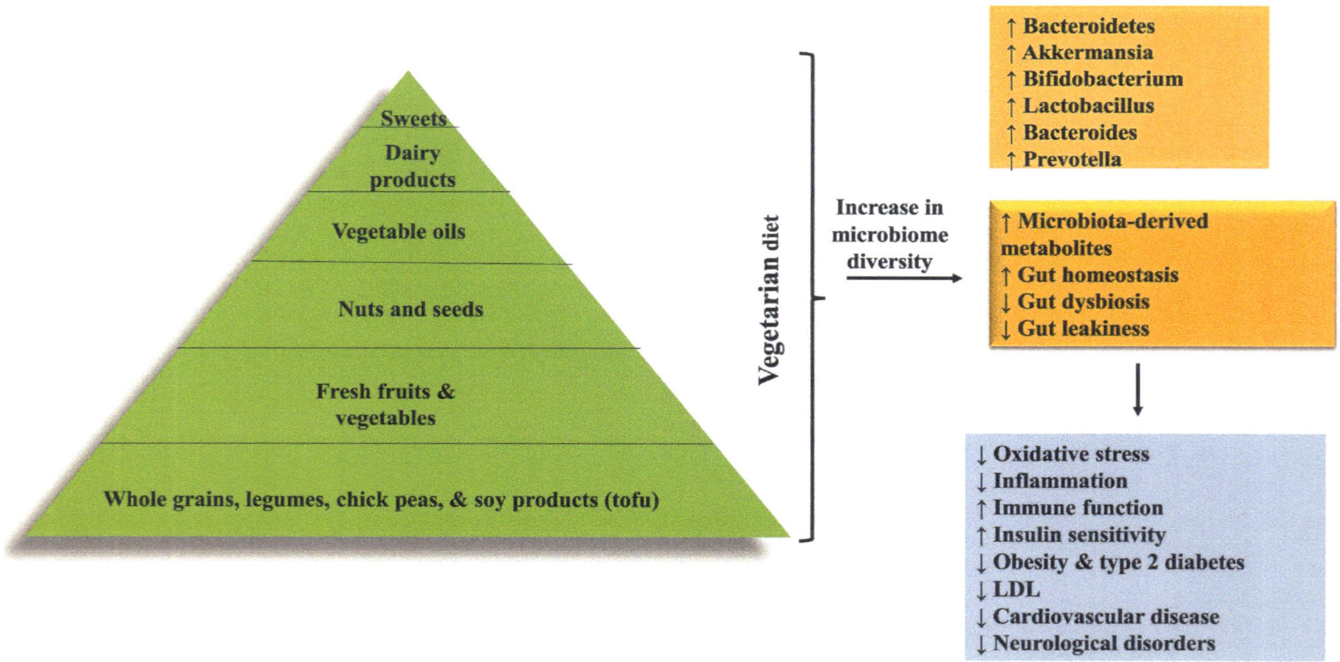

Fig. 5 Composition of vegetarian diet and its effects on microbiota.

Score in a recent study that included 1475 participants.[93] Thus, long-term consumption of a vegetarian diet promotes its effects through multiple pathways, including better glycemic control, lower inflammatory activity, and altered neurotransmitter metabolism via dietary intake or intestinal activity. In addition, a vegetarian diet inhibits oxidative stress and retards inflammation due to the maintenance of antioxidant enzymes in neural and nonneural cells.[94]

7 Effects of long-term consumption of the ketogenic diet

The ketogenic diet is composed of 80%–90% fat, with carbohydrates and proteins constituting the remainder of the intake. This diet provides sufficient protein for growth, but insufficient amounts of carbohydrates for the body's metabolic needs. The ketogenic diet phenocopies several biochemical characteristics of fasting including reduced insulin level, Forkhead box O (FoxO) signaling, inhibition of the mammalian target of rapamycin (mTOR), and activation of AMP-activated protein kinase (AMPK).[95] Recent studies suggest the ketogenic diet significantly impacts gut microbiota, but with mixed results. For example, a mouse study noted that the ketogenic diet increased beneficial species of gut microbiota including *Akkermansia muciniphila* and *Lactobacillus* bacteria, which are capable of producing SCFAs that may provide the host with beneficial health outcomes.[96] However, these changes occur at the expense of reduced overall microbial diversity, possibly due to the minimized carbohydrate intake, which can disrupt other beneficial microbes.[97] Human studies also point out some potential negative effects on the gut microbiota. For example, overweight and obese subjects following the ketogenic diet for 8 weeks had a significantly reduced amount of a beneficial bacterium, *Bifidobacterium*, in their colon and decreased plasma levels of SCFAs.[98]

The ketogenic diet provides energy mainly from the oxidation of fatty acids in mitochondria as opposed to glucose, which is the main energy source following consumption of a normal diet. Following prolonged consumption of the ketogenic diet, fatty acids are oxidized at a high rate, which results in an overproduction of acetyl-CoA. The overproduction of acetyl-CoA then leads to the synthesis of ketone bodies (β-hydroxybutyrate, acetoacetate, and acetone) primarily in the liver.[99] The accumulation of ketone bodies in the bloodstream causes ketosis, a survival mechanism, which is activated during prolonged fasting, starvation, or lack of carbohydrate ingestion. Ketone bodies readily cross the blood-brain barrier either by simple diffusion (acetone) or with the aid of monocarboxylic transporters (β-hydroxybutyrate, acetoacetate), whose expression is related to the level of ketosis.[100] During prolonged ketosis, the brain is capable of utilizing ketone bodies as an alternate fuel, thus reducing its requirement for glucose.[101]

Studies on the effect of a ketogenic diet on gut microbiota in mice indicate that in four days a ketogenic diet produces significant changes in gut bacterial taxonomy. Two species of bacteria, *Akkermansia* and *Parabacteriodes*, are significantly increased. The increase of these two bacterial species in the gut lead to a decrease in production of γ-glutamyl transpeptidase by the gut microbiome. This enzyme catalyzes the transfer of γ-glutamyl functional groups from molecules such as glutathione to an acceptor that may be an amino acid-generating glutamate.[102] Moreover, it is also shown that there is a decrease in a subset of ketogenic γ-glutatamylated (GG) amino acids (i.e., γ-glutamyl-leucine) both in the gut and blood. GG amino acids are supposed to have transport properties across the blood-brain barrier, different from non-γ-glutamylated forms.[103] This property is involved in glutamate and GABA biosynthesis.[104]

The consumption of a ketogenic diet by mice produces a variety of neurovascular improvements strictly linked to a lower risk of developing AD. The beneficial effects of a ketogenic diet may be related to gut microbiota composition, which shows an increase in levels of *A. muciniphila* and *Lactobacillus*. Furthermore, the ketogenic diet not only decreases the overall microbial alpha diversity due to the low carbohydrate (complex carbohydrate) content of diet, but also produces high levels of SCFAs to promote host metabolism, which may provide energy.[97]

8 Conclusion

The gut microbiota is the central regulator of mammalian food intake by processing nutrients into absorbable metabolites. They also produce vitamins and essential metabolites that are not synthesized by the host. Changes in the composition of the gut microbiota mediate important repercussions on the regulation of host biochemical pathways and metabolic networks. Many bacterial metabolite end products of the gut microbiota (tryptophan, indole, hydrogen peroxide, SCFAs, choline, phosphatidylcholine, L-carnitine, betaine, trimethylamine, γ-butyrobetaine, and crotonobetaine) play crucial roles in host metabolic homeostasis, immunological processes, and neurobiology, and underline the fundamental importance of this extended genome in human health and disease. Long-term consumption of Western diet promotes alpha diversity by decreasing levels of *Firmicutes* and *Actinobacteria* and increasing the abundance of bile-tolerant microorganisms (*Alistipes*, *Bilophila*, and *Bacteroides*). This microbiota composition produces detrimental effects on human health. It is suggested that long-term consumption of a Western diet results in induction of oxidative stress and development of low-grade inflammation associated with visceral diseases (obesity, type 2 diabetes, MetS, heart disease, osteoporosis, rheumatoid arthritis, and cancer) and neurological disorders (stroke, AD, and depression). In contrast, long-term consumption of the Mediterranean diet reduces obesity by increasing the Firmicutes-to-Bacteroidetes ratio. Thus, the Mediterranean diet acts by reducing the detrimental effects of oxidative stress and low-grade neuroinflammation, normalizing mitochondrial dysfunction, improving endothelial function, and increasing cognitive function. Similarly, a vegetarian diet consumption increases lactic acid bacteria, such as *Ruminococcus*, *Eubacterium rectale*, and *Roseburia*, and reduces *Clostridium* and *Enterococcus* species. Presence of polyphenols in the Mediterranean and vegetarian diets increases levels ofBacteroidetes and *Akkermansia*, *Bifidobacteria*, *Lactobacillus*, *Bacteroides*,and *Prevotella* while reducing Firmicutes and the Firmicutes-to-Bacteroidetes ratio in the gut. This composition deceases oxidative stress and reduces low-grade inflammation, both of which are associated with chronic visceral and neurological disorders.

References

1. Clarke G, Sandhu KV, Griffin BT, Dinan TG, Cryan JF, Hyland NP. Gut reactions: breaking down xenobiotic-microbiome interactions. *Pharmacol Rev.* 2019;71:198–224.
2. den Besten G, Bleeker A, Gerding A, et al. Short-chain fatty acids protect against high-fat diet-induced obesity via a PPARgamma-dependent switch from lipogenesis to fat oxidation. *Diabetes.* 2015;64:2398–2408.
3. Gensollen T, Iyer SS, Kasper DL, Blumberg RS. How colonization by microbiota in early life shapes the immune system. *Science.* 2016;352:539–544.
4. Baumler AJ, Sperandio V. Interactions between the microbiota and pathogenic bacteria in the gut. *Nature.* 2016;535:85–93.
5. Natividad JMM, Verdu EF. Modulation of intestinal barrier by intestinal microbiota: pathological and therapeutic implications. *Pharmacol Res.* 2013;69:42–51.
6. Yang J-Y, Kweon M-N. The gut microbiota: a key regulator of metabolic diseases. *BMB Rep.* 2016;49:536–541.
7. Ardissone AN, de la Cruz DM, Davis-Richardson AG, et al. Meconium microbiome analysis identifies bacteria correlated with premature birth. *PLoS One.* 2014;9, e90784.
8. Dominguez-Bello MG, Costello EK, Contreras M, et al. Delivery mode shapes the acquisition and structure of the initial microbiota across multiple body habitats in newborns. *Proc Natl Acad Sci U S A.* 2010;107:11971–11975.
9. Albenberg LG, Wu GD. Diet and the intestinal microbiome: associations, functions, and implications for health and disease. *Gastroenterology.* 2014;146:1564–1572.
10. Chwalba A, Otto-Buczkowska E. Participation of the microbiome in the pathogenesis of diabetes mellitus. *Clin Diabetol.* 2017;6:178–181.

11. Ley RE, Hamady M, Lozupone C, et al. Evolution of mammals and their gut microbes. *Science.* 2008;320:1647–1651.
12. Faith JJ, Guruge JL, Charbonneau M, et al. The long-term stability of the human gut microbiota. *Science.* 2013;341:1237439.
13. Ottman N, Smidt H, de Vos WM, Belzer C. The function of our microbiota: who is out there and what do they do? *Front Cell Infect Microbiol.* 2012;2:104.
14. Candela M, Biagi E, Brigidi P, O'Toole PW, De Vos WM. Maintenance of a healthy trajectory of the intestinal microbiome during aging: a dietary approach. *Mech Ageing Dev.* 2014;136:70–75.
15. Miele L, Giorgio V, Alberelli MA, De Candia E, Gasbarrini A, Grieco A. Impact of gut microbiota on obesity, diabetes, and cardiovascular disease risk. *Curr Cardiol Rep.* 2015;17:120.
16. Degruttola AK, Low D, Mizoguchi A, Mizoguchi E. Current understanding of dysbiosis in disease in human and animal models. *Inflamm Bowel Dis.* 2016;22:1137–1150.
17. McIntyre CW, Harrison LE, Eldehni MT, et al. Circulating endotoxemia: a novel factor in systemic inflammation and cardiovascular disease in chronic kidney disease. *Clin J Am Soc Nephrol.* 2011;6:133–141.
18. Harris K, Kassis A, Major G, Chou CJ. Is the gut microbiota a new factor contributing to obesity and its metabolic disorders? *J Obes.* 2012;2012:879151.
19. Cândido FG, Valente XF, Grześkowiak LM, Moreira AP, Rocha DM, Alfenas R. Impact of dietary fat on gut microbiota and low-grade systemic inflammation: mechanisms and clinical implications on obesity. *Int J Food Sci Nutr.* 2017;69:125–143.
20. Lundin A, Bok CM, Aronsson L, et al. Gut flora, toll-like receptors and nuclear receptors: a tripartite communication that tunes innate immunity in large intestine. *Cell Microbiol.* 2008;10:1093–1103.
21. Chacon MR, Lozano-Bartolome J, Portero-Otin M, et al. The gut mycobiome composition is linked to carotid atherosclerosis. *Benef Microbes.* 2017;9:1–14.
22. Akira S, Uematsu S, Takeuchi O. Pathogen recognition and innate immunity. *Cell.* 2006;124:783–801.
23. Neves AL, Coelho J, Couto L, Leite-Moreira A, Roncon-Albuquerque Jr R. Metabolic endotoxemia: a molecular link between obesity and cardiovascular risk. *J Mol Endocrinol.* 2013;51:R51–R64.
24. Belkaid Y, Hand TW. Role of the microbiota in immunity and inflammation. *Cell.* 2014;157:121–141.
25. Noverr MC, Huffnagle GB. Does the microbiota regulate immune responses outside the gut? *Trends Microbiol.* 2004;12:562–568.
26. Ding Y, Subramanian S, Montes VN, et al. Toll-like receptor 4 deficiency decreases atherosclerosis but does not protect against inflammation in obese low-density lipoprotein receptor-deficient mice. *Arterioscler Thromb Vasc Biol.* 2012;32:1596–1604.
27. Edfeldt K, Swedenborg J, Hansson GK, Yan ZQ. Expression of toll-like receptors in human atherosclerotic lesions: a possible pathway for plaque activation. *Circulation.* 2002;105:1158–1161.
28. Wu GD, Chen J, Hoffmann C, et al. Linking long-term dietary patterns with gut microbial enterotypes. *Science.* 2011;334:105–108.
29. Cenit MC, Sanz Y, Codoñer-Franch P. Influence of gut microbiota on neuropsychiatric disorders. *World J Gastroenterol.* 2017;23:5486–5498.
30. Quigley EMM. Microbiota-brain-gut axis and neurodegenerative diseases. *Curr Neurol Neurosci Rep.* 2017;17:94.
31. Li X, Watanabe K, Kimura I. Gut microbiota dysbiosis drives and implies novel therapeutic strategies for diabetes mellitus and related metabolic diseases. *Front Immunol.* 2017;8:1882.
32. Farooqui AA. *Insulin Resistance as a Risk Factor in Visceral and Neurological Disorders.* San Diego, CA: Academic Press an Imprint of Elsevier; 2020.

33. Backhed F, Ley RE, Sonnenburg JL, Peterson DA, Gordon JI. Host-bacterial mutualism in the human intestine. *Science.* 2005;307:1915–1920.
34. Sonnenburg ED, Sonnenburg JL. Starving our microbial self: the deleterious consequences of a diet deficient in microbiota-accessible carbohydrates. *Cell Metab.* 2014;20:779–786.
35. Koeth RA, Wang Z, Levison BS, et al. Intestinal microbiota metabolism of L-carnitine, a nutrient in red meat, promotes atherosclerosis. *Nat Med.* 2013;19:576–585.
36. Wang Z, Klipfell E, Bennett BJ, et al. Gut flora metabolism of phosphatidylcholine promotes cardiovascular disease. *Nature.* 2011;472:57–63.
37. Jaworska K, Huc T, Samborowska E, et al. Hypertension in rats is associated with an increased permeability of the colon to TMA, a gut bacteria metabolite. *PLoS One.* 2017;12, e0189310.
38. Nicholson JK, Holmes E, Kinross J, et al. Host–gut microbiota metabolic interactions. *Science.* 2012;336:1262–1267.
39. Brown JM, Hazen SL. Microbial modulation of cardiovascular disease. *Nat Rev Microbiol.* 2018;16:171–181.
40. Macfarlane GT, Macfarlane S. Fermentation in the human large intestine: its physiologic consequences and the potential contribution of prebiotics. *J Clin Gastroenterol.* 2011;45(Suppl):S120–S127.
41. Canfora EE, Jocken JW, Blaak EE. Short-chain fatty acids in control of body weight and insulin sensitivity. *Nat Rev Endocrinol.* 2015;11:577–591.
42. Kasubuchi M, Hasegawa S, Hiramatsu T, Ichimura A, Kimura I. Dietary gut microbial metabolites, short-chain fatty acids, and host metabolic regulation. *Nutrients.* 2015;7:2839–2849.
43. Aoyama M, Kotani J, Usami M. Butyrate and propionate induced activated or non-activated neutrophil apoptosis via HDAC inhibitor activity but without activating GPR-41/GPR-43 pathways. *Nutrition.* 2010;26:653–661.
44. Dashwood RH, Myzak MC, Ho E. Dietary HDAC inhibitors: time to rethink weak ligands in cancer chemoprevention? *Carcinogenesis.* 2006;27:344–349.
45. Gao F, Lv YW, Long J, et al. Butyrate improves the metabolic disorder and gut microbiome dysbiosis in mice induced by a high-fat diet. *Front Pharmacol.* 2019;10:1040.
46. De Vadder F, Kovatcheva-Datchary P, Goncalves D, et al. Microbiota-generated metabolites promote metabolic benefits via gut–brain neural circuits. *Cell.* 2014;156:84–96.
47. Donohoe DR, Garge N, Zhang X, et al. The microbiome and butyrate regulate energy metabolism and autophagy in the mammalian colon. *Cell Metab.* 2011;13:517–526.
48. Schiattarella GG, Sannino A, Toscano E, et al. Gut microbe-generated metabolite trimethylamine-N-oxide as cardiovascular risk biomarker: a systematic review and doseresponse metaanalysis. *Eur Heart J.* 2017;38:2948–2956.
49. Trøseid M, Ueland T, Hov JR, et al. Microbiota-dependent metabolite trimethylamine-N-oxide is associated with disease severity and survival of patients with chronic heart failure. *Intern Med.* 2015;277:717–726.
50. Wang Z, Zhao Y. Gut microbiota derived metabolites in cardiovascular health and disease. *Protein Cell.* 2018;9:416–431.
51. Zhu W, Gregory JC, Org E, et al. Gut microbial metabolite TMAO enhances platelet hyperreactivity and thrombosis risk. *Cell.* 2016;165:111–124.
52. Haro C, Montes-Borrego M, Rangel-Zúñiga OA, et al. Two healthy diets modulate gut microbial community improving insulin sensitivity in a human obese population. *J Clin Endocrinol Metab.* 2016;101:233–242.
53. Turnbaugh PJ, Bäckhed F, Fulton L, Gordon JI. Diet-induced obesity is linked to marked but reversible alterations in the mouse distal gut microbiome. *Cell Host Microbe.* 2008;3:213–223.
54. Murphy EF, Cotter PD, Healy S, et al. Composition and energy harvesting capacity of the gut microbiota: relationship to diet, obesity and time in mouse models. *Gut.* 2010;59:1635–1642.

55. Ravussin Y, Koren O, Spor A, et al. Responses of gut microbiota to diet composition and weight loss in lean and obese mice. *Obesity*. 2012;20:738–747.
56. Cani PD, Possemiers S, Van de Wiele T, et al. Changes in gut microbiota control inflammation in obese mice through a mechanism involving GLP-2-driven improvement of gut permeability. *Gut*. 2009;58:1091–1103.
57. Vajro P, Paolella G, Fasano A. Microbiota and gut-liver axis: their influences on obesity and obesity-related liver disease. *J Pediatr Gastroenterol Nutr*. 2013;56:461–468.
58. Annalisa N, Alessio T, Claudette TD, Erald V, Antonino dL, Nicola DD. Gut microbioma population: an indicator really sensible to any change in age, diet, metabolic syndrome, and life-style. *Mediators Inflamm*. 2014;2014:901308.
59. Guillemot-Legris O, Masquelier J, Everard A, Cani PD, Alhouayek M, Muccioli GG. High-fat diet feeding differentially affects the development of inflammation in the central nervous system. *J Neuroinflammation*. 2016;13:206.
60. Minihane AM, Vinoy S, Russell WR, et al. Low-grade inflammation, diet composition and health: current research evidence and its translation. *Br J Nutr*. 2015;114:999–1012.
61. Siri PW, Krauss RM. Influence of dietary carbohydrate and fat on LDL and HDL particle distributions. *Curr Atheroscler Rep*. 2005;7:455–459.
62. Feng R, Du S, Chen Y, et al. High carbohydrate intake from starchy foods is positively associated with metabolic disorders: a cohort study from a Chinese population. *Sci Rep*. 2015;5:16919.
63. Farooqui AA. *Metabolic Syndrome—An Important Risk Factor for Stroke, Alzheimer Disease, and Depression*. New York: Springer; 2013.
64. McNeilly AD, Stewart CA, Sutherland C, Balfour DJ. High fat feeding is associated with stimulation of the hypothalamic-pituitary-adrenal axis and reduced anxiety in the rat. *Psychoneuroendocrinology*. 2015;52:272–280.
65. Sofi F, Cesari F, Abbate R, Gensini GF, Casini A. Adherence to Mediterranean diet and health status: meta-analysis. *BMJ*. 2008;337:a1344.
66. Farooqui AA. *Phytochemical, Signal Transduction, and Neurological Disorders*. New York: Springer; 2012.
67. Sofi F, Macchi C, Abbate R, Gensini GF, Casini A. Mediterranean diet and health. *Biofactors*. 2013;39:335–342.
68. Farooqui AA. *Molecular Mechanisms of Dementia*. San Diego, CA: Academic Press An Imprint of Elsevier; 2019.
69. Haro C, Rangel-Zúñiga OA, Alcalá-Díaz JF, et al. Intestinal microbiota is influenced by gender and body mass index. *PLoS One*. 2016;11(5), e0154090.
70. Haro C, García-Carpintero S, Rangel-Zúñiga OA, et al. Consumption of two healthy dietary patterns restored microbiota dysbiosis in obese patients with metabolic dysfunction. *Mol Nutr Food Res*. 2017;61:1700300.
71. Devkota S, Devkota S, Wang Y, et al. Dietary-fat-induced taurocholic acid promotes pathobiont expansion and colitis in Il10−/− mice. *Nature*. 2012;487:104–108.
72. David LA, Maurice CF, Carmody RN, et al. Diet rapidly and reproducibly alters the human gut microbiome. *Nature*. 2014;505:559–563.
73. Hamaker BR, Tuncil YE. A perspective on the complexity of dietary fiber structures and their potential effect on the gut microbiota. *J Mol Biol*. 2014;426:3838–3850.
74. El Kaoutari A, Armougom F, Gordon JI, Raoult D, Henrissat B. The abundance and variety of carbohydrate-active enzymes in the human gut microbiota. *Nat Rev Microbiol*. 2013;11:497–504.
75. De Filippo C, Cavalieri D, Di Paola M, et al. Impact of diet in shaping gut microbiota revealed by a comparative study in children from Europe and rural Africa. *Proc Natl Acad Sci U S A*. 2010;107:14691–14696.
76. Sonnenburg JL, Sonnenburg ED. Starving our microbial self: the deleterious consequences of a diet deficient in microbiota-accessible carbohydrates. *Science*. 2019;366:eaaw9255.

77. Lopez-Legarrea P, Fuller NR, Zulet MA, Martinez JA, Caterson ID. The influence of Mediterranean, carbohydrate and high protein diets on gut microbiota composition in the treatment of obesity and associated inflammatory state. *Asia Pac J Clin Nutr.* 2014;23:360–368.
78. AlEssa HB, Malik VS, Yuan C, et al. Dietary patterns and cardiometabolic and endocrine plasma biomarkers in US women. *Am J Clin Nutr.* 2017;105:432–441.
79. Mendez MA, Popkin BM, Jakszyn P, et al. Adherence to a Mediterranean diet is associated with reduced 3-year incidence of obesity. *J Nutr.* 2006;136:2934–2938.
80. Larsen N, Vogensen FK, van den Berg FW, et al. Gut microbiota in human adults with type 2 diabetes differs from non-diabetic adults. *PLoS One.* 2010;5:e9085.
81. Turnbaugh PJ, Ley RE, Mahowald MA, Magrini V, Mardis ER, Gordon JI. An obesity-associated gut microbiome with increased capacity for energy harvest. *Nature.* 2006;444:1027–1031.
82. Gutiérrez-Díaz I, Fernández-Navarro T, Sánchez B, Margolles A, González S. Mediterranean diet and faecal microbiota: a transversal study. *Food Funct.* 2016;7:2347–2356.
83. Farooqui AA. *Neurochemical Aspects of Alzheimer Disease: Risk Factors, Pathogenesis, Biomarkers, and Treatment Strategies.* Academic Press; 2017.
84. Tomova A, Bukovsky I, Rembert E, et al. The effects of vegetarian and vegan diets on gut microbiota. *Front Nutr.* 2019;6:47.
85. Lyu H, Xu T, Brotman D, et al. Overtreatment in the United States. *PLoS One.* 2017;12:e0181970.
86. Rizzo NS, Jaceldo-Siegl K, Sabate J, Fraser GE. Nutrient profiles of vegetarian and nonvegetarian dietary patterns. *J Acad Nutr Diet.* 2013;113:1610–1619.
87. Stabler SP. Vitamin B12 deficiency. *N Engl J Med.* 2013;368:149–160.
88. Köbe T, Witte AV, Schnelle A, et al. Vitamin B-12 concentration, memory performance, and hippocampal structure in patients with mild cognitive impairment. *Am J Clin Nutr.* 2016;103:1045–1054.
89. Ganguly P, Alam SF. Role of homocysteine in the development of cardiovascular disease. *Nutr J.* 2015;14:6.
90. McCaddon A, Regland B, Hudson P, Davies G. Functional vitamin B12 deficiency and Alzheimer disease. *Neurology.* 2002;58:1395–1399.
91. Moore E, Mander A, Ames D, Carne R, Sanders K, Watters D. Cognitive impairment and vitamin B12: a review. *Int Psychogeriatr.* 2012;24:541–556.
92. Spence JD. Metabolic vitamin B12 deficiency: a missed opportunity to prevent dementia and stroke. *Nutr Res.* 2016;36:109–116.
93. Clarys P, Deliens T, Huybrechts I, et al. Comparison of nutritional quality of the vegan, vegetarian, semi-vegetarian, pesco-vegetarian and omnivorous diet. *Nutrients.* 2014;6:1318–1332.
94. Farooqui AA, Farooqui T. Effects of mediterranean diet components on neurodegenerative diseases. In: Farooqui T, Farooqui AA, eds. *Role of the Mediterranean Diet in the Brain and Neurodegenerative Diseases.* Academic Press; 2018:1–16.
95. Veech RL, Bradshaw PC, Clarke K, Curtis W, Pawlosky R, King MT. Ketone bodies mimic the life span extending properties of caloric restriction. *IUBMB Life.* 2017;69:305–314.
96. Ma D, Wang AC, Parikh I, et al. Ketogenic diet enhances neurovascular function with altered gut microbiome in young healthy mice. *Sci Rep.* 2018;8:6670.
97. Swidsinski A, Dorffel Y, Loening-Baucke V, et al. Reduced mass and diversity of the colonic microbiome in patients with multiple sclerosis and their improvement with ketogenic diet. *Front Microbiol.* 2017;8:1141.
98. Brinkworth GD, Noakes M, Clifton PM, Bird AR. Comparative effects of very low-carbohydrate, high-fat and high-carbohydrate, low-fat weight-loss diets on bowel habit and faecal short-chain fatty acids and bacterial populations. *Br J Nutr.* 2009;101:1493–1502.

99. Freeman J, Veggiotti P, Lanzi G, et al. The ketogenic diet: from molecular mechanisms to clinical effects. *Epilepsy Res.* 2006;68:145–180.
100. Pierre K, Pellerin L. Monocarboxylate transporters in the central nervous system: distribution, regulation and function. *J Neurochem.* 2005;94:1–14.
101. LaManna JC, Salem N, Puchowicz M, et al. Ketones suppress brain glucose consumption. *Adv Exp Med Biol.* 2009;645:301–306.
102. Whitfield JB. γ Glutamyl transferase. *Crit Rev Clin Lab Sci.* 2001;38:263–355.
103. Pica A, Chi MC, Chen YY, d'Ischia M, Lin LL, Merlino A. The maturation mechanism of γ-glutamyl transpeptidases: insights from the crystal structure of a precursor mimic of the enzyme from bacillus licheniformis and from site-directed mutagenesis studies. *Biochim Biophys Acta.* 2016;1864:195–203.
104. Hertz L. The glutamate-glutamine (GABA) cycle: importance of late postnatal development and potential reciprocal interactions between biosynthesis and degradation. *Front Endocrinol (Lausanne).* 2013;4–59.

CHAPTER 3

Importance of fiber in human diet: Contribution of microbiota in human health

Akhlaq A. Farooqui
Department of Molecular and Cellular Biochemistry, The Ohio State University, Columbus, OH, United States

1 Introduction

Dietary fiber occurs in foods of plant origin. Chemically, dietary fiber mostly comprises carbohydrate polymers (ten or more monomeric units) that resist hydrolysis by digestive enzymes in the mammalian small intestine, but can be fermented by large intestinal microbiota.Bioactive metabolites generated by microbiota produce beneficial physiological effects in humans and rodents.[1,2] Fiber entering the large intestine is hydrolyzed by gut microbiota or may be resistant to fermentation, passing through the digestive tract relatively unchanged. As there are no reports on the deficiency state of fiber, this component of diet is not considered a nutrient. Among Americans, the mean intake of fiber ranged from 11 to 19 g/day for individuals aged two years and older in 2009–10. However, women and men need 25 and 38 g fiber/day, respectively.[3,4]

Dietary fiber is classified on the basis of solubility in water, microbial fermentation in the large intestine, and viscosity. Soluble dietary fiber includes pectin, inulin, psyllium, some gums, and β-glucans, and polysaccharides, whereas insoluble dietary fiber includes cellulose, hemicelluloses, lignin, and resistance starch[1,5,6] (Fig. 1). Soluble fiber is found in cucumbers, blueberries, beans, and nuts. Soluble fiber tends to be fermentable. It can be dissolved into a gel-like texture, helping to slow down digestion. This helps one to feel full for a longer time and is important for controlling weight. Fiber stimulates the liver to produce more bile, thus helping digestion. Insoluble fiber also facilitates digested food to pass through the intestines more quickly, thus contributing to "regularity" and reducing the length of time the gut lining is in contact with any potentially harmful substances contained in the digested food. Soluble dietary fiber not only produces

Gut Microbiota in Neurologic and Visceral Diseases
https://doi.org/10.1016/B978-0-12-821039-0.00010-1

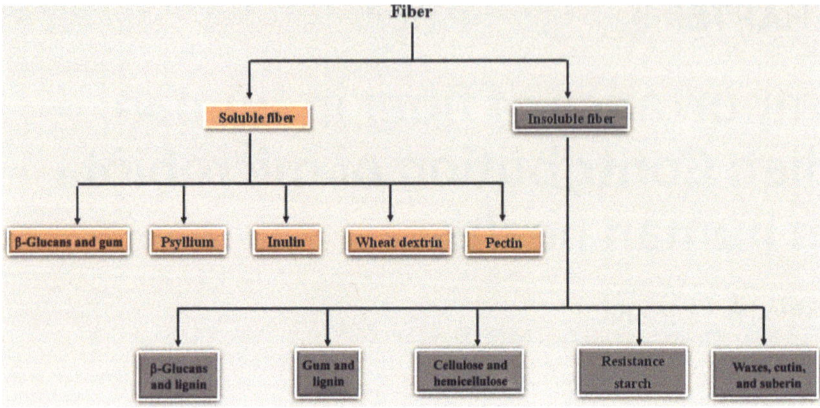

Fig. 1 Sources of soluble and insoluble fibers in the diet.

diarrhea due to increase in osmotic pressure in the colon, but also increases fecal output.[7] Insoluble fiber is found in foods like dark green leafy vegetables, green beans, celery, and carrots. Insoluble fiber does not dissolve at all and helps add bulk to feces. This helps food to move through the gastrointestinal tract more quickly for healthy elimination. Insoluble fiber generally has low-to-moderate fermentability and may cause mucosal damage due to mechanical stimulation of the intestinal mucosa. A severe dietary problem with dietary fiber is that it interferes with mineral absorption because of its binding or bulking characteristics.[8] Insoluble dietary fiber contains a relatively high level of phytate, which is known to bind calcium and magnesium, making them insoluble in the gastrointestinal tract. Thus, level of phytate in fiber may be a key regulatory factor in mineral absorption, as well as the diluting effect on gastrointestinal tract content due to fiber's bulking characteristics.[7,9]

2 Beneficial effects of dietary fiber consumption on human health

It is well known that consumption of fiber retards excess dietary energy intake by a number of mechanisms including: (1) displacement of energy-dense foods from the diet; (2) inhibition of absorption of macronutrients at the intestinal surface; (3) an altered luminal environment; and (4) an increased gastric distension.[10] In general, soluble fiber is completely fermented and has a greater viscosity than insoluble fibers. Prebiotic dietary fiber is not classified in terms of solubility or viscosity, but is defined by the resistance to digestion and absorption in the small intestine, partial or complete fermentation by microbiota in the large intestine, and the ability to stimulate growth of selective microbiota. Only two such dietary fibers fit all of these criteria. They are considered as prebiotics: inulin and trans-galacto-oligosaccharides. Fiber not only helps in providing a feeling of fullness, but also promotes digestive health and laxation. Increasing total dietary fiber has been shown to not only reduce body fat, improve glycemic response, and reduce blood pressure, but also to lower triglycerides and low-density lipoprotein cholesterol (LDL-C).[11–15]

Dietary fiber not only exerts a physiological influence throughout the digestive tract by structuring digesta (relevant to satiety and control of food intake) and modulating digestion processes (relevant to control of circulating glucose and lipid levels), but it also acts as a prime substrate for microbial fermentation (relevant to control of laxation and reduced risk of colon cancer among other chronic diseases). Foods that are rich in dietary fiber release glucose more slowly into the blood, which is not only relevant to the prevention of visceral disorders (obesity, type 2 diabetes, metabolic syndrome, and intestinal and rectal cancers) but also neurological disorders (stroke, Alzheimer's disease (AD), and Parkinson's disease (PD)).[16–18] Dietary fiber increases the gastrointestinal tract biomass, changes the composition of gut flora, and may decrease the risk of metabolic disorders like dyslipidemia, hypercholesterolemia, and hyperglycemia, as well as substantially influence immune-based actions. The consumption of dietary fiber is also known to decrease the oxidation of lipids, which in turn is associated with favorable effect on blood pressure (BP), serum lipid profile, insulin sensitivity, and decrease in inflammation.[19–22]

Dietary fiber produces antiinflammatory effects via the decrease in the production of proinflammatory cytokines (IL-6 and TNFα), inhibition of proinflammatory enzymes (cyclooxygenase 2, COX-2), and induction of nitric oxide synthase (iNOS) gene expression.[23] The molecular mechanisms associated with beneficial effects of dietary fiber on metabolic health are not well established. However, it is proposed that fiber mediated changes in intestinal viscosity, nutrient absorption, rate of passage, production of short-chain fatty acids (SCFAs), and production of gut hormones are linked not only with decrease in body weight, insulin resistance, and type 2 diabetes but also with marked reduction in cardiovascular disease. Dietary fiber produces satiation, a process, defined as the feeling of fullness during the consumption of a meal, which leads to meal termination.[24] Satiation is controlled by numerous factors, including macronutrient composition, digestion, gastric emptying, and nutrient absorption, which together influence postprandial satiety responses. Collective evidence suggests that a major function of satiation is to retard overconsumption during individual meals, thereby averting deleterious consequences from incomplete digestion as well as excessive disturbances in circulating levels of glucose and other nutrients. Signals for satiation arise from multiple sites in the gastrointestinal system, including the stomach, proximal small intestine, distal small intestine, colon, and pancreas.[24] Satiation is modulated by number of hormones. The glucagon-like peptides GLP-1 and GLP-2 are produced in enteroendocrine L cells of the small and large intestine and secreted in a nutrient-dependent manner. GLP-1 regulates nutrient assimilation via inhibition of gastric emptying and food intake. GLP-1 controls blood glucose following nutrient absorption via stimulation of glucose-dependent insulin secretion, insulin biosynthesis, islet proliferation, neogenesis, and inhibition of glucagon secretion. GLP-1 is an insulinotropic hormone. GLP-1 inhibits glucagon secretion and lowers blood glucose in normal subjects and in patients with type 2 diabetes. Collective evidence suggests that GLP-1 is secreted by the L cells of the ileum and proximal colon in a nutrient-dependent manner, enhances satiety, and reduces food intake when administered to normal subjects.[25] Leptin acts as an afferent signal in a negative-feedback loop by inhibiting appetite and regulating adiposity.[26] Leptin also increases insulin sensitivity in normal and diabetic murine models.[27] Studies on feeding animals fermentable fiber (such as inulin, lactitol, resistant starch, and fructooligosaccharides) indicate that consumption of fiber not only results in an increase in concentration of plasma GLP-1, proximal colon GLP-1, and plasma PYY, but also in less weight gain and a decrease in levels of ghrelin and leptin.[28, 29] Converging evidence suggests that dietary fiber not only improves constipation, reduces serum cholesterol levels, and excretion of carcinogens into feces but also increases satiety, reduces energy intake, and promotes longevity (Fig. 2).

3 Gut microbiota and human health

The human gut microbiota community has been classified into at least seven phyla, of which four are the most predominant (usually ~98% of the total population). These are the Firmicutes (58%–88%), Bacteroidetes (8.5%–28%), Proteobacteria (0.1%–8%), and Actinobacteria (2.5%–5%).[30–32] Microbiota community profiling from feces has shown that as many as 60% of bacterial species have not yet been identified.[33] For this reason, many studies report at higher taxonomy than the level of species, due to the lack of representative sequences in prokaryote databases.[34, 35] Overall,

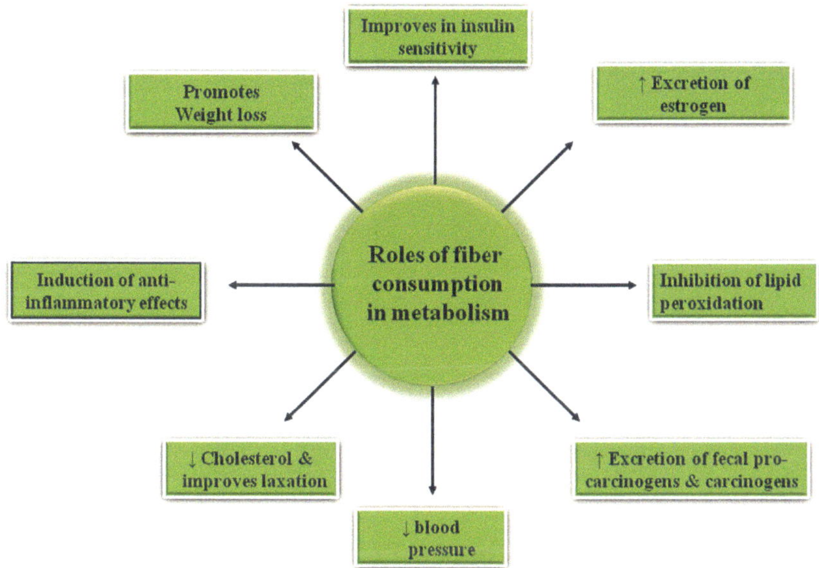

Fig. 2 Beneficial effects of dietary fibers on metabolic processes.

the predominant bacterial groups in the microbiome are Gram-positive Firmicutes and Gram-negative Bacteroidetes.[36] Recently, it has been shown that microbiota can effectively be subdivided into different enterotypes, each enriched by particular bacterial genera, but that all seem to share high functional uniformity (Arumugam). The ratio between Bacteroidetes and Firmicutes phylum remains remarkably stable over time.[37]

Environmental factors, diet, medications, age, geographic factors, surgical interventions, and genes related to the host immune system and metabolism modulate the population of gut microbiota.[38–40] Gut microbiota composition exerts protective and neurotoxic effects not only by compromising systemic physiology (by modulating the multidirectional pathways) but also by altering the homeostasis of indigenous gut microbiota-derived metabolites.[41, 42] Elevation or reduction in microbiota composition result in dysbiosis, a process that contributes to the release of lipopolysaccharide (LPS) from Gram-negative bacteria. This immunogenic endotoxin produces chronic immune responses associated with inflammation.

4 Effects of gut microbiota-derived metabolite in host tissues

Fermentation of fiber (undigested carbohydrates and proteins) by gut microbiota produces a wide range of bioactive molecules. Major

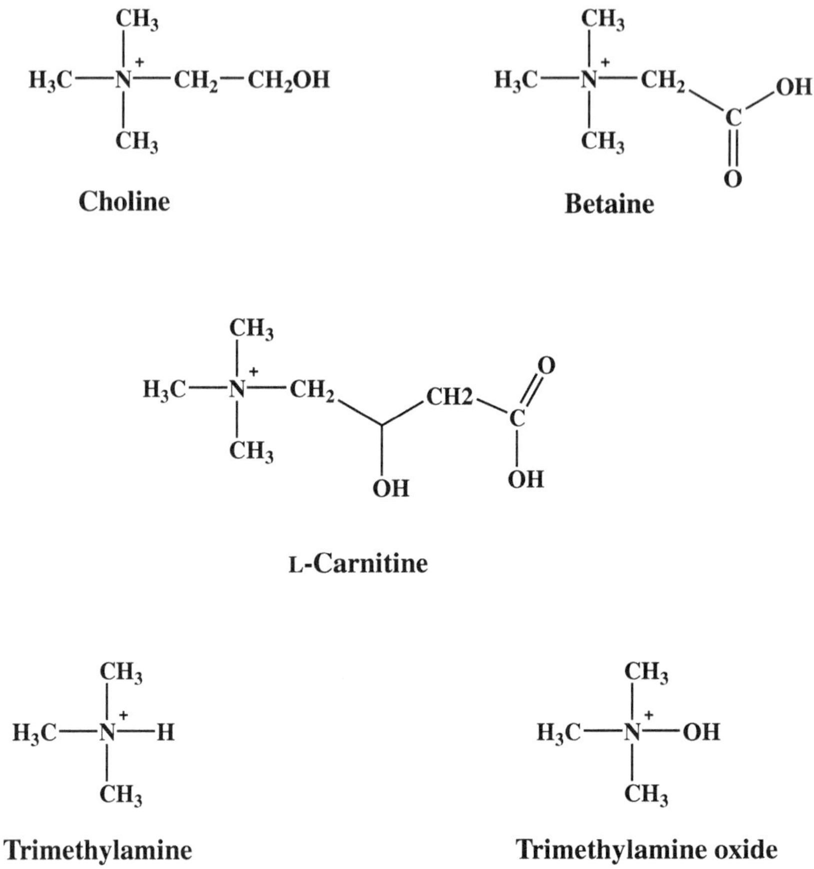

Fig. 3 Chemical structures of choline, betaine, carnitine, trimethylamine, and trimethylamine oxide.

microbiota-derived bioactive molecules include SCFAs (acetic, propionic, and butyric acids), choline, phosphatidylcholine, L-carnitine, betaine, and trimethylamine (TMA) (Fig. 3), while minor bioactive molecules are tryptophan, indole, hydrogen peroxide, γ-butyrobetaine, and crotono-betaine. These bioactive products enter the circulation and modulate circulatory system homeostasis.[43, 44] In addition, fermentation also generates gases (H_2, CO_2, H_2S, and methane).[45] To enter the circulation, gut microbiota-derived bioactive molecules pass through the intestinal barrier. The integrity and permeability of the intestinal barrier depend on numerous factors, including intestinal blood flow.[46] These microbiota-derived bioactive molecules not only play crucial roles in host metabolic homeostasis, immunological processes, and neurobiology, but also contribute to human visceral (insulin resistance, type 2 diabetes, metabolic syndrome, and heart disease) and neurological disorders (PD, AD, multiple sclerosis, stroke, and autism).[47,48] The molecular mechanisms associated with the induction of obesity, insulin resistance, type 2 diabetes, metabolic syndrome, and heart disease are regulated by host food intake behavior, energy harvest, energy expenditure, and fat storage. In contrast, gut microbiota communicates with the central nervous system (CNS) not only through the gut-brain axis and vagus nerve, but also through the immune system, neuroendocrine pathways, and microbiota-derived metabolites.[16, 18, 49] This bidirectional communication not only modulates gut functions (nausea, satiety, and pain) but also neurotransmission. Alterations in neurotransmission are related to the pathogenesis of neurological and neuropsychiatric disorders.[16, 18, 49]

5 Role of microbiota-derived short-chain fatty acids

Microbial fermentation of fiber in the large intestine by microbiota results in the synthesis of SCFAs, including acetic, propionic, and butyric acids, according to the following equations:

$$59\ C_6H_{12}O_6 + 38\ H_2O \rightarrow 60\ \text{Acetate} + 22\ \text{Propionate} + 18\ \text{Butyrate} + 96\ CO_2 + 256\ H^+$$

The majority of SCFAs (up to 95%) are rapidly absorbed by the colonocytes, resulting in decreasing concentrations from the proximal to distal colon and small intestine.[50, 51] Only a minor fraction of SCFAs (about 5%) are excreted in feces. Among SCFAs, acetate is the most abundant found in the circulation and has been shown to cross the blood–brain barrier.[52, 53] SCFAs play distinct physiological roles in host metabolism. They not only shape the gut environment and influence the physiology of the colon, but they also can be used as energy sources by host cells.[54] This production of SCFAs by microbiota reduces circulating cholesterol levels and produces anticarcinogenic effects mediated by interference from numerous regulators of the cell cycle such as β-catenin, p53, p21, Bax, and caspase 3 genes in chronic inflammation.[55, 56] SCFAs are regarded as mediators in the communication between the intestinal microbiome and the immune system. The signal they produce is transferred, among others, in immune cells via free fatty acid receptors (FFARs), which belong to the family of G protein-coupled receptors (GPCRs). SCFAs produce both obesogenic and antiobesogenic effects by regulating glucose, lipid metabolism, and insulin sensitivity.[57] SCFA receptors are also expressed at the mRNA level in various insulin-sensitive tissues such as the adipose tissue,[58] skeletal muscle, liver,[59] and pancreatic beta cells,[60] which illustrates their broad metabolic role. SCFAs (butyrate and propionate) are known to regulate gene expression not only by inhibiting histone deacetylase (HDAC)[61] but also by inducing conformational changes in the active site of HDAC.

Among SCFAs, acetate is the most abundant in circulation and has been shown to cross the blood-brain barrier.[52, 53] Acetate affects host energy and substrate metabolism via secretion of the gut hormones like glucagon-like peptide-1 and peptide YY, thereby affecting appetite via a reduction in whole-body lipolysis, systemic pro-inflammatory cytokine levels, and an increase in energy expenditure and fat oxidation. Acetate can also be converted to acetyl-CoA and incorporated into the tricarboxylic acid (TCA) cycle in various peripheral tissues.[62] Acetate may also increase fatty acid synthesis through epigenetic mechanisms such as histone acetylation.[63] Additionally, acetate may activate other receptors important for blood pressure regulation including olfactory receptors 51E2 (Olfr51E2) and 78 (Olfr78) in renal tissue.[64] Propionate is utilized locally through conversion into glucose by intestinal gluconeogenesis[65] or diffuses into the portal vein to be utilized as a substrate for hepatic gluconeogenesis.[66]

At the molecular level, SCFAs act by increasing the levels of phosphorylated-adenosine 5′-monophosphate-activated protein kinase (p-AMPK) and facilitating expression of glucose transporter member 4 (GLUT4) in the adipose tissue. This observation suggests that oral supplements of butyrate may decrease or delay the onset of cardiovascular disease.[67] Collectively, these studies suggest that fiber-derived acetate, propionate, and butyrate are important metabolites that not only contribute to energy production by gut microbiota but also are associated with intestinal and liver gluconeogenesis, de novo lipid synthesis, and protein synthesis through distinct mechanisms.[65, 68] In addition, in the intestinal tract acetate can also be used as a precursor for cholesterol synthesis.

6 Role of microbiota-derived trimethylamine and trimethylamine-*N*-oxide

Phosphatidylcholine (PtdCho) is a phospholipid integral to cell membranes that is present in higher-fat foods. Gut microbiota releases choline from dietary PtdCho where it is then metabolized to trimethylamine (TMA). TMA is transported to the liver via the portal vein where it is oxidized by flavin monooxygenase-3 to trimethylamine-*N*-oxide (TMAO)[43, 44, 69] (Fig. 4).

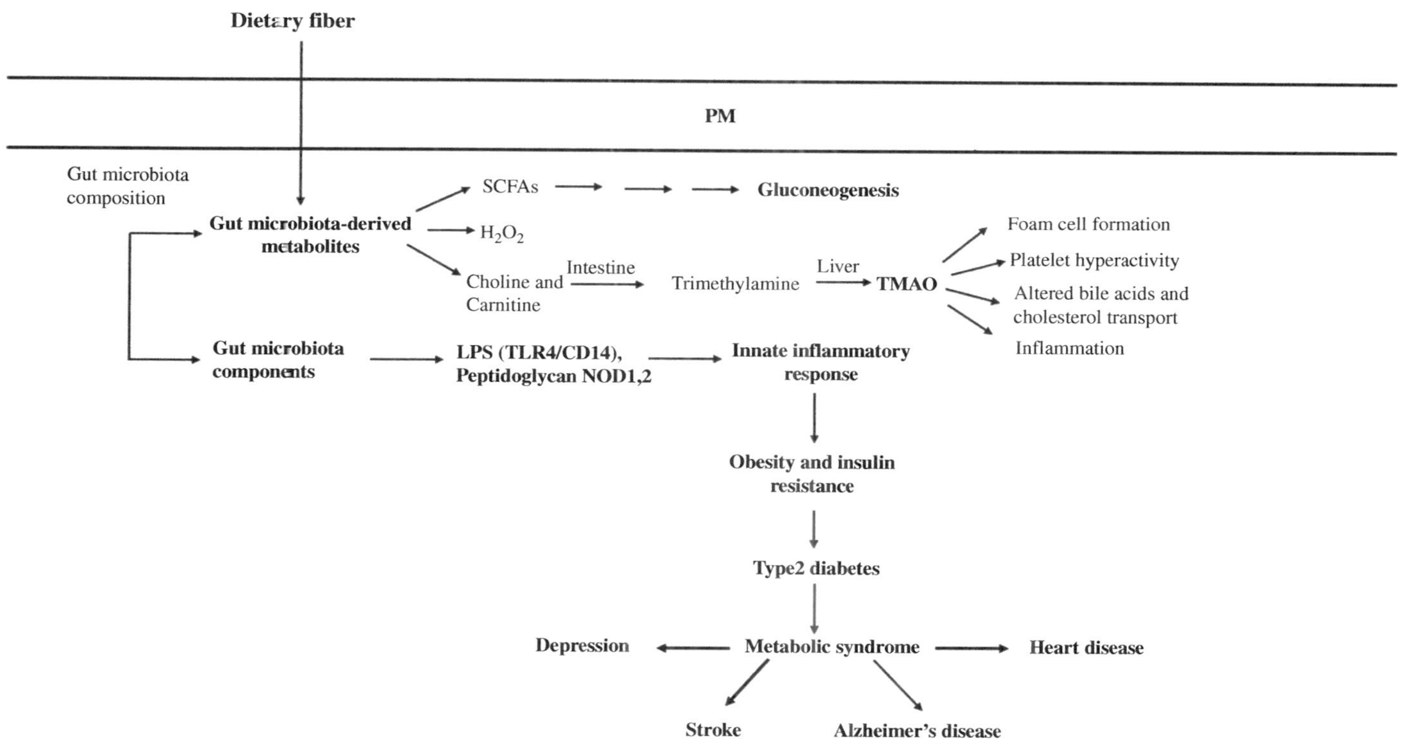

Fig. 4 Effect of fiber and microbiota-derived metabolites on human health. *LPS*, lipopolysaccharides; *NLRs*, NOD-like receptors, *NOD1* and *NOD2* (intracellular pattern recognition receptors); *PM*, plasma membrane; *TLR4*, toll-like receptor4; *TMAO*, trimethylamine oxide.

It is reported that increased levels of plasma TMAO, choline, and betaine have dose-dependent relationships with the presence of cardiovascular disease in a cohort of 1876 men and women, after controlling for established risk factors and medication use.[43, 44] This evidence suggests that TMAO induces pro-atherogenic effects.[43, 44] While the microbiota is necessary for methylamine production (GF mice do not excrete TMA), dietary interventions can modulate the metabolic outcomes.[70] Elevated plasma TMAO also is a signature biomarker of a high-meat diet as compared to low-meat and vegetarian diets in humans.[71] Circulating TMAO levels are associated with prevalence of cardiovascular disease and can independently predict incident risk for major adverse cardiac events, including myocardial infarction, stroke, or death after adjustment for traditional cardiac risk factors and renal function.[44, 72] Molecular mechanisms through which TMAOs promote atherosclerosis and thrombosis have been studied at the molecular level. In vascular smooth muscle cells and endothelial cells, TMAO activates MAPK and nuclear factor-κB (NF-κB) signaling, leading to inflammatory gene expression and endothelial cell adhesion of leukocytes.[73] Meanwhile, TMAO can also activate the NLRP3 inflammasome.[74–76] In vivo, TMAOs have been reported to increase scavenger receptor, CD36, and SR-A1 expression, leading to more uptake of modified LDL for macrophages to form foam cells.[44] Alternatively, TMAO decreases expression of two key enzymes, CYP7A1 and CYP27A1, essential for bile acid biosynthesis and multiple bile acid transporters (OATP1, OATP4, MRP2, and NTCP) in the liver, which decreases bile acid pool and results in decreased reverse cholesterol efflux.[43] Moreover, TMAO increases endoplasmic reticulum calcium release in platelet cells, leading to platelet aggregation and thrombosis.[77] Decreasing the consumption of TMA precursors in the diet and inhibiting TMAO production protect from harmful effects of microbiota. Recently, metabolism of TMAO has been linked to insulin sensitivity and glucose metabolism.[78] High levels of TMAO induce vascular inflammation and prothrombosis by exaggerating platelet hyperresponsiveness to multiple agonists, and can be implicated in the pathogenesis of obesity and type 2 diabetes.[77] Accumulating evidence suggests that the intestinal microbiota plays an important role in the pathophysiology and outcome of stroke. Ischemic stroke alters the composition of the intestinal microbiota. Conversely, the intestinal microbiota can modulate the outcome of stroke and play a role in the pathogenesis of stroke. Clinical and experimental studies indicate that the gut microbiota is associated with risk factors for stroke such as hypertension, diabetes, and obesity.[79] A recent study has also indicated that cerebrospinal fluid TMAO levels are increased in AD patients, and this increase in TMAO levels is associated with increase in AD pathology and neurodegeneration (Fig. 4). TMAO levels may play a role in AD pathology by promoting cerebrovascular disease, as vascular risk factors are common in AD patients, and cerebrovascular pathology frequently coexists with AD pathology.[80]

7 Role of tryptophan and its metabolites in metabolism

The phenolic compounds phenol, p-cresol, and indole are the major metabolites of bacterial fermentation of the aromatic amino acids (tyrosine, phenylalanine, and tryptophan). These metabolites are largely and rapidly absorbed by the colonic mucosa cells and are excreted in urine after sulfate or glucuronide conjugation in the mucosa or the liver.[81] In healthy subjects, these compounds do not accumulate in the body. Therefore, their urinary elimination is often considered a reliable estimate of their production in the colon.[82] In mammalian cells, tryptophan is primarily catabolized (95%) through the kynurenine pathway. About 5% of tryptophan is used by microbiota through the indole pathway and a small fraction of tryptophan is converted to serotonin and melatonin via the serotonin pathway, mainly in the gastrointestinal tract (Fig. 5). It is becoming increasingly evident that changes in the microbiota modulate the host immune system by modulating tryptophan metabolism. Moreover, tryptophan and endogenous metabolites of tryptophan (kynurenines, serotonin, and melatonin) and bacterial tryptophan metabolites (indole, indolic acid, skatole, and tryptamine) mediate profound effects on gut microbial composition, microbial metabolism, the host's immune system, the host-microbiome interface, and host immune system-intestinal microbiota interactions.[67] It is proposed that tryptophan metabolites and bacterial metabolites, including indole, indole propionic acid, indole acetic acid, skatole, and tryptamine, interact with the aryl hydrocarbon receptor (AhR) to produce intestinal immunity and immune homeostasis.[83] Additional factors such as aging, stress, probiotics, and diseases (spondyloarthritis, irritable bowel syndrome, inflammatory bowel disease, colorectal cancer), which are associated with variability in tryptophan metabolism, are also modulated by tryptophan—microbiome—immune system interactions in the gut.

8 Conclusion

Dietary fiber is the nondigestible form of carbohydrates and lignin. Dietary fiber occurs in plants and is fermented by microbiota in the rectum.

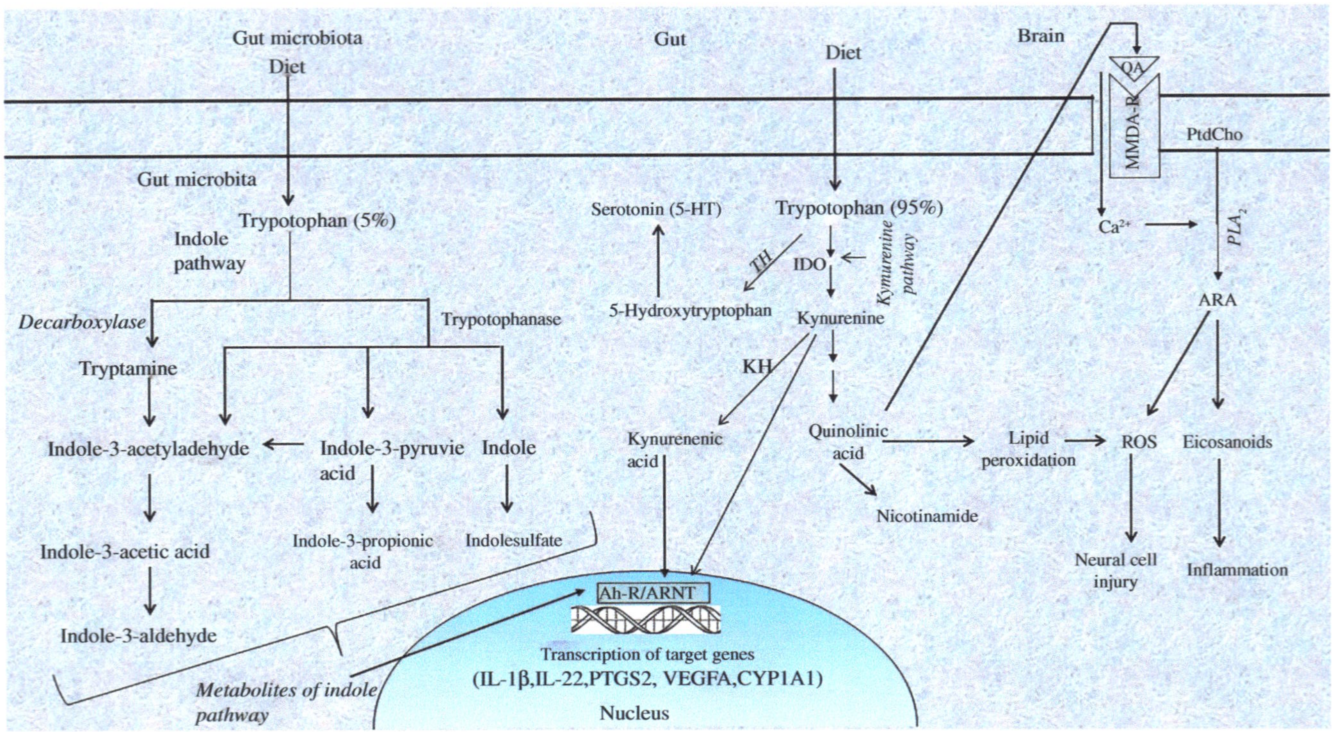

Fig. 5 Metabolism of tryptophan by kynurenine and indole pathways. *AhR*, aryl hydrocarbon receptor; *ARA*, arachidonic acid; *ARNT*, binding partner of AhR; *cPLA2*, cytosolic phospholipase A_2; *IDO*, indolamine 2,3-dioxygenase; *IL-22*, interleukin-22; *IL-6*, interleukin-6; *KMO*, kynurenine 3-monooxygenase; *NMDA-R*, *N*-methyl-D-aspartate receptor; *PtdCho*, phosphatidylcholine; *Ptgs2*, prostaglandin-endoperoxide synthase 2; *ROS*, reactive oxygen species; *TH*, tryptophan hydroxylase; *VEGF*, vascular endothelial growth factor. Quinolenic acid (QA) not only generates ROS but also stimulates NMDA receptors and promotes the generation of eicosanoids and platelet-activating factor. These products induce oxidative stress and facilitate neuroinflammation.

Consumption of a high-fiber diet produces beneficial effects on human health through many potential mechanisms. Soluble, viscous fiber modulates the absorption of food from the small intestine because of the formation of gels that attenuate postprandial blood glucose and lipid rises. The formation of gels also slows gastric emptying, maintaining levels of satiety and contributing towards less weight gain. Dietary fiber also increases fecal bulking and viscosity and there is less contact time between potential carcinogens and mucosal cells. In addition, dietary fiber increases the binding between bile acids and carcinogens. Consumption of dietary fiber not only increases the amount of estrogen excreted in the feces due to an inhibition of estrogen absorption in the intestine, but also promotes healthy lipid profiles, glucose tolerance, and ensures normal gastrointestinal function. Microbiota produce SCFAs (acetate, propionate, and butyrate) via fermentation of fiber. SCFAs produce beneficial effects on host health. Butyrate is the major energy source for colonocytes. Propionate is largely taken up by the liver. Acetate enters the peripheral circulation to be metabolized by peripheral tissues. Propionate, a gluconeogenerator, inhibits cholesterol synthesis. Collective evidence suggests that dietary fiber and specific types of microbiota act in several different ways: (1) they improve laxation by increasing bulk and reducing transit time of feces through the bowel; (2) they increase excretion of bile acid, estrogen, and fecal procarcinogens and carcinogens by interacting with them; (3) they lower serum cholesterol; (4) they slow glucose absorption and improve insulin sensitivity; (5) they lower blood pressure; (6) they promote weight loss; and (7) they produce anti-inflammatory effects.[84]

References

1. Papathanasopoulos A, Camilleri M. Dietary fiber supplements: effects in obesity and metabolic syndrome and relationship to gastrointestinal functions. *Gastroenterology.* 2010;138:65–72. e1–2.
2. Raninen K, Lappi J, Mykkänen H, Poutanen K. Dietary fiber type reflects physiological functionality: comparison of grain fiber, inulin, and polydextrose. *Nutr Rev.* 2011;69:9–21.
3. Institute of Medicine (U.S.) Panel on Macronutrients. Institute of Medicine (U.S.). Standing committee on the scientific evaluation of dietary reference intakes. In: *Dietary Reference Intakes for Energy, Carbohydrate, Fiber, Fat, Fatty Acids, Cholesterol, Protein, and Amino Acids.* Washington, DC: National Academies Press; 2005:339.
4. U.S. Department of Agriculture, Agricultural Research Service. *Nutrient Intakes from Food: Mean Amounts Consumed per Individual, by Gender and Age, What We Eat in America, NHANES 2009–2010*; 2012. Available online http://www.ars.usda.gov/ba/bhnrc/fsrg. Accessed 15 February 2014.
5. Slavin J. Fiber and prebiotics: mechanisms and health benefits. *Nutrients.* 2013;5:1417–1435.

6. Williams BA, Mikkelsen D, Flanagan BM, Gidley MJ. "Dietary fibre": moving beyond the "soluble/insoluble" classification for monogastric nutrition, with an emphasis on humans and pigs. *J Anim Sci Biotechnol.* 2019;10:45.
7. Ishikawa F, Takayama H, Matsumoto K. Effect of β 1-4 linked galactooligosaccharides on human fecal microflora. *Bifidus.* 1995;9:5–18.
8. Cassidy MM, Fitzpatrick LR, Vahouny GV. The effect of fiber in the postweaning diet on nutritional and intestinal morphological indices in the rats. In: Vahouny CV, Kritchevsky D, eds. *Dietary Fiber, Basic and Clinical Aspects.* New York: Plenum Press; 1986:229–252.
9. McGuire S. U.S. department of agriculture and U.S. department of health and human services, dietary guidelines for Americans, 2010. *Adv Nutr.* 2011;2:293–294.
10. Heaton KW. Food fibre as an obstacle to energy intake. *Lancet.* 1973;ii:1418–1421.
11. Brown L, Rosner B, Willett WW, Sacks FM. Cholesterol-lowering effects of dietary fiber: a meta-analysis. *Am J Clin Nutr.* 1999;69:30–42.
12. Chandalia M, Garg A, Lutjohann D, von Bergmann K, Grundy SM, Brinkley LJ. Beneficial effects of high dietary fiber intake in patients with type 2 diabetes mellitus. *N Engl J Med.* 2000;342:1392–1398.
13. Evans CEL, Greenwood DC, Threapleton DE, et al. Effects of dietary fibre type on blood pressure: a systematic review and meta-analysis of randomized controlled trials of healthy individuals. *J Hypertens.* 2015;33:897–911.
14. Jenkins DJA, Kendall CWC, Vuksan V, et al. Soluble fiber intake at a dose approved by the US Food and Drug Administration for a claim of health benefits: serum lipid risk factors for cardiovascular disease assessed in a randomized controlled crossover trial. *Am J Clin Nutr.* 2002;75:834–839.
15. Tucker LA, Thomas KS. Increasing total fiber intake reduces risk of weight and fat gains in women. *J Nutr.* 2009;139:576–581.
16. Kobayashi Y, Sugahara H, Shimada K, et al. Therapeutic potential of bifidobacterium breve strain A1 for preventing cognitive impairment in Alzheimer's disease. *Sci Rep.* 2017;7, 13510.
17. Nugent AP. Health properties of resistant starch. *Nutr Bull.* 2005;30:27–54.
18. Quigley EMM. Microbiota-brain-gut axis and neurodegenerative diseases. *Curr Neurol Neurosci Rep.* 2017;17:94.
19. Ajani UA, Ford ES, Mokdad AH. Dietary fiber and C-reactive protein: findings from national health and nutrition examination survey data. *J Nutr.* 2004;134:1181–1185.
20. Lattimer JM, Haub MD. Effects of dietary fiber and its components on metabolic health. *Nutrients.* 2010;2:1266–1289.
21. Ma Y, Griffith J, Chasan-Tuber L. Association between dietary fiber and serum C-reactive protein. *Am J Clin Nutr.* 2007;83:760–766.
22. Turner ND, Lupton JR. Dietary fiber. *Adv Nutr.* 2011;2:151–152.
23. Slavin JL. Dietary fiber and body weight. *Nutrition.* 2005;21:411–418.
24. Schroeder N, Marquart LF, Gallaher DD. The role of viscosity and fermentability of dietary fibers on satiety- and adiposity-related hormones in rats. *Nutrients.* 2013;5:2093–2113.
25. Bosch G, Verbrugghe A, Hesta M, et al. The effects of dietary fibre type on satiety-related hormones and voluntary food intake in dogs. *Br J Nutr.* 2009;102:318–325.
26. Frühbeck G, Gómez-Ambrosi J. Control of body weight: a physiologic and transgenic perspective. *Diabetologia.* 2003;46:143–172.
27. Frühbeck G, Salvador J. Relation between leptin and the regulation of glucose metabolism. *Diabetologia.* 2000;43:3–12.
28. Brubaker PL. The glucagon-like peptides: pleiotropic regulators of nutrient homeostasis. *Ann NY Acad Sci.* 2006;1070:10–26.
29. Velázquez M, Davies C, Marett R, Slavin JL, Feirtag JM. Effect of oligosaccharides and fibre substitutes on short-chain fatty acid production by human faecal microflora. *Anaerobe.* 2006;6:87–92.

30. Donaldson GP, Lee SM, Mazmanian SK. Gut biogeography of the bacterial microbiota. *Nat Rev Microbiol.* 2016;14:20–32.
31. Peterson DA, Frank DN, Pace NR, Gordon JI. Metagenomic approaches for defining the pathogenesis of inflammatory bowel diseases. *Cell Host Microbe.* 2008;3:417–427.
32. Wu GD, Bushmanc FD, Lewis JD. Diet, the human gut microbiota, and IBD. *Anaerobe.* 2013;24:117–120.
33. Qin J, Li R, Raes J, et al. A human gut microbial gene catalogue established by metagenomic sequencing. *Nature.* 2010;464:59–65.
34. Gevers D, Cohan FM, Lawrence JG, et al. Opinion: re-evaluating prokaryotic species. *Nat Rev Microbiol.* 2005;3:733–739.
35. Sunagawa S, Mende DR, Zeller G, et al. Metagenomic species profiling using universal phylogenetic marker genes. *Nat Methods.* 2013;10:1196–1199.
36. Walker AW, Ince J, Duncan SH, et al. Dominant and diet-responsive groups of bacteria within the human colonic microbiota. *ISME J.* 2011;5:220–230.
37. Faith JJ, Guruge JL, Charbonneau M, et al. The long-term stability of the human gut microbiota. *Science.* 2013;341, 1237439.
38. Candela M, Biagi E, Brigidi P, O'Toole PW, De Vos WM. Maintenance of a healthy trajectory of the intestinal microbiome during aging: a dietary approach. *Mech Ageing Dev.* 2014;136:70–75.
39. Gensollen T, Iyer SS, Kasper DL, Blumberg RS. How colonization by microbiota in early life shapes the immune system. *Science.* 2016;352:539–544.
40. Ottman N, Smidt H, de Vos WM, Belzer C. The function of our microbiota: who is out there and what do they do? *Front Cell Infect Microbiol.* 2012;2:104.
41. Chwalba A, Otto-Buczkowska E. Participation of the microbiome in the pathogenesis of diabetes mellitus. *Clin Diabetology.* 2017;6:178–181.
42. Miele L, Giorgio V, Alberelli MA, De Candia E, Gasbarrini A, Grieco A. Impact of gut microbiota on obesity, diabetes, and cardiovascular disease risk. *Curr Cardiol Rep.* 2015;17:120.
43. Koeth RA, Wang Z, Levison BS, et al. Intestinal microbiota metabolism of L-carnitine, a nutrient in red meat, promotes atherosclerosis. *Nat Med.* 2013;19:576–585.
44. Wang Z, Klipfell E, Bennett BJ, et al. Gut flora metabolism of phosphatidylcholine promotes cardiovascular disease. *Nature.* 2011;472:57–63.
45. Holscher HD. Dietary fiber and prebiotics and the gastrointestinal microbiota. *Gut Microbes.* 2017;8:172–184.
46. Jaworska K, Huc T, Samborowska E, et al. Hypertension in rats is associated with an increased permeability of the colon to TMA, a gut bacteria metabolite. *PLoS One.* 2017;12, e0189310.
47. Brown JM, Hazen SL. Microbial modulation of cardiovascular disease. *Nat Rev Microbiol.* 2018;16:171–181.
48. Nicholson JK, Holmes E, Kinross J, et al. Host–gut microbiota metabolic interactions. *Science.* 2012;336:1262–1267.
49. Cenit MC, Sanz Y, Codoñer-Franch P. Influence of gut microbiota on neuropsychiatric disorders. *World J Gastroenterol.* 2017;23:5486–5498.
50. Karaki S, Tazoe H, Hayashi H, et al. Expression of the short-chain fatty acid receptor, GPR43, in the human colon. *J Mol Histol.* 2008;39:135–142.
51. Nohr MK, Pedersen MH, Gille A, et al. GPR41/FFAR3 and GPR43/FFAR2 as cosensors for short-chain fatty acids in enteroendocrine cells vs. FFAR3 in enteric neurons and FFAR2 in enteric leukocytes. *Endocrinology.* 2013;154:3552–3564.
52. Frost G, Sleeth ML, Sahuri-Arisoylu M, et al. The short-chain fatty acid acetate reduces appetite via a central homeostatic mechanism. *Nat Commun.* 2014;5:3611.
53. Perry RJ, Peng L, Barry NA, et al. Acetate mediates a microbiome–brain–β- cell axis to promote metabolic syndrome. *Nature.* 2016;534:213–217.

54. Macfarlane GT, Macfarlane S. Fermentation in the human large intestine: its physiologic consequences and the potential contribution of prebiotics. *J Clin Gastroenterol.* 2011;45(Suppl):S120–S127.
55. den Besten G, van Eunen K, Groen AK, Venema K, Reijngoud DJ, Bakker BM. The role of short-chain fatty acids in the interplay between diet, gut microbiota, and host energy metabolism. *J Lipid Res.* 2013;54:2325–2340.
56. Young GP, Hu Y, Le Leu RK, Nyskohus L. Dietary fibre and colorectal cancer: a model for environment—gene interactions. *Mol Nutr Food Res.* 2005;49:571–584.
57. Canfora EE, Jocken JW, Blaak EE. Short-chain fatty acids in control of body weight and insulin sensitivity. *Nat Rev Endocrinol.* 2015;11:577–591.
58. Le Poul E, Loison C, Struyf S, et al. Functional characterization of human receptors for short chain fatty acids and their role in polymorphonuclear cell activation. *J Biol Chem.* 2003;278:25481–25489.
59. Brown AJ, Goldsworthy SM, Barnes AA, et al. The orphan G protein-coupled receptors GPR41 and GPR43 are activated by propionate and other short chain carboxylic acids. *J Biol Chem.* 2003;278:11312–11319.
60. Tang C, Ahmed K, Gille A, et al. Loss of FFA2 and FFA3 increases insulin secretion and improves glucose tolerance in type 2 diabetes. *Nat Med.* 2015;21:173–177.
61. Kasubuchi M, Hasegawa S, Hiramatsu T, Ichimura A, Kimura I. Dietary gut microbial metabolites, short-chain fatty acids, and host metabolic regulation. *Nutrients.* 2015;7:2839–2849.
62. Mittendorfer B, Sidossis LS, Walser E, Chinkes DL, Wolfe RR. Regional acetate kinetics and oxidation in human volunteers. *Am J Physiol.* 1998;274:E978–E983.
63. Gao X, Lin S-H, Ren F, et al. Acetate functions as an epigenetic metabolite to promote lipid synthesis under hypoxia. *Nat Commun.* 2016;7, 11960.
64. Pluznick JL, Protzko RJ, Gevorgyan H, et al. Olfactory receptor responding to gut microbiota-derived signals plays a role in renin secretion and blood pressure regulation. *Proc Natl Acad Sci U S A.* 2013;110:4410–4415.
65. De Vadder F, Kovatcheva-Datchary P, Goncalves D, et al. Microbiota-generated metabolites promote metabolic benefits via gut-brain neural circuits. *Cell.* 2014;156:84–96.
66. Cummings JH, Pomare EW, Branch HWJ, Naylor E, Macfarlane GT. Short chain fatty acids in human large intestine, portal, hepatic and venous blood. *Gut.* 1987;28:1221–1227.
67. Gao J, Xu K, Liu H, et al. Impact of the gut microbiota on intestinal immunity mediated by tryptophan metabolism. *Front Cell Infect Microbiol.* 2018;8:13.
68. Donohoe DR, Garge N, Zhang X, et al. The microbiome and butyrate regulate energy metabolism and autophagy in the mammalian colon. *Cell Metab.* 2011;13:517–526.
69. Harris K, Kassis A, Major G, Chou CJ. Is the gut microbiota a new factor contributing to obesity and its metabolic disorders? *J Obes.* 2012;2012, 879151.
70. Dumas ME, Barton RH, Toye A, et al. Metabolic profiling reveals a contribution of gut microbiota to fatty liver phenotype in insulin-resistant mice. *Proc Natl Acad Sci U S A.* 2006;103(33):12511–12516.
71. Stella C, Beckwith-Hall B, Cloarec O, et al. Susceptibility of human metabolic phenotypes to dietary modulation. *J Proteome Res.* 2006;5:2780–2788.
72. Tang WH, Wang Z, Levison BS, et al. Intestinal microbial metabolism of phosphatidylcholine and cardiovascular risk. *N Engl J Med.* 2013;368:1575–1584.
73. Seldin MM, Meng Y, Qi H, et al. Trimethylamine **N**-oxide promotes vascular inflammation through signaling of mitogen-activated protein kinase and nuclear factor-kappaB. *J Am Heart Assoc.* 2016;5. pii: e002767.
74. Boini KM, Hussain T, Li PL, Koka S. Trimethylamine-N-oxide instigates NLRP3 inflammasome activation and endothelial dysfunction. *Cell Physiol Biochem.* 2017;44:152–162.
75. Chen ML, Zhu XH, Ran L, Lang HD, Yi L, Mi MT. Trimethylamine-N-oxide induces vascular inflammation by activating the NLRP3 inflammasome through the SIRT3-SOD2-mtROS signaling pathway. *J Am Heart Assoc.* 2017;6. pii: e006347.

76. Sun X, Jiao X, Ma Y, et al. Trimethylamine **N**-oxide induces inflammation and endothelial dysfunction in human umbilical vein endothelial cells via activating ROS-TXNIP-NLRP3 inflammasome. *Biochem Biophys Res Commun*. 2016;481:63–70.
77. Zhu W, Gregory JC, Org E, et al. Gut microbial metabolite TMAO enhances platelet hyperreactivity and thrombosis risk. *Cell*. 2016;165:111–124.
78. Wang Z, Zhao Y. Gut microbiota derived metabolites in cardiovascular health and disease. *Protein Cell*. 2018;9:416–431.
79. Winek K, Dirnagl U, Meisel A. The gut microbiome as therapeutic target in central nervous system diseases: implications for stroke. *Neurotherapeutics*. 2016;13:762–774.
80. Vogt NM, Romano KA, Darst BF, et al. The gut microbiota-derived metabolite trimethylamine N-oxide is elevated in Alzheimer's disease. *Alzheimers Res Ther*. 2018;10:124.
81. Roediger W, Babidge W. Human colonocyte detoxification. *Gut*. 1997;41:731–734.
82. Smith EA, Macfarlane GT. Enumeration of human colonic bacteria producing phenolic and indolic compounds: effects of pH, carbohydrate availability and retention time on dissimilatory aromatic amino acid metabolism. *J Appl Bacteriol*. 1996;81:288–302.
83. Sun M, Ma N, He T, Johnston LJ, Ma X. Tryptophan (Trp) modulates gut homeostasis via aryl hydrocarbon receptor (AhR). *Crit Rev Food Sci Nutr*. 2019;29:1–9.
84. Yang J-Y, Kweon M-N. The gut microbiota: a key regulator of metabolic diseases. *BMB Rep*. 2016;49:536–541.

CHAPTER 4

Importance of fermented foods on human health

Akhlaq A. Farooqui
Department of Molecular and Cellular Biochemistry, The Ohio State University, Columbus, OH, United States

1 Introduction

Foods and beverages produced through controlled edible microbial growth are called fermented foods. The conversion of food components through controlled microbial growth involves microbial enzymatic action. This process not only provides satisfying flavors, aromas, and textures to foods, but also results in health benefits to humans.[1] Edible bacteria used in production of fermented foods include lactic acid bacteria, *Propionibacterium*, *Acetobacter*, yeast, molds, and *Bacillus* species. Examples of commonly used fermented foods include alcoholic beverages (beers and wines), vinegars, pickled vegetables, sausages, cheeses, yogurts, vegetable protein amino acid/peptide sauces and pastes, kefir, kombucha, sauerkraut, tempeh, nattō, miso, kimchi, meat, fish, and sourdough breads.[2,3] In the absence of oxygen, edible microbial fermentation converts sugars into simple acids, alcohols, and carbon dioxide. This reaction requires proper environmental conditions such as temperature, pH, and moisture content. Furthermore, biotransformation reactions (removal of glycol-side residues) during fermentation promote health-benefitting activities, facilitate the removal of antinutrients (e.g., phytic acid, bitter-tasting phenolic compounds) from fermented foods, and deliver probiotics.[4] Thus, fermented food provides several benefits over nonfermented food. Fermentation not only enhances the presence of many nutrients (e.g., linoleic acid, bioactive peptides) in fermented foods, but also promotes the removal of toxic or undesirable food constituents (e.g., phytic acid, bitter-tasting phenolic compounds). In addition, fermentation extends the shelf life and safety of fermented foods. It is estimated that fermented foods and beverages comprise anywhere from 5%–40% of the human diet in some populations.[5] The consumption of fermented foods produces significant positive improvements in balancing intestinal permeability and barrier function and is linked with mood, behavior, energy, weight, food

Gut Microbiota in Neurologic and Visceral Diseases
https://doi.org/10.1016/B978-0-12-821039-0.00012-5

cravings, hormone balance, immunity, and overall wellness. Based on this information, fermented foods are considered "functional foods" or "nutraceuticals" due to their putative health benefits.[6]

Due to cultural and socioeconomic differences, traditional methods for microbial fermentation vary from one country to another and may increase the risk of contamination by foodborne pathogens. Food safety issues are mainly due to improper fermentation conditions or procedures rather than the fermented foods.[3] Therefore, attention should be given to standardize the protocols and microorganisms used in fermentation for improved food safety and enhanced health benefits. Fermented foods are often contaminated with toxins and pathogenic microbes. Mycotoxins (aflatoxins, fumonisins, sterigmatocystin, nivalenol, deoxynivalenol, zearalenone, ochratoxin, and alternariol), bacterial toxins (shiga toxin and botulinum), biogenic amines, and cyanogenic glycosides are the common toxins found in fermented foods in addition to pathogenic microbes.[7] Among fermented foods, milk products and meat sausages are extremely vulnerable to contamination.

2 Beneficial effects of fermented foods

The intake of fermented foods is gaining increasing attention due to their health-promoting benefits (Fig. 1). Among them, fermented dairy foods have been associated with obesity prevention and reduction of the risk of metabolic disorders, cardiovascular and immune-related diseases, and

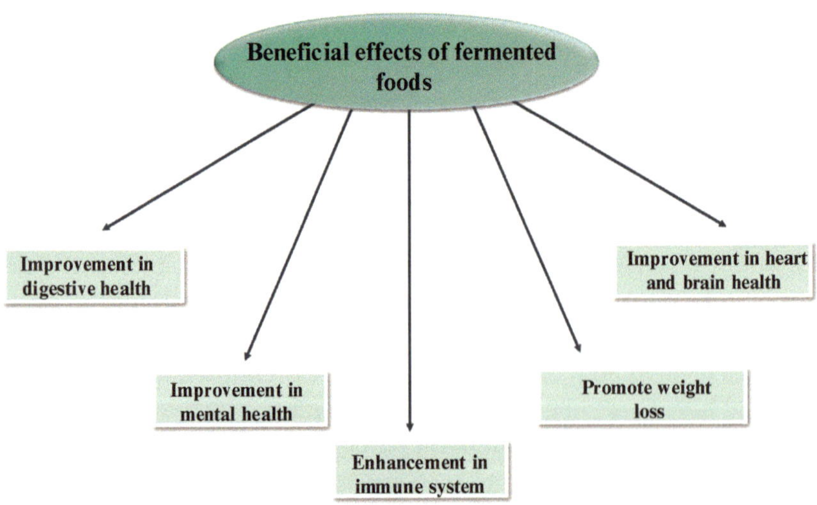

Fig. 1 Beneficial effects of fermented foods.

cognitive decline.[7–10] Fermented foods induce health benefits by providing the consumer with both easily metabolizable nutrients and beneficial microorganisms.

Apart from their content of fatty acids, vitamins, and minerals, fermented products contain bioactive peptides and living microorganisms that modulate immune responses and impact intestinal microbiota composition and functionality.[11,12] Bioactive peptides are generally short peptides (3–20 amino acids) derived from proteins that can exert biological activities over and above their expected nutritional value.[13] These peptides are often functionally inactive within the native proteins and must be released by proteolysis (in vivo digestion, in vitro enzymatic hydrolysis, or bacterial fermentation) to achieve their specific "bioactive" roles. Many of these food-derived peptides demonstrate antihypertensive, antiinflammatory, antidiabetic, and antioxidant properties under experimental conditions (Fig. 2).[14,15] While some studies have observed the effects of single peptides, many others have examined protein hydrolysates composed of a mixture of diverse bioactive peptides.[16,17] Given their food-based sources and a perceived lack of serious side effects, bioactive peptides and peptide-rich protein hydrolysates can potentially provide a better alternative to synthetic pharmaceuticals for the prevention and treatment of chronic illnesses that affect an increasing number of people.

The consumption of fermented foods affects human gut microbiota, a complex and dynamic system represented by trillions of microorganisms that plays an important role in the maintenance of health. Indeed, recent

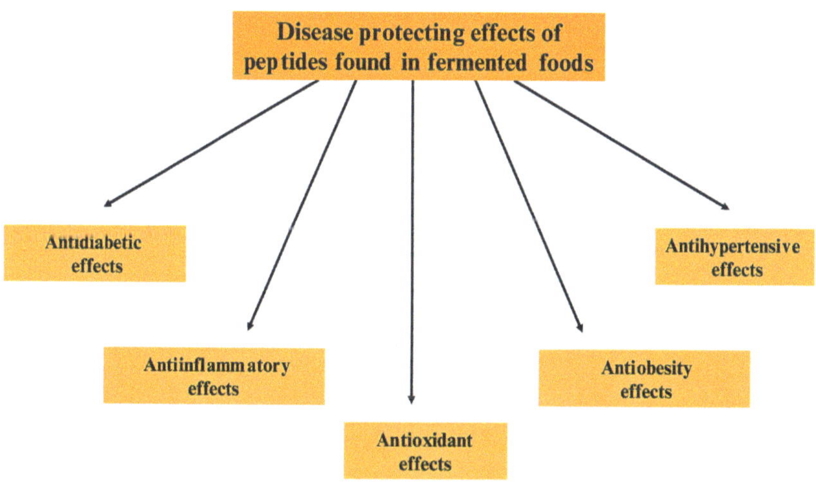

Fig. 2 Disease-protecting effects of peptides found in fermented foods.

studies have consistently identified disease-specific microbiota signatures in different health disorders.[18] The microbiota of healthy adults is represented mainly by anaerobic bacteria from the Firmicutes and Bacteroidetes phyla.[19] Under physiological conditions, the gut microbiota co-exist in mutualistic symbiosis with the host, contributing to the body's homeostatic control through the regulation of immune, endocrine, and neural pathways and functions.[20] The communication between the hypothalamic-pituitary-adrenal (HPA) axis of the brain and microbiota involves several mechanisms such as expression of immune-modulatory cytokines and chemokines and vagal nerve signaling. The generation of neuroactive metabolites and endocrine secretions (e.g., glucocorticoids, neuropeptides, etc.) or their receptors[21] play important roles in the signaling process.[22] These mediators play an important role in cognitive and emotional outcomes later in life.[23,24] Gut microbiota produce and release a variety of neuroactive factors (metabolites) including dopamine, gamma-aminobutyric acid (GABA), 5-hydroxytryptamine (5-HT), and short-chain fatty acids (SCFAs) (acetate, butyrate, propionate) (Fig. 3). These metabolites not only impact carbohydrate metabolism[25] but are also taken up by the circulatory system and utilized by the citric acid cycle; propionate is incorporated into glucose metabolism.[26,27] In addition, SCFAs have been shown to affect transit time through modulation of gut motility and to affect insulin sensitivity and energy expenditure, giving them a link to metabolic syndrome (MetS).[28] Butyrate acts by promoting AMP-activated protein kinase (AMPK) activation and glucose transporter4 (GLUT4) expression in the adipose tissue,

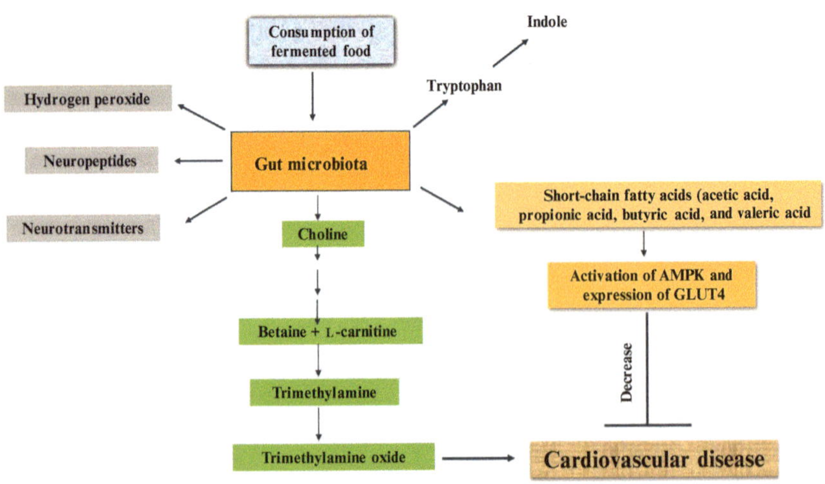

Fig. 3 Fermented foods and their effects on gut microbiota-derived metabolites.

improving cardiovascular disease (CVD)-related metabolic disorder, resisting high-fat diet-induced gut microbiome dysbiosis, and promoting resolvin E1 and lipoxin biosynthesis. All of this supports the view that oral supplements of the natural fermentation product (butyrate) can potentially be used to prevent CVD. Similarly, lactic acid bacteria have been reported to synthesize vitamins and minerals, produce biologically active peptides, and remove some harmful nutrients. Among biologically active peptides, conjugated linoleic acids (CLAs) help lower blood pressure, exopolysaccharides exhibit prebiotic effects, bacteriocins show antimicrobial effects, sphingolipids induce anticarcinogenic and antimicrobial effects, and bioactive peptides exhibit antioxidant, antimicrobial, opioid antagonist, and antiallergenic effects. As a result, fermented foods provide many health benefits such as antioxidant, antimicrobial, antifungal, antiinflammatory, antidiabetic, and antiatherosclerotic activity.[29] SCFAs also improve immunity against pathogens, blood-brain barrier integrity, provision of energy substrates, and regulation of critical functions of the intestine.[30] In addition, gut microbiota also releases bioactive neuropeptide such as peptide YY (PYY), glucagon-like peptide-1 (GLP-1) GLP-2, and neuropeptide Y (NYP).[31] These peptides, for example, PYY, are transported from the gut through blood to the brain where they exert their impact and effect on satiety.[32] Satiation-signaling peptides arise primarily from the gastrointestinal tract, but most of them are also synthesized within the brain.[33] By synthesizing and releasing these bioactive metabolites, microbiota may modulate brain development, inflammatory processes, immune system, pain, brain function, and behavior.[34] Converging evidence suggests that fermented food-mediated changes in composition of gut microbiota may have a strong impact on hosts' behavior through different modalities, including interaction through the nervous system, mutual crosstalk with the immune and the endocrine systems, and finally with direct synthesis and managing of neuro-chemicals.[35,36]

An imbalance in the gut equilibrium (dysbiosis)[37,38] is linked with several pathologies (inflammatory bowel disease, celiac disease, obesity, metabolic disorder, etc.).[39] Gut dysbiosis not only affects gut permeability but also blood-brain barrier integrity.[40] These processes contribute to hippocampal dysfunction and neurological disorders.[26,41]

3 Importance of fermented foods on human health

Many fermented foods and beverages are commercially available these days. Examples include pickled vegetables, sausages, cheeses, yogurts, sauerkraut, tempeh, nattō, miso, kimchi, fermented meat and fish, vegetable protein amino acid/peptide, and sauces and pastes. Examples of fermented beverages are alcoholic beverages (beers and wines), vinegars, kefir, and kombucha.[3] The production of foods such as yogurt and cultured milk, wine and beer, sauerkraut and kimchi, and fermented sausage were initially valued because of their improved shelf life, safety, and organoleptic properties. However, in recent years it is becoming increasingly evident that fermented foods and beverages also contain enhanced nutritional and functional properties due to transformation of substrates and formation of bioactive or bioavailable end-products. Many fermented foods also contain living microorganisms that may act as probiotics.

3.1 Consumption of kefir

Kefir is a traditional fermented milk beverage that is typically manufactured by inoculating a kefir grain, a cauliflower-like exopolysaccharide (EPS) matrix containing a symbiotic community of bacteria and yeast[42] into milk and incubating it at room temperature for approximately 24 h. This results in a beverage that has a sour taste or yogurt-like taste.[43,44] This flavor can vary depending on the microbial composition of the grain that is used.[45] High-throughput sequencing investigations have demonstrated that kefir grains are typically dominated by the bacterial genus *Lactobacillus* and the fungal phylum Ascomycota.[45] In contrast, kefir milk is dominated by the bacterial genera *Lactobacillus*, *Lactococcus*, *Acetobacter*, and *Leuconostoc* and the fungal genera *Kazachstania*, *Kluyveromyces*, *Naumovozyma*, and *Saccharomyces*.[45,46] Consumption of kefir not only results in numerous effects (antimicrobial, immunomodulatory, antiallergenic, antitumoral, antidiabetic, antiinflammatory, antimutagenic, and antioxidative activities),[47–50] but is also associated with the alleviation of symptoms of lactose intolerance and the reduction of cholesterol.[51] These effects are due to the production of organic acids, bacteriocins, carbon dioxide, hydrogen peroxide, ethanol, and diacetyl.[52] In addition, production of bioactive peptide from casein stimulates the immune system in animal models,[53] while kefiran reduced ovalbumin-induced cytokine production in a murine asthma model.[54]

Based on this information, it is suggested that the consumption of kefir provides benefits in cardiometabolic diseases including hypertension, vascular endothelial dysfunction, dyslipidemia, and insulin resistance.[55] To this end, mechanisms of actions of kefir in cardiometabolic diseases include recruitment of endothelial progenitor cells, improvement of the balance vagal/sympathetic nervous system, diminution of excessive generation of reactive oxygen species, inhibition of angiotensin converting enzymes, an antiinflammatory cytokine profile, and alteration of the intestinal microbiota.

3.2 Consumption of dairy products (yogurt and cheese)

Lactic acid bacteria are present in yogurt and cheese. These bacteria release bioactive metabolites (exopolysaccharides and peptides). These metabolites have the potential to exert a wide range of metabolic and regulatory functions. In particular, peptides derived from fermented dairy products are likely to exert greater cardiometabolic and antiinflammatory effects than nonfermented dairy. It has been hypothesized that lactic acid bacteria-derived bioactive peptides have the potential to protect the host against cardiometabolic diseases such as CVD, type 2 diabetes, and MetS through antimicrobial actions and to effect changes in gene expression of glucose regulatory and antiinflammatory signaling pathways. The peptides released through fermentation may explain some of the health effects of fermented dairy products on cardiometabolic diseases. MetS is a risk factor for type 2 diabetes, CVD, stroke, Alzheimer's disease, and depression.[56,57] Dairy products, such as cheese, do not exert negative effects on blood lipids as predicted solely by the content of saturated fat. Calcium and other bioactive components may modify the effects on low-density lipoprotein (LDL) cholesterol and triglycerides. Apart from supplying valuable dairy nutrients, yogurt may also exert beneficial probiotic effects. The consumption of yogurt, and other dairy products, in observational studies is associated with a reduced risk of weight gain and obesity as well as of CVD, and these findings are, in part, supported by randomized trials.

3.3 Consumption of kombucha

Kombucha is a traditional beverage of Manchurian origins. It is usually prepared from the fermentation of black or green tea (sweetened with 5%–8% of sugar) by a symbiotic microbial consortium that is mainly composed of white sugar, acetic acid bacteria, and osmophilic yeasts.[58] This symbiotic culture of bacteria and yeast is called SCOBY. During fermentation, yeast converts sucrose to ethanol, organic acids, and carbon dioxide. The microbial and metabolite composition of kombucha varies according to the exact composition of the SCOBY, the type and concentration of tea and sugar,[59,60] levels of oxygen, fermentation time (10–20 days),[61] temperature (range of 20–30 °C),[61] and storage duration.[59] The low pH of kombucha, owing mainly to the production of a high concentration of acetic acid, has been shown to prevent the growth of pathogenic bacteria such as *Helicobacter pylori*, *Escherichia coli*, *Salmonella typhimurium*, and *Campylobacter jejuni*.[62]

Studies on the composition of kombucha have indicated that kombucha is rich in organic acids (acetic, glucuronic, gluconic acids), vitamins, and tea polyphenols. The pH of kombucha is low, avoiding bacterial contamination.[63–65] The beverage has a refreshing taste and regular consumption improves digestion.[66] Due to the presence of green and black tea, kombucha is enriched in polyphenols, flavonoids, and saponins.[67,68] In vitro or animal models (in vivo) have indicated that kombucha not only produces antimicrobial activity against foodborne and human pathogens, but also has antiinflammatory, hypocholesterolemic, antiproliferative, and hypoglycemic activities.[65,69]

3.4 Consumption of sauerkraut

Sauerkraut is sour cabbage that is prepared from the fermentation of fresh shredded cabbage and mixed with 2.3%–3.0% salt. The mixture is left to undergo spontaneous natural fermentation generally involving *Leuconostoc* spp., *Lactobacillus* spp., and *Pediococcus* spp.[70] The pH of the final product ranges from 3.5 to 3.8.[71] At this pH, the cabbage or other vegetables will be preserved for a long period of time.[2] Sauerkraut brine is an important by-product of the cabbage fermentation industry and can be used as a substance for the production of carotenoids by *Rhodotorula rubra* or for β-glucosidase production by *Candida wickerhamii* for commercial applications.[72]

Treatment of Wistar rats with sauerkraut juices increased the activity of glutathione S-transferase (GST) and NAD(*P*)H:quinone oxidoreductase 1 (NQO1), key liver and kidney detoxifying enzymes.[73] Certain lactic acid bacteria present in sauerkraut also produce conjugated linoleic acid,[74] contributing to anticarcinogenic and antiatherosclerotic activity in animals.[75] Furthermore, *Lactobacillus plantarum* P2 isolated from sauerkraut significantly promoted the induction and expression of TNF-α and IL-12 and prevented adhesion and invasion of Caco-2 cells by *Salmonella enteritidis*.[76] Sauerkraut contains glucosinolate breakdown products including kaempferol (a flavonoid), isothiocyanates, indole-3-carbinol, goitrin, allyl cyanide, and nitriles.[77] The importance of these phytochemicals in human health is unclear, however, kaempferol has been shown to have radical scavenging activity, to protect from oxidative damage, and to attenuate cytokine-mediated reactive oxygen species in vitro.[78] Isothiocyanates have been shown to have antimicrobial properties, preventing the growth of a range of species, including *Escherichia coli*, *Clostridioides difficile*, *Campylobacter jejuni*, and *Clostridium perfringens*.[79]

3.5 Consumption of tempeh

Tempeh is a traditional Indonesian food that is produced by fermenting boiled and dehulled soybeans with a starter culture of *Rhizopus oligosporus* fungal species at room temperature for 35–37 h.[80,81] This process results in production of a soft white cake with a chewy texture and mushroom-like flavor. The microbial composition of tempeh varies according to variations in production.[82] Tempeh contains lactic acid bacteria,[83] *Enterococcus faecium*,[83] and *Rhizopus* filamentous fungi. Fermentation of soybeans by *R. oligosporus* results in reduction in concentrations of protease inhibitors, phytic acid, and phenols.[84] It is well known that tempeh preparations reduce the serum concentrations of total cholesterol, LDLs, and triglycerides if consumed in place of animal protein. In addition, tempeh preparation produces antinutritional factors that are high in raw soybeans.[85] The consumption of tempeh decreases cholesterol levels, oxidative stress, and appetite while improving bone health. Tempeh also contains prebiotics, which may improve digestive health and reduce inflammation. Based on this information, it is suggested that tempeh may attenuate the effects of diabetes mellitus, blood pressure, cardiac disorders, and cancer-related issues.[7] Tempeh has gained considerable attention due to its greater nutritional value as compared to nonfermented foods.

3.6 Consumption of miso

Miso is a traditional Japanese paste of fermented soybean used to make miso soup. Miso is made by fermenting soybeans with Koji produced from *Aspergillus oryzae*, although *Saccharomyces cerevisiae* and lactic acid bacteria may additionally be used. As with other fermented soy foods, miso production varies greatly in terms of ingredients, temperature and fermentation time, salt concentration, and the strain of *A. oryzae* used. Miso is not only enriched in essential minerals, but is also a good source of various B vitamins, vitamins E and K, and folic acid. As a fermented food, miso provides the gut with beneficial bacteria that help humans to stay healthy, vibrant, and happy; good gut health is known to be linked to overall mental and physical wellness.

It is well known that fermented soy foods such as miso, tempeh, and nattō contain considerable amounts of isoflavones.[86,87] The three isoflavones of genistein, daidzein, and glycitein and their respective glycosides account for approximately 50%, 40%, and 10%, respectively, of the total isoflavone content of soybeans.[88]

Many studies have indicated that miso intake decreases stomach cancer risk in Japanese men.[89,90]

3.7 Consumption of nattō

Nattō is a cheese-like food made from fermented soybeans with *Bacillus subtilis*. It has been consumed as a traditional food in Asian countries for more than 2000 years. Nattō consumption is believed to be a significant contributor to the longevity of the Japanese population. Recent studies have demonstrated that a high nattō intake is associated with decreased risk of total CVD mortality and, in particular, a decrease in risk of mortality from ischemic heart diseases.[91] Beneficial effects of nattō on heart disease are due to the presence of an enzyme called nattokinase (EC 3.4.21.62), a profibrinolytic serine protease with 275 amino acid residues (molecular weight of approximately 28 kDa) (Fig. 4). Nattokinase not only has potent fibrinolytic/antithrombotic activity[92,93] but also has antihypertensive,[94,95] antiatherosclerotic,[96,97] lipid-lowering,[96,98] antiplatelet/anticoagulant,[99] and neuroprotective effects.[100] An Egyptian study also found high levels of histamine in fermented fish.[101,102] All these pharmacologic actions of nattokinase have relevance to the prevention and treatment of CVD. Indeed, nattokinase supplementation has shown to enhance markers of fibrinolysis and anticoagulation and to decrease blood pressure and atherosclerosis in human subjects.[95,96] It should be noted that compared to other fibrinolytic enzymes (urokinase, t-PA, and streprokinase), nattokinase has no side effects,

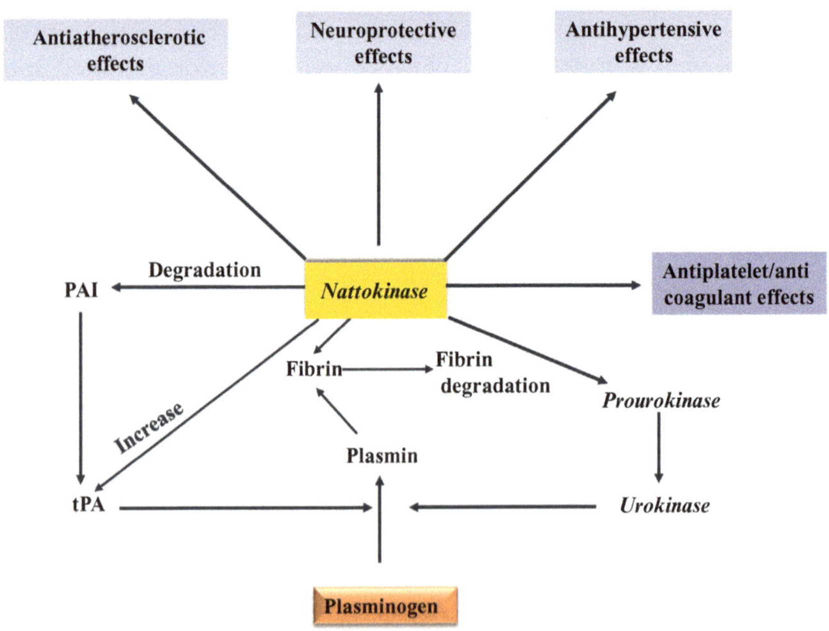

Fig. 4 Molecular mechanism of action of nattokinase on disease-related processes. Nattokinase dissolves blood clots by directly hydrolyzing fibrin and plasmin substrate. It converts endogenous prourokinase to urokinase (uPA). It also degrades plasminogen activator inhibitor (PAI-1) and increases the level of tissue plasminogen activator (t-PA).

is inexpensive, and has a long lifetime. As such, it has the potential to be used as a drug for treating CVD and a functional food additive.

3.8 Consumption of kimchi

Kimchi, one of the most popular Korean traditional foods, is prepared by fermenting baechu cabbage together with other vegetables and lactic acid bacteria (*Lactobacillus brevis*, *Lactobacillus fermentum*, *L. plantarum*, *Leuconostoc mesenteroides*, *Weissella confusa* and *Pediococcus pentosaceus*) with functional potential.[103] Other ingredients of kimchi are brine, red pepper powder, and fermented fish sauce, and other spices such as garlic, green onion, ginger, and some seaweed. Many types of ingredients are added to kimchi to enhance its taste, flavor, nutritional value, texture, and so on. A number of bacteria are involved in the fermentation of kimchi, but lactic acid bacteria are the dominant species in the fermentation process. Kimchi possesses antiinflammatory, antibacterial, antioxidant, anticancer, antiobesity, anticonstipation, antiaging, and probiotic properties as well as promotes colorectal health and lowers cholesterol[103] (Fig. 5).

Several studies have indicated that consumption of kimchi affects gut microbiota. A study on diet-induced obesity murine models showed that *L. plantarum* HAC01, isolated from kimchi, results in greater *Adlercreutzia* and

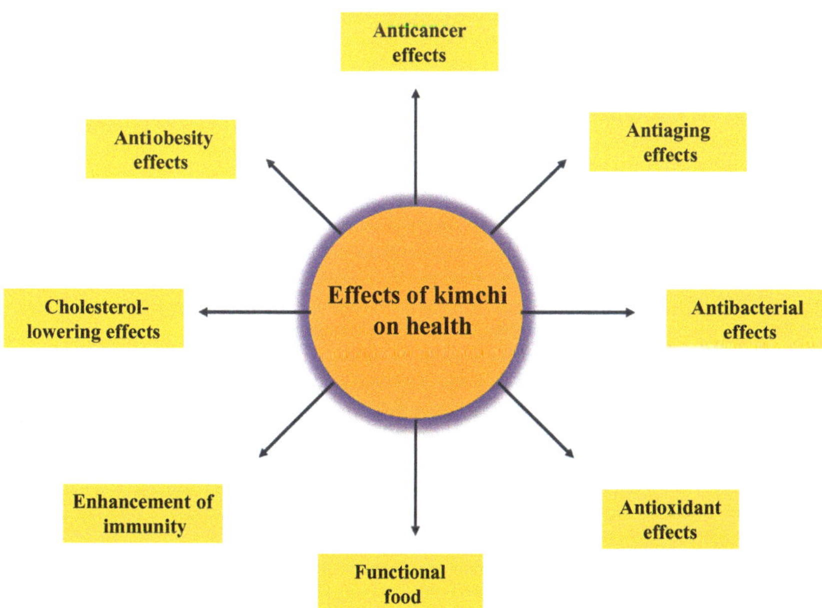

Fig. 5 Health benefits of kimchi.

lesser *Bacteroides*, *Mucispirillum*, and *Ruminococcus* proportions compared to control mice.[104] In humans, a nonrandomized study of six people in South Korea showed that consumption of 300 g/day kimchi for 4 weeks increased stool concentrations of *Lactobacillus* and *Leuconostoc* and decreased stool pH as compared to 60 g/day of kimchi.[105] In a randomized controlled trial of fermented compared to fresh (unfermented) kimchi in 24 women with obesity, women who were randomized to receive 180 g/day fermented kimchi for 8 weeks showed a decrease in *Blautia* abundance and an increase in *Prevotella* and *Bacteroides* abundance compared to baseline. Both groups (fermented and unfermented kimchi) experienced increases in *Proteobacteria* and *Actinobacteria* abundance.[106] Another, human study demonstrated that consumption of kimchi fermented for 8 weeks promoted changes in expression of genes related to metabolic pathways and immunity.[106] In a mouse colitis model, *Lactobacillus paracasei* LS2, a strain isolated from kimchi, decreased cytokine production, myeloperoxidase activity, and the number of macrophages and neutrophils in the lamina propria lymphocytes, suggesting a potential antiinflammatory effect.[107] The anticarcinogenic properties of kimchi can be attributed to the inhibition of gastric cancer cell growth.[108] Notably, as kimchi comprises a variety of ingredients, its impact on the gut microbiota and health is thought to result from a synergic effect of the microorganisms it contains as well as the nutrient content (e.g., phytochemicals, fiber, vitamins) of the foods used in the preparation.[109,110]

4 Adverse effects of fermented foods

In parts of Asia, fermented fish sauce is widely consumed. A study performed in the Chaoshan area of China indicated that there is an increased risk of squamous cell carcinoma of the esophagus in habitual consumers of fermented fish sauce.[111] Another Chinese study showed that *N*-nitroso compounds and genotoxins present before and after nitrosation appear to be responsible for cancer risk.[112]

5 Conclusion

Fermented foods and beverages are products made by microbial organisms and enzymatic conversions of major and minor food components. In addition to their commonly recognized effects on digestive health (e.g., dysbiosis) and well being, there is now strong evidence for the impact of fermented foods and beverages (e.g., yogurt, pickles, bread, kefir, beers, wines, mead) on general health, namely, their effects on the gut microbiota balance and brain functionality. Fermented foods and beverages are popular because they enhance the shelf life, safety, functionality, and sensory characteristics of food and have great nutritional value (production of vitamins and peptides) and chronic disease-delaying effects. As discussed, many fermented foods contain live microorganisms that may improve gastrointestinal health and provide other health benefits, including lowering the risk of type 2 diabetes, MetS, and CVD. The number of organisms in fermented foods can vary significantly depending on how products are manufactured and processed as well as conditions and duration of storage. This chapter describes studies on commonly consumed fermented foods and beverages, including cultured dairy products, cheese, fermented sausage, fermented vegetables, soy-fermented foods, and fermented cereal products. Although few specific recommendations and claim legislations for what constitutes a relevant dose exist, the studies from this survey revealed that many fermented foods and beverages are good sources of live bacteria that may provide benefits in human health.

References

1. Marco ML, Heeney D, Binda S, Cifelli CJ, Cotter PD, et al. Health benefits of fermented foods: microbiota and beyond. *Curr Opin Biotechnol.* 2017;44:94–102.
2. Steinkraus KH. Classification of fermented foods: worldwide review of household fermentation techniques. *Food Control.* 1997;8:311–317.
3. Steinkraus KH. In: Moselio S, ed. *Encyclopedia of Microbiology.* 3rd ed. Oxford: Academic Press; 2009:45–53.
4. van Hylckama Vlieg JET, Veiga P, Zhang C, Derrien M, Zhao L. Impact of microbial transformation of food on health—from fermented foods to fermentation in the gastro-intestinal tract. *Curr Opin Biotechnol.* 2011;22:211–219.
5. Borresen EC, Henderson AJ, Kumar A, Weir TL, Ryan EP. Fermented foods: patented approaches and formulations for nutritional supplementation and health promotion. *Recent Pat Food Nutr Agric.* 2012;4:134–140.
6. Aslam H, Green J, Jacka FN, et al. Fermented foods, the gut and mental health: a mechanistic overview with implications for depression and anxiety. *Nutr Neurosci.* 2018;11:1–13.
7. Sivamaruthi BS, Kesika P, Chaiyasut C. Toxins in fermented foods: prevalence and preventions-a mini review. *Toxins (Basel).* 2018;11. pii: E4.
8. Guo J, Astrup A, Lovegrove JA, Gijsbers L, Givens DI, Soedamah-Muthu SS. Milk and dairy consumption and risk of cardiovascular diseases and all-cause mortality: dose–response meta-analysis of prospective cohort studies. *Eur J Epidemiol.* 2017;32:269–287.
9. Kok CR, Hutkins R. Yogurt and other fermented foods as sources of health-promoting bacteria. *Nutr Rev.* 2018;76:4–15.
10. Salas-Salvadó J, Guasch-Ferré M, Díaz-López A, Babio N. Yogurt and diabetes: overview of recent observational studies. *J Nutr.* 2017;147:1452S–1461S.
11. Chakrabarti S, Jahandideh F, Wu J. Food-derived bioactive peptides on inflammation and oxidative stress. *Biomed Res Int.* 2014;2014, 608979.

12. Severyn CJ, Bhatt AS. With probiotics, resistance is not always futile. *Cell Host Microbe.* 2018;24:334–336.
13. Kitts DD, Weiler K. Bioactive proteins and peptides from food sources. Applications of bioprocesses used in isolation and recovery. *Curr Pharm Des.* 2003;9:1309–1323.
14. Erdmann K, Cheung BWY, Schröder H. The possible roles of food-derived bioactive peptides in reducing the risk of cardiovascular disease. *J Nutr Biochem.* 2008;19:643–654.
15. Sousa GTD, Lira FS, Rosa JC, et al. Dietary whey protein lessens several risk factors for metabolic diseases: a review. *Lipids Health Dis.* 2012;11:67.
16. Izumi H, Ishizuka S, Inafune A, et al. α-lactalbumin hydrolysate stimulates glucagon-like peptide-2 secretion and small intestinal growth in suckling rats. *J Nutr.* 2009;139:1322–1327.
17. Sato K, Egashira Y, Ono S, et al. Identification of a hepatoprotective peptide in wheat gluten hydrolysate against D-galactosamine-induced acute hepatitis in rats. *J Agric Food Chem.* 2013;61:6304–6310.
18. Duvallet C, Gibbons SM, Gurry T, Irizarry RA, Alm EJ. Meta-analysis of gut microbiome studies identifies disease-specific and shared responses. *Nat Commun.* 2017;8:1784.
19. Eckburg PB, Lauber CL, Costello EK, et al. Diversity of the human intestinal microbial flora. *Science.* 2005;308:1635–1638.
20. Romani-Perez M, Agusti A, Sanz Y. Innovation in microbiome-based strategies for promoting metabolic health. *Curr Opin Clin Nutr Metab Care.* 2017;20:484–491.
21. Moloney RD, Desbonnet L, Clarke G, Dinan TG, Cryan JF. The microbiome: stress, health and disease. *Mamm Genome.* 2014;25:49–74.
22. Bravo JA, Forsythe P, Chew MV, et al. Ingestion of lactobacillus strain regulates emotional behavior and central gaba receptor expression in a mouse via the vagus nerve. *Proc Natl Acad Sci U S A.* 2011;108:16050–16055.
23. Farooqui AA. *Metabolic Syndrome—An Important Risk Factor for Stroke, Alzheimer Disease, and Depression.* New York: Springer; 2013.
24. Francis H, Stevenson R. The longer-term impacts of Western diet on human cognition and the brain. *Appetite.* 2013;63:119–128.
25. COX LM, Weiner HL. Microbiota signaling pathways that influence neurologic disease. *Neurotherapeutics.* 2018;15:135–145.
26. Farooqui AA. *High Calorie Diet and the Human Brain.* Springer, International Publishing Switzerland; 2015.
27. Jeffery IB, O'Toole PW. Diet-microbiota interactions and their implications for healthy living. *Nutrients.* 2013;5:234–252.
28. Gao Z, Yin J, Zhang J, et al. Butyrate improves insulin sensitivity and increases energy expenditure in mice. *Diabetes.* 2009;58:1509–1517.
29. Şanlier N, Gökcen BB, Sezgin AC. Health benefits of fermented foods. *Crit Rev Food Sci Nutr.* 2019;59:506–527.
30. Tomova A, Bukovsky I, Rembert E, et al. The effects of vegetarian and vegan diets on gut microbiota. *Front Nutr.* 2019;6:47.
31. Holzer P, Farzi A, Hassan AM, Zenz G, Jačan A, Reichmann F. Visceral inflammation and immune activation stress the brain. *Front Immunol.* 2017;8:1613.
32. Romijn JA, Corssmit EP, Havekes LM, Pijl H. Gut-brain axis. *Curr Opin Clin Nutr Metab Care.* 2008;11:518–521.
33. Cummings DE, Overduin J. Gastrointestinal regulation of food intake. *J Clin Invest.* 2007;117:13–23.
34. Holzer P, Farzi A. Neuropeptides and the microbiota-gut-brain axis. *Adv Exp Med Biol.* 2014;817:195–219.
35. Foster JA, Rinaman L, Cryan JF. Neurobiology of stress & the gut-brain axis: regulation by the microbiome. *Neurobiol Stress.* 2017;7:124–136.

36. Lyte M. Microbial endocrinology in the microbiome-gut-brain axis: how bacterial production and utilization of neurochemicals influence behavior. *PLoS Pathog.* 2013;9:e1003726.
37. Panebianco C, Potenza A, Andriulli A, Pazienza V. Exploring the microbiota to better understand gastrointestinal cancers physiology. *Clin Chem Lab Med.* 2018;56:1400–1412.
38. Petersen C, Round JL. Defining dysbiosis and its influence on host immunity and disease. *Cell Microbiol.* 2014;16:1024–1033.
39. Schippa S, Conte MP. Dysbiotic events in gut microbiota: impact on human health. *Nutrients.* 2014;6:5786–5805.
40. Braniste V, Al-Asmakh M, Kowal C, et al. The gut microbiota influences blood-brain barrier permeability in mice. *Sci Transl Med.* 2014;6:263ra158.
41. Garcia AL, Udeh A, Kalahasty K, Hackam AS. A growing field: the regulation of axonal regeneration by Wnt signaling. *Neural Regen Res.* 2018;13:43–52.
42. Nielsen B, Gürakan GC, Ünlü G. Kefir: a multifaceted fermented dairy product. *Probiotics Antimicrob Proteins.* 2014;6:123–135.
43. Bourrie BC, Willing BP, Cotter PD. The microbiota and health promoting characteristics of the fermented beverage kefir. *Front Microbiol.* 2016;7:647.
44. Güzel-Seydim ZB, Seydim AC, Greene AK, Bodine AB. Determination of organic acids and volatile flavor substances in kefir during fermentation. *J Food Compos Anal.* 2000;13:35–43.
45. Marsh AJ, O'Sullivan O, Hill C, Ross RP, Cotter PD. Sequencing-based analysis of the bacterial and fungal composition of kefir grains and milks from multiple sources. *PLoS One.* 2013;8:e69371.
46. Korsak N, Taminiau B, Leclercq M, et al. Short communication: evaluation of the microbiota of kefir samples using metagenetic analysis targeting the 16S and 26S ribosomal DNA fragments. *J Dairy Sci.* 2015;98:3684–3689.
47. Hadisaputro S, Djokomoeljanto RRJ, Judiono, Soesatyo MHNE. The effects of oral plain kefir supplementation on proinflammatory cytokine properties of the hyperglycemia Wistar rats induced by streptozotocin. *Acta Med Indones.* 2012;44:100–104.
48. Khoury N, El-Hayek S, Tarras O, El-Sabban M, El-Sibai M, Rizk S. Kefir exhibits antiproliferative and proapoptotic effects on colon adenocarcinoma cells with no significant effects on cell migration and invasion. *Int J Oncol.* 2014;45:2117–2127.
49. Lee MY, Ahn K-S, Kwon O-K, et al. Anti-inflammatory and anti-allergic effects of kefir in a mouse asthma model. *Immunobiology.* 2007;22:647–654.
50. Rodrigues KL, Caputo LR, Carvalho JC, Evangelista J, Schneedorf JM. Antimicrobial and healing activity of kefir and kefiran extract. *Int J Antimicrob Agents.* 2005;25:404–408.
51. Liu J-R, Wang S-Y, Chen M-J, Chen H-L, Yueh P-Y, Lin C-W. Hypocholesterolaemic effects of milk-kefir and soyamilk-kefir in cholesterol-fed hamsters. *Br J Nutr.* 2006;95:939–946.
52. De Oliveira Leite AM, Miguel MA, Peixoto RS, Rosado AS, Silva JT, Paschoalin VM. Microbiological, technological and therapeutic properties of kefir: a natural probiotic beverage. *Braz J Microbiol.* 2013;44:341–349.
53. Matar C, Valdéz J, Perdigón G, Leblanc J, Leblanc J. Immunomodulating effects of peptidic fractions issued from milk fermented with *Lactobacillus helveticus*. *J Dairy Sci.* 2002;85:2733–2742.
54. Kwon O-K, Ahn K-S, Lee M-Y, et al. Inhibitory effect of kefiran on ovalbumin-induced lung inflammation in a murine model of asthma. *Arch Pharm Res.* 2008;31:1590–1596.
55. Pimenta FS, Luaces-Regueira M, Ton AM, et al. Mechanisms of action of kefir in chronic cardiovascular and metabolic diseases. *Cell Physiol Biochem.* 2018;48:1901–1914.

56. Farooqui AA. *Insulin Resistance as a Risk Factor in Visceral and Neurological Disorders.* San Diego, CA: Academic Press, An Imprint of Elsevier; 2020.
57. Farooqui AA. *Metabolic Syndrome.* New York: Springer Science Media; 2013.
58. Jarrell J, Cal T, Bennett JW. The Kombucha consortia of yeasts and bacteria. *Mycologist.* 2000;14:166–170.
59. Fu C, Yan F, Cao Z, Xie F, Lin J. Antioxidant activities of kombucha prepared from three different substrates and changes in content of probiotics during storage. *Food Sci Technol.* 2014;34:123–126.
60. Gaggìa F, Baffoni L, Galiano M, et al. Kombucha beverage from green, black and rooibos teas: a comparative study looking at microbiology, chemistry and antioxidant activity. *Nutrients.* 2018;11:1.
61. Loncar E, Djuric M, Malbaša R, Kolarov L, Klašnja M. Influence of working conditions upon kombucha conducted fermentation of black tea. *Food Bioprod Process.* 2006;84:186–192.
62. Sreeramulu G, Zhu Y, Knol W. Kombucha fermentation and its antimicrobial activity. *J Agric Food Chem.* 2000;48:2589–2594.
63. Dufresne C, Farnworth E. Tea, kombucha, and health: a review. *Food Res Int.* 2000;33:409–421.
64. Jayabalan R, Marimuthu S, Swaminathan K. Changes in content of organic acids and tea polyphenols during kombucha tea fermentation. *Food Chem.* 2007;102:392–398.
65. Jayabalan R, Malbaša RV, Lončar ES, Vitas JS, Sathishkumar M. A review on kombucha tea-microbiology, composition, fermentation, beneficial effects, toxicity, and tea fungus. *Compr Rev Food Sci Food Saf.* 2014;13:538–550.
66. Greenwalt CJ, Steinkraus KH, Ledford RA. Kombucha, the fermented tea: microbiology, composition, and claimed health effects. *J Food Prot.* 2000;63:976–981.
67. Chu SC, Chen C. Effects of origins and fermentation time on the antioxidant activities of kombucha. *Food Chem.* 2006;98:502–507.
68. Valduga AT, Gonçalves IL, Magri E, Delalibera Finzer JR. Chemistry, pharmacology and new trends in traditional functional and medicinal beverages. *Food Res Int.* 2019;478–503.
69. Villarreal-Soto SA, Beaufort S, Bouajila J, Souchard JP, Tailandier P. Understanding kombucha tea fermentation: a review. *J Food Sci.* 2018;83:580–588.
70. Swain MR, Anandharaj M, Ray RC, Rani RP. Fermented fruits and vegetables of Asia: a potential source of probiotics. *Biotechnol Res Int.* 2014;2014, 250424.
71. Gardner NJ, Savard T, Obermeier P, Caldwell G, Champagne CP. Selection and characterization of mixed starter cultures for lactic acid fermentation of carrot, cabbage, beet and onion vegetable mixtures. *Int J Food Microbiol.* 2001;64:261–275.
72. Yoon KY, Woodams EE, Hang YD. Production of probiotic cabbage juice by lactic acid bacteria. *Bioresour Technol.* 2006;97:1427–1430.
73. Krajka-Kuzniak V, Szaefer H, Bartoszek A, Baer-Dubowska W. Modulation of rat hepatic and kidney phase II enzymes by cabbage juices: comparison with the effects of indole-3-carbinol and phenethyl isothiocyanate. *Br J Nutr.* 2011;105:816–826.
74. Zeng Z, Lin J, Gong D. Identification of lactic acid bacterial strains with high conjugated linoleic acid-producing ability from natural sauerkraut fermentations. *J Food Sci.* 2009;74:M154–M158.
75. Den Hartigh LJ. Conjugated linoleic acid effects on cancer, obesity, and atherosclerosis: a review of pre-clinical and human trials with current perspectives. *Nutrients.* 2019;11:370.
76. Feng J, Liu P, Yang X, Zhao X. Screening of immunomodulatory and adhesive Lactobacillus with antagonistic activities against Salmonella from fermented vegetables. *World J Microbiol Biotechnol.* 2015;31:1947–1954.

77. Tolonen M, Taipale M, Viander B, Pihlava J-M, Korhonen H, Ryhänen E-L. Plant-derived biomolecules in fermented cabbage. *J Agric Food Chem.* 2002;50:6798–6803.
78. Devi KP, Malar DS, Nabavi SF, et al. Kaempferol and inflammation: from chemistry to medicine. *Pharmacol Res.* 2015;99:1–10.
79. Romeo L, Iori R, Rollin P, Bramanti P, Mazzon E. Isothiocyanates: an overview of their antimicrobial activity against human infections. *Molecules.* 2018;23:624.
80. Cao Z-H, Green-Johnson JM, Buckley ND, Lin Q-Y. Bioactivity of soy-based fermented foods: a review. *Biotechnol Adv.* 2019;37:223–238.
81. Jayachandran M, Xu B. An insight into the health benefits of fermented soy products. *Food Chem.* 2019;271:362–371.
82. Nout M, Kiers J. Tempe fermentation, innovation and functionality: update into the third millenium. *J Appl Microbiol.* 2005;98:789–805.
83. Moreno M, Tee L, De Vuyst L, et al. Microbial analysis of Malaysian tempeh, and characterization of two bacteriocins produced by isolates of *Enterococcus faecium*. *J Appl Microbiol.* 2002;92:147–157.
84. Abu-Salem FM, Mohamed R, Gibriel A, Rasmy NMH. Levels of some antinutritional factors in tempeh produced from some legumes and jojobas seeds. *Int Sch Sci Res Innov.* 2014;8:296–301.
85. Sutardi, Buckle K. Characterization of extra- and intracellular phytases from *Rhizopus oligosporus* used in tempeh production. *Int J Food Microbiol.* 1988;6:67–79.
86. Nakajima N, Nozaki N, Ishihara K, Ishikawa A, Tsuji H. Analysis of isoflavone content in tempeh, a fermented soybean, and preparation of a new isoflavone-enriched tempeh. *J Biosci Bioeng.* 2005;100:685–687.
87. Rowland I, Faughnan M, Hoey L, Wahala K, Williamson G, Cassidy A. Bioavailability of phyto-oestrogens. *Br J Nutr.* 2003;89:S45–S58.
88. Murphy PA, Barua K, Hauck CC. Solvent extraction selection in the determination of isoflavones in soy foods. *J Chromatogr B Anal Technol Biomed Life Sci.* 2002;777:129–138.
89. Hoshiyama Y, Sasaba T. A case-control study of single and multiple stomach cancers in Saitama prefecture, Japan. *Jpn J Cancer Res.* 1992;83:937–943.
90. Khan MMH, Goto R, Kobayashi K, et al. Dietary habits and cancer mortality among middle aged and older Japanese living in hokkaido, Japan by cancer site and sex. *Asian Pac J Cancer Prev.* 2004;5:58–65.
91. Nagata C, Wada K, Tamura T, et al. Dietary soy and natto intake and cardiovascular disease mortality in Japanese adults: the Takayama study. *Am J Clin Nutr.* 2017;105:426–431.
92. Sumi H, Hamada H, Tsushima H, Mihara H, Muraki H. A novel fibrinolytic enzyme (nattokinase) in the vegetable cheese Natto; a typical and popular soybean food in the Japanese diet. *Experientia.* 1987;43:1110–1111.
93. Sumi H, Hamada H, Nakanishi K, Hiratani H. Enhancement of the fibrinolytic activity in plasma by oral administration of nattokinases. *Acta Haematol.* 1990;84:139–143.
94. Fujita M, Ohnishi K, Takaoka S, Ogasawara K, Fukuyama R, Nakamuta H. Antihypertensive effects of continuous oral administration of nattokinase and its fragments in spontaneously hypertensive rats. *Biol Pharm Bull.* 2011;34:1696–1701.
95. Kim JY, Gum SN, Paik JK, et al. Effects of nattokinase on blood pressure: a randomized, controlled trial. *Hypertens Res.* 2008;31:1583–1588.
96. Ren N, Chen H, Li Y, McGowan E, Lin Y. A clinical study on the effect of nattokinase on carotid artery atherosclerosis and hyperlipidaemia. *Nat Med J China.* 2017;97:2038–2042.
97. Suzuki Y, Kondo K, Matsumoto Y, et al. Dietary supplementation of fermented soybean, natto, suppresses intimal thickening and modulates the lysis of mural thrombi after endothelial injury in rat femoral artery. *Life Sci.* 2003;73:1289–1298.

98. Duan Z, Jiang X, Jiang H, Zhang S, Dong M, Zhao X. Study on the antioxidative activity and effects on experimental hyperlipidemia of natto extract. *Acta Nutr Sin.* 2003;26:296–299.
99. Jang JY, Kim TS, Cai J, et al. Nattokinase improves blood flow by inhibiting platelet aggregation and thrombus formation. *Lab Anim Res.* 2013;29:221–225.
100. Fadl N, Ahmed H, Booles H, Sayed A. Serrapeptase and nattokinase intervention for relieving Alzheimer's disease pathophysiology in rat model. *Hum Exp Toxicol.* 2013;32:721–735.
101. Ji H, Yu L, Liu K, et al. Mechanisms of nattokinase in protection of cerebral ischemia. *Eur J Pharmacol.* 2014;745:144–151.
102. Rabie MA, Elsaidy S, el-Badawy AA, Siliha H, Malcata FX. Biogenic amine contents in selected Egyptian fermented foods as determined by ion-exchange chromatography. *J Food Prot.* 2011;74:681–685.
103. Patra JK, Das G, Paramithiotis S, Shin HS. Kimchi and other widely consumed traditional fermented foods of Korea: a review. *Front Microbiol.* 2016;7:1493.
104. Park S, Ji Y, Jung HY, et al. *Lactobacillus plantarum* HAC01 regulates gut microbiota and adipose tissue accumulation in a diet-induced obesity murine model. *Appl Microbiol Biotechnol.* 2017;101:1605–1614.
105. Kil J-H, Jung K-O, Lee H-S, Hwang I-K, Kim Y-J, Park K-Y. Effects of kimchi on stomach and colon health of *Helicobacter pylori*-infected volunteers. *Prev Nutr Food Sci.* 2004;9:161–166.
106. Han K, Bose S, Wang J-H, et al. Contrasting effects of fresh and fermented kimchi consumption on gut microbiota composition and gene expression related to metabolic syndrome in obese Korean women. *Mol Nutr Food Res.* 2015;59:1004–1008.
107. Park J-S, Joe I, Rhee PD, Jeong C-S. A lactic acid bacterium isolated from kimchi ameliorates intestinal inflammation in DSS-induced colitis. *J Microbiol.* 2017;55:304–310.
108. Park K-B, Kim S-G, Oh C-H, Jeon J-I, Oh S-H. Gastric cancer cell growth inhibitory effects of cabbage kimchi by fermentation and storage conditions. *Korean J Food Nutr.* 2014;27:692–698.
109. Kim J, Bang J, Beuchat LR, Kim H, Ryu J-H. Controlled fermentation of kimchi using naturally occurring antimicrobial agents. *Food Microbiol.* 2012;32:20–31.
110. Sim KH, Han YS. Effect of red pepper seed on kimchi antioxidant activity during fermentation. *Food Sci Biotechnol.* 2008;17:295–301.
111. Ke L, Yu P, Zhang ZX. Novel epidemiologic evidence for the association between fermented fish sauce and esophageal cancer in South China. *Int J Cancer.* 2002;99:424–426.
112. Chen CS, Pignatelli B, Malaveille C, et al. Levels of direct-acting mutagens, total *N*-nitroso compounds in nitrosated fermented fish products, consumed in a high-risk area for gastric cancer in southern China. *Mutat Res.* 1992;265:211–221.

CHAPTER 5

Contribution of gut microbiota in the pathogenesis of amyotrophic lateral sclerosis and Huntington's disease

Akhlaq A. Farooqui
Department of Molecular and Cellular Biochemistry, The Ohio State University, Columbus, OH, United States

1 Introduction

Amyotrophic lateral sclerosis (ALS) and Huntington's disease (HD) are neurodegenerative diseases characterized by misfolding, aggregation, and accumulation of TDP43 and huntingtin (HTT), respectively, as well as neurodegeneration in the brain.[1] ALS is a fatal motor neuron disease typified by the degeneration of both upper and lower motor neurons. Loss of these neurons causes muscle weakness, spasticity, atrophy, paralysis, and premature death.[2] Common symptoms of ALS are muscle tenderness, cramping, twitching, and muscle impairment. Later in the advance stage of ALS, patients experience dysphagia (swallowing difficulty), dysarthria (speech difficulty), and dyspnea (breathing difficulty).[3,4] Until recently, ALS was classified primarily within the neuromuscular domain, although new imaging and neuropathological data have indicated the involvement of the nonmotor neuraxis in disease pathology.[5,6] It is estimated that by 2040 the number of ALS patients will increase to 400,000 worldwide.[7] ALS is more common in men than in women. Causes of ALS remain unknown. However, it is suggested that mitochondrial dysfunction, excitotoxicity, oxidative stress, protein aggregation, altered cholinergic synapse, neuroinflammation, production of pro-inflammatory cytokines, problems with autophagy, and altered RNA metabolism are closely associated with the pathogenesis of ALS.[8–10] Approximately 10% of ALS patients have a family history of disease, whereas the remainder (90%) of cases are classified as sporadic.[11]

Gut Microbiota in Neurologic and Visceral Diseases
https://doi.org/10.1016/B978-0-12-821039-0.00004-6

2 Pathogenesis of familial ALS

Many genes have been implicated in the pathogenesis of familial ALS (Table 1). Among 30 genes that have been implicated in ALS, genes encoding for Cu/Zn superoxide dismutase (*SOD1*), *TARDBP* (the gene encoding 43 kDa transactive response), and *TDP-43* (gene for DNA-binding protein) have been extensively investigated for mutation.[15,16] SOD1 is an enzyme that catalyzes the dismutation of superoxide radicals and protects the cell against reactive oxygen species. Mutant *SOD1* leads to aggregation of SOD1 protein in motor neurons within the central nervous system (CNS).[17] SOD1 mutant mice have been reported to show length-dependent axonopathy with synaptic degeneration[18] and a decrease in synaptophysin-positive presynaptic bouton in the remaining motor neurons.[19] In addition to changes in SOD1, alterations also occur in metabolic proteins including proteins associated with glycolysis, β-oxidation, and mitochondrial metabolism in $SOD1^{G93A}$ in the mouse spinal cord well before onset of ALS.[17] These changes are accompanied by a decrease in maximal respiration and glycolytic flux in $SOD1^{G93A}$ dermal fibroblasts and an increase in hydrogen peroxide and lipid hydroperoxide production in mitochondria from sciatic nerve and gastrocnemius muscle fibers at the end stage of the disease.

Mutation and aggregation of another DNA-/RNA-binding protein called TDP-43 is observed in familial ALS and frontotemporal lobar degeneration.[20] In ALS this protein modulates RNA splicing and micro RNA biogenesis.[21] Transgenic animals with mutated human TDP-43 exhibit not only reduction in levels of synaptophysin in the brain but also cognitive dysfunction due to impairment in motor function, as evidenced by behavioral tests.[22,23] FUS is another DNA/RNA-binding protein; it has 526 amino

Table 1 Genes associated with the pathogenesis of familial ALS.

Gene	Reference
MATR3	12
CHCHD1O	12
TBK1	12
TUBA4A	12
NEK1	12
C21 or f2	12
CCNF	12
SOD1	
PFN1	13
TDP-43	14

acids. This protein has a similar structure and functions as TDP-43.[24] The mutation in nuclear localization signal (NLS) of FUS leads to increased cytoplasmic FUS position, which induces the aggregation of FUS mutants, as pathogenesis of ALS proceeds.[25] The normal physiological function of FUS is unclear. However, known roles include transcriptional control,[26] RNA processing through splicing regulation of pre-mRNAs,[27] and DNA repair.[28,29] In addition, mutations in FUS can also be linked to synaptic dysfunctions. Overexpression of mutated FUS disrupts formation of pre-synaptic active zones, consequently reducing the synaptic transmission with decreased quantal size.[30] In addition, there are other factors such as oligogenic nature of the disease that may contribute to the heritability of ALS. The difficulty in uncovering the genetic determinants of ALS suggests the disease has a complex genetic architecture that may consist of combinations of gene variants differing in frequency and noxiousness. Interest in the potential oligogenic nature of ALS has recently increased. The number of ALS patients reported to have more than one ALS risk variant varies depending on the study, but has been reported to be between 1% and 4%.[31,32]

3 Involvement of the kynurenine pathway in the pathogenesis of sporadic ALS

Despite intensive research over many decades, the etiology of sporadic ALS remains unknown. However, it has been hypothesized that sporadic ALS may be caused by complex interactions between genetic susceptibility and environmental factors. It is also reported that induction of oxidative stress can be linked to the degeneration of skeletal muscles and motor neurons in ALS.[33–35] Recent investigations have also indicated that the kynurenine pathway (KP) is associated with neurodegeneration in ALS.[36] Inflammatory cytokines such as IFN-γ are known to activate the KP.[37,38] This pathway interconnects several processes such as neuroinflammation, excitotoxicity, oxidative stress, immune system activation, and dysregulation of energy metabolism with the pathogenesis of ALS. In mammals, more than 95% of dietary tryptophan is metabolized through the KP, ultimately leading to the production of nicotinamide adenosine dinucleotide (NAD^+). A number of metabolites of this pathway are neuroactive and can modulate activity of several glutamate receptors.[10,39] Among metabolites of the KP, pathway 3-hydroxykynurenine (3-HK) and quinolinic acid (QUIN) are neurotoxins, whereas kynurenic acid (KYNA) is a neuroprotectant. Neurotoxic metabolites 3-HK and QUIN can cross the blood-brain barrier (BBB) and promote the pathogenesis of ALS. Some KP metabolites including kynurenine (KYN), 3HK, and anthranilic acid can be actively transported by the large neutral amino acid carrier system; others enter cells via passive diffusion.[40] This suggests that KP metabolite-mediated inflammation of peripheral tissues can be translocated to the brain and spinal cord, where it can alter immune regulation and elevate neurotoxicity.[41] An excess of these metabolites may result in excessive depolarization, calcium influx, and neurotoxicity, leading to the speculation that an overproduction or suppressed removal of QUIN may be an important factor in ALS. These KP metabolites are quite stable. They can be estimated in body fluids and tissue. It is interesting to note that tryptophan (Trp) metabolites have profound effects on gut microbial composition, microbial metabolism, the host's immune system, the host-microbiome interface, and host immune system-intestinal microbiota interactions.[42] The aryl hydrocarbon receptor (AhR) mediates the regulation of intestinal immunity by Trp metabolites (as ligands of AhR), which is beneficial for immune homeostasis. Among Trp metabolites, AhR ligands consist of endogenous metabolites, including KYN, KYNA, xanthurenic acid, and cinnabarinic acid, and bacterial metabolites, including indole, indole propionic acid, indole acetic acid, skatole, and tryptamine.[42] Based on this information, it is proposed that Trp metabolites, 3-HK, and QUIN may be a relevant set of biomarkers for ALS.[36] Other possible mechanisms of neurodegeneration in ALS include mitochondrial dysfunction,[34] induction of endoplasmic reticulum (ER) stress, KP of tryptophan degradation, axonal deterioration, changes in neurofilaments, deficiencies in factors related to growth, and deposition of toxic ubiquitinated neuronal inclusions (Fig. 1).[34]

As stated, most of the aforementioned mechanisms are interconnected, and interactions among excitotoxicity, oxidative stress, and neuroinflammation along with mitochondrial dysfunction may play a major role in pathogenesis of ALS (Fig. 1).[43,44] Collectively, these studies indicate that glutamate-induced excitotoxicity and mitochondrial dysfunction generate oxidative stress in ALS. Enhanced oxidative alterations to the SOD1 protein may lead to protein misfolding and degradation and destruction of motor neurons in ALS, leading to muscle weakness.[45]

4 Contribution of gut microbiota in the pathogenesis of amyotrophic lateral sclerosis

The gut is considered "the second brain," as it contains roughly the same number of neurons as the spinal cord. The gut-neuron axis is a bidirectional communication system. Intestinal functions are largely controlled by the

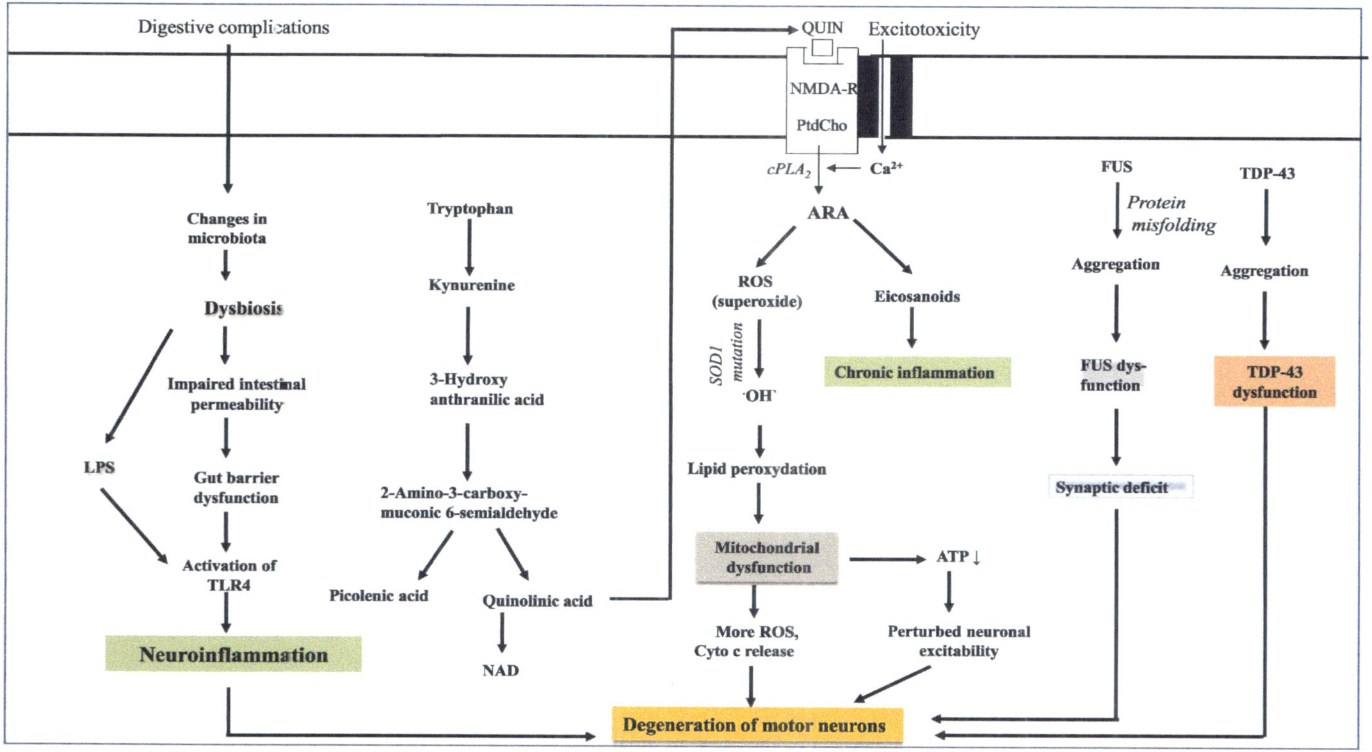

Fig. 1 Contribution of microbiota and excitotoxicity in the pathogenesis of ALS. Plasma membrane (PM); *N*-methyl-D-aspartate receptor (NMDA-R); glutamate (Glu); phosphatidylcholine (PtdCho); cytosolic phospholipase A_2 ($cPLA_2$); arachidonic acid (ARA); superoxide dismutase 1 (SOD1); *fused in sarcoma gene* (*FUS*); TAR DNA binding protein, TDP-43; lipopolysaccharide (LPS); toll-like receptor 4 (TLR4).

autonomic and enteric nervous systems.[46] The integrative function of gut-secreted hormones, endocrine factors, or bacterial products plays an important role in the physiology and pathophysiology of neurological and psychiatric diseases.[47] Many ALS patients show digestive problems (delay in gastric emptying and colonic transit times), increase in metabolic rate, intestinal barrier dysfunction, and weight loss.[48] These findings are related to alterations in microbiota composition and gut microbiome dysbiosis. Specifically, ALS patients have a greater relative abundance of the phylum Firmicutes and lesser abundance of Bacteroidetes. Importantly, a larger Firmicutes-to-Bacteroidetes ratio has been proposed as an indicator of poor gastrointestinal health, suggesting that dysbiosis of gut microbiota may contribute to ALS etiology and progression.[49,50] Some ALS patients share genetic and pathological mechanisms with frontotemporal dementia (FTD). These patients show abnormal aggregates of TAR DNA-binding protein 43 (TDP-43). Blood endotoxin (LPS) levels are increased in ALS patients,[51] possibly as a result of gut inflammation and microbiome changes.[52] It is suggested that increased endotoxin levels plus aggregable TDP-43 along with increased levels of proinflammatory cytokines in cerebrospinal fluid (CSF) may contribute to neurodegeneration in ALS and FTD.[53] Studies on mouse models of ALS have revealed that ALS damages tight junction structure and increases permeability in the intestine versus controls along with a shifted microbiome profile, including decreased levels of butyrate-producing bacteria. Dietary supplementation with butyrate relieves symptoms of ALS such as weight loss and increases survival time.[54] A small study indicated that ALS patients show decreased levels of butyrate-producing bacteria compared to healthy subjects. Butyrate is the preferred energy source of intestinal epithelial cells (Fig. 2).

5 Molecular mechanisms contributing to neurodegeneration in Huntington's disease

HD is a lethal autosomal dominant and progressive neurodegenerative disorder characterized by motor, cognitive, and behavioral impairment. Symptoms of HD include cognitive dysfunction (e.g., organizational deficit, lack of attention, and motor skill learning deficits), motor impairment (e.g., chorea, rigidity, gait abnormalities, and bradykinesia), sleep disturbance, and weight loss.[55] HD is caused by an elongated CAG repeat (36 repeats or more) on the short arm of chromosome 4p16.3 in the huntingtin gene. In HD, the huntingtin protein is expressed ubiquitously in somatic as well as brain and spinal cord tissues with profound degeneration in the striatum and

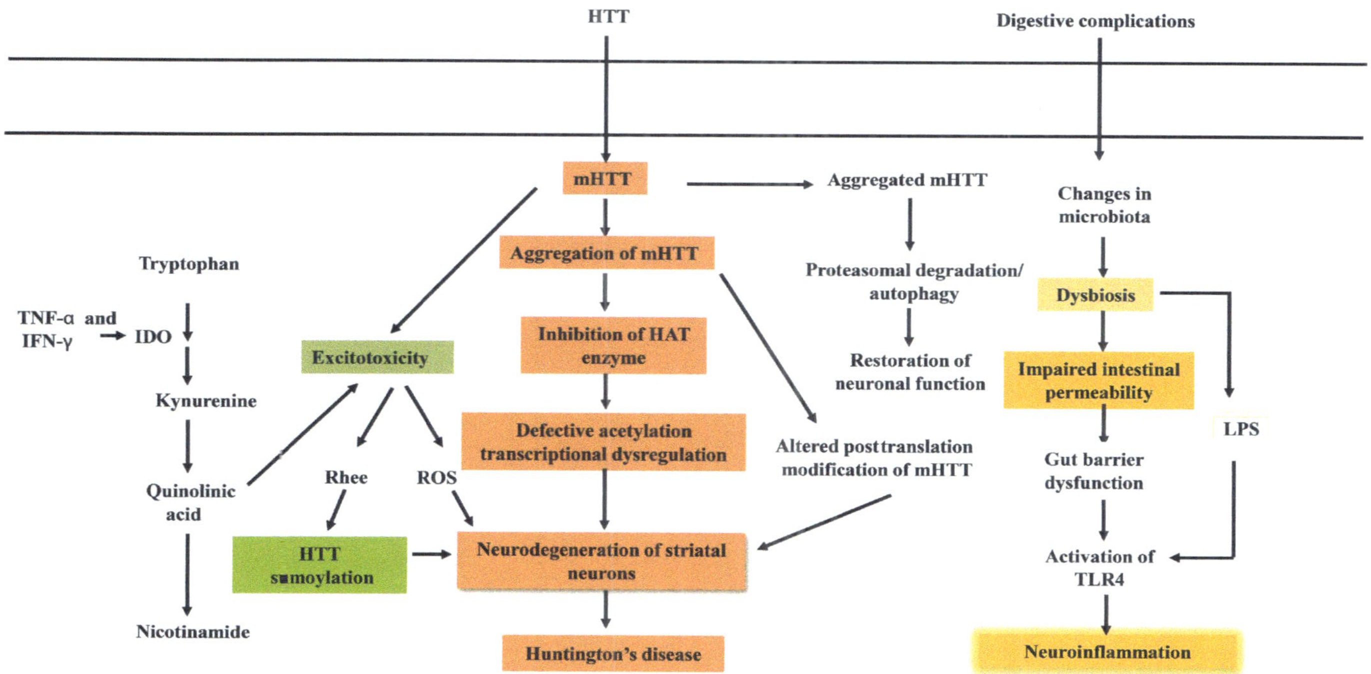

Fig. 2 Contribution of microbiota and excitotoxicity in the pathogenesis of Huntington's disease. Huntingtin (HTT); mutated Huntingtin (mHTT); reactive oxygen species (ROS); lipopolysaccharide (LPS); toll-like receptor 4 (TLR4); indoleamine 2,3-dioxygenase (IDO).

the cerebral cortex. Several studies have indicated that the large huntingtin protein (359 kDa) is cleaved to produce a shorter N-terminal fragment containing the polyglutamine expansion. The polyglutamine expansion causes the protein fragment to misfold and form aggregates (inclusions) of huntingtin to produce mutant huntingtin (mHTT) intracellularly.[56] This protein not only has ability to undergo phosphorylation, acetylation, and palmitoylation, but it also contains several domains that can interact with other macromolecules.[57] Mutant mHTT holds an unusual polyglutamine sequence, corresponding to the expanded CAG repeat, which is known to be toxic in nature and results in neuronal cell death or dysfunction. Neurons of the striatal region are particularly prone to this mutant protein; however, HD has been documented as a disorder of whole brain and body. Abnormality of huntingtin protein leads to neuronal death through several mechanisms including direct effect of the mutant protein exon 1 and tendency of the mutant protein to form aggregates with direct effect on axonal transport, protein homeostasis, and mitochondrial function.[58,59] Despite the disease being caused by a single gene, there is a major variability in the neuropathology as well as major heterogeneity in the symptom profiles observed in HD patients.[60] mHTT aggregates are localized both in the nucleus and cytosol.[61] mHTT activates the glutamate receptor to produce excitotoxicity, which is partially responsible for neurodegeneration in HD. However, studies on correlation among mHTT aggregates, regional neurodegeneration, and HD progression indicate that this correlation is weak.[62] Therefore, it is not known whether the inclusions are genuinely harmful or beneficial. However, it has been suggested that mHTT may cause neurodegeneration through transcriptional dysregulation resulting in multiple cellular dysfunctions such as induction of oxidative stress, impairment of proteasome function, interference with cellular trafficking, axonal transport defects, reduction in neurotrophic transcription (BDNF), induction of mitochondrial dysfunction, excitotoxicity, and neuroinflammation along with elevated levels of interleukins and tumor necrosis factor (Fig. 3).[63–66] It is reported that neuron-bearing mHTT inclusions undergo a rapid aging process leading to neuronal cell death via apoptosis as well as necrosis.[67] It is reported that mHTT monomers are positive for caspase-3 and die quickly, but for those with mHTT aggregates, cells experience a delayed cell death,[67] which supports the argument that high-ordered aggregates or inclusions may be neuroprotective.

Recent studies have also indicated that perturbation of the KP and its metabolites are closely associated with the pathogenesis of HD.[68] As stated, KP generates three neuroactive metabolites, the neurotoxins 3-HK and

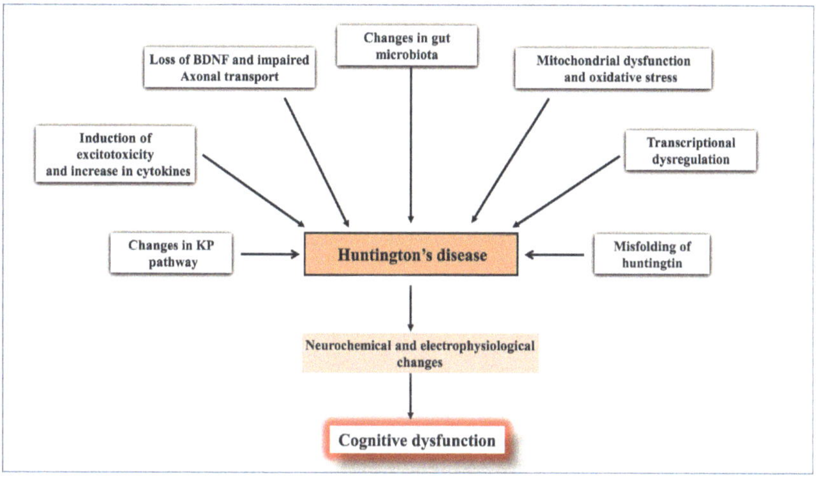

Fig. 3 Factors contributing to the pathogenesis of Huntington's disease.

QUIN, and the neuroprotectant KYNA. In model systems in vitro and in vivo, 3-HK and QUIN have been shown to cause neurodegeneration via a combination of excitotoxic mechanisms and oxidative stress. Recent studies with HD patient samples and in HD model systems support the hypothesis that a shift away from the synthesis of KYNA and toward the formation of 3-HK and QUIN may trigger the neuropathological characteristics of HD. The enzyme kynurenine 3-monooxygenase (KMO) is located at a critical branching point in the KP and inhibition of this enzyme by either pharmacological or genetic means shifts the flux in the pathway toward the formation of KYNA. This treatment ameliorates disease-relevant phenotypes in HD models.[68] In addition, L-tryptophan and its metabolites play crucial roles in the balance between intestinal immune tolerance and gut microbiota maintenance.[42] Thus, endogenous L-tryptophan metabolites (kynurenines, serotonin, and melatonin), and bacterial Trp metabolites (indole, indolic acid, skatole, and tryptamine) are known to produce profound effects on gut microbial composition, microbial metabolism, and the host's immune system through the involvement of AhR. The activation of AhR by Trp metabolites is known to induce the expression of downstream cytokines such as interleukin-22 (IL-22) and interleukin-17 (IL-17), thereby regulating intestinal homeostasis and immunity.[42,69] In addition, interactions between dietary Trp metabolites and AhR also result in microglial activation and TGFα and VEGF-B production by modulating the transcriptional program of astrocytes. These processes are closely associated with microglial control of astrocytes and suppression of neuroinflammation in the brain.[70]

6 Contribution of gut microbiota in Huntington's disease

HD patients show alterations in gut function due to abnormalities in gut microbiota, which modulate brain function through immunological, neuroendocrine, and direct neural mechanisms. Little information is available on microbiota composition in HD patients. However, studies on microbiota in transgenic mouse models of HD have indicated that there are significant differences in microbiota composition in HD compared to WT littermate control mice at 12 weeks of age, which coincides with onset of a motor coordination deficit tested on the rotarod, but prior to onset of the hind-limb clasping phenotype. Specifically, it is observed that there is an increase in Bacteriodetes and a proportional decrease in Firmicutes in the gut microbiome of HD patients.[71] Observed gut dysbiosis coincides with impairment in body weight gain despite greater food intake as well as motor deficits at 12 weeks of age. Gut dysbiosis is not only associated with a change in the gut microenvironment, but also with greater fecal water content in HD mice at 12 weeks of age.[71]

7 Conclusion

ALS and HD are neurodegenerative diseases characterized by the accumulation and aggregation of TDP43 and mHTT, respectively, as well as neurodegeneration in the brain. The vast majority of ALS cases are sporadic. However, about 10% of ALS cases are familial. In the last decade, enormous progress has been made in the etiology of ALS, with the identification of ALS-causing mutations in new genes as well as key molecular players involved in the disease's origin or progression. It has been proposed that the KP, the major biochemical pathway for tryptophan metabolism, is dysregulated in ALS. In particular, the neuroactive intermediate QUIN has been shown to accumulate with a concomitant decrease in other neuroprotective and immunomodulatory KP metabolites.

HD is a lethal autosomal dominant and progressive neurodegenerative disorder characterized by motor, cognitive, and behavioral impairment. Onset of HD symptoms emerges usually at 35–45 years of age and varies considerably. HD leads to severe brain atrophy and death, with a clinical course that spans 15–20 years. In HD, neurotoxicity is mediated by misfolded mHTT in its soluble form, and/or in aggregates, and/or in the process of aggregation. A wide range of potential mechanisms for neurotoxicity have been proposed, including caspase activation, dysregulation of transcriptional pathways, increased production of reactive oxygen species, and inhibition of proteasome activity. Despite the identification of the mHTT gene as critical for the pathogenesis of HD, effective therapies for the disease remain elusive.

References

1. Farooqui AA. *Neurochemical Aspects of Neurotraumatic and Neurodegenerative Diseases.* New York: Springer; 2010.
2. Zarei S, Carr K, Reiley L, Diaz K, Guerra O, et al. A comprehensive review of amyotrophic lateral sclerosis. *Surg Neurol Int.* 2015;6:171.
3. Kori M, Aydın B, Unal S, Arga KY, Kazan D. Metabolic biomarkers and neurodegeneration: a pathway enrichment analysis of Alzheimer's disease, Parkinson's disease, and amyotrophic lateral sclerosis. *Omics.* 2016;20:645–661.
4. Marin B, Desport JC, Kajeu P, et al. Alteration of nutritional status at diagnosis is a prognostic factor for survival of amyotrophic lateral sclerosis patients. *J Neurol Neurosurg Psychiatry.* 2011;82:628–634.
5. Eisen A, Weber M. The motor cortex and amyotrophic lateral sclerosis. *Muscle Nerve.* 2001;24:564.
6. Mochizuki Y, Mizutani T, Takasu T. Amyotrophic lateral sclerosis with marked neurological asymmetry: clinicopathological study. *Acta Neuropathol.* 1995;90:44.
7. Arthur KC, Calvo A, Price TR, Geiger JT, Chiò A, Traynor BJ. Projected increase in amyotrophic lateral sclerosis from 2015 to 2040. *Nat Commun.* 2016;7:12408.
8. Chen S, Zhang X, Song L, Le W. Autophagy dysregulation in amyotrophic lateral sclerosis. *Brain Pathol.* 2012;22:110–116.
9. Droppelmann CA, Campos-Melo D, Ishtiaq M, Volkening K, Strong MJ. RNA metabolism in ALS: when normal processes become pathological. *Amyotroph Lateral Scler Frontotemporal Degener.* 2014;15:321–326.
10. Rothstein JD. Current hypotheses for the underlying biology of amyotrophic lateral sclerosis. *Ann Neurol.* 2009;65(Suppl. 1):S3–S9.
11. Talbott EO, Malek AM, Lacomis D. The epidemiology of amyotrophic lateral sclerosis. *Handb Clin Neurol.* 2016;138:225–238.
12. Chia R, Chio A, Traynor BJ. Novel genes associated with amyotrophic lateral sclerosis: diagnostic and clinical implications. *Lancet Neurol.* 2018;17:94–102.
13. Yang C, Danielson EW, Qiao T, et al. Mutant PFN1 causes ALS phenotypes and progressive motor neuron degeneration in mice by a gain of toxicity. *Proc Natl Acad Sci U S A.* 2016;113:E6209–E6218.
14. Xu Z, Yang C. TDP-43-the key to understanding amyotrophic lateral sclerosis. *Rare Dis.* 2014;2:e944443.
15. Maurel C, Dangoumau A, Marouillat S, et al. Causative genes in amyotrophic lateral sclerosis and protein degradation pathways: a link to neurodegeneration. *Mol Neurobiol.* 2018;55:6480–6499.
16. Pansarasa O, Bordoni M, Diamanti L, et al. SOD1 in amyotrophic lateral sclerosis: ambivalent behavior connected to the disease. *Int J Mol Sci.* 2018;19:1345.
17. Pharaoh G, Sataranatarajan K, Street K, et al. Metabolic and stress response changes precede disease onset in the spinal cord of mutant SOD1 ALS mice. *Front Neurosci.* 2019;13:487.
18. Tallon C, Russell KA, Sakhalkar S, Andrapallayal N, Farah MH. Length-dependent axo-terminal degeneration at the neuromuscular synapses of type II muscle in SOD1 mice. *Neuroscience.* 2016;312:179–189.

19. Zang DW, Lopes EC, Cheema SS. Loss of synaptophysin-positive boutons on lumbar motor neurons innervating the medial gastrocnemius muscle of the SOD1G93A G1H transgenic mouse model of ALS. *J Neurosci Res.* 2005;79:694–699.
20. Neumann M, Sampathu DM, Kwong LK, et al. Ubiquitinated TDP-43 in frontotemporal lobar degeneration and amyotrophic lateral sclerosis. *Science.* 2006;314:130–133.
21. Xu ZS. Does a loss of TDP-43 function cause neurodegeneration? *Mol Neurodegener.* 2012;7:27.
22. Handley EE, Pitman KA, Dawkins E, et al. Synapse dysfunction of layer V pyramidal neurons precedes neurodegeneration in a mouse model of TDP-43 proteinopathies. *Cereb Cortex.* 2017;27:3630–3647.
23. Medina DX, Orr ME, Oddo S. Accumulation of C-terminal fragments of transactive response DNA-binding protein 43 leads to synaptic loss and cognitive deficits in human TDP-43 transgenic mice. *Neurobiol Aging.* 2014;35:79–87.
24. Nolan M, Talbot K, Ansorge O. Pathogenesis of FUS-associated ALS and FTD: insights from rodent models. *Acta Neuropathol Commun.* 2016;4:99.
25. Da Cruz S, Cleveland DW. Understanding the role of TDP-43 and FUS/TLS in ALS and beyond. *Curr Opin Neurobiol.* 2011;21:904–919.
26. Tan AY, Riley TR, Coady T, Bussemaker HJ, Manley JL. TLS/FUS (translocated in liposarcoma/fused in sarcoma) regulates target gene transcription via single-stranded DNA response elements. *Proc Natl Acad Sci U S A.* 2012;109:6030–6035.
27. Lagier-Tourenne C, Polymenidou M, Hutt KR, et al. Divergent roles of ALS-linked proteins FUS/TLS and TDP-43 intersect in processing long pre-mRNAs. *Nat Neurosci.* 2012;15:1488–1497.
28. Mastrocola AS, Kim SH, Trinh AT, Rodenkirch LA, Tibbetts RS. The RNA-binding protein fused in sarcoma (FUS) functions downstream of poly(ADP-ribose) polymerase (PARP) in response to DNA damage. *J Biol Chem.* 2013;288:24731–24741.
29. Wang WY, Pan L, Su SC, et al. Interaction of FUS and HDAC1 regulates DNA damage response and repair in neurons. *Nat Neurosci.* 2013;16:1383–1391.
30. Machamer JB, Collins SE, Lloyd TE. The ALS gene FUS regulates synaptic transmission at the Drosophila neuromuscular junction. *Hum Mol Genet.* 2014;23:3810–3822.
31. Cady J, Allred P, Bali T, et al. Amyotrophic lateral sclerosis onset is influenced by the burden of rare variants in known amyotrophic lateral sclerosis genes. *Ann Neurol.* 2015;77:100–113.
32. Morgan S, Shatunov A, Sproviero W, et al. A comprehensive analysis of rare genetic variation in amyotrophic lateral sclerosis in the UK. *Brain.* 2017;140:1611–1618.
33. Ohta Y, Nomura E, Shang J, Feng T, Huang Y, et al. Enhanced oxidative stress and the treatment by edaravone in mice model of amyotrophic lateral sclerosis. *J Neurosci Res.* 2018;97.
34. Sreedharan J, Brown Jr RH. Amyotrophic lateral sclerosis: problems and prospects. *Ann Neurol.* 2013;74:309–316.
35. Wang Z, Bai Z, Qin X, Cheng Y. Aberrations in oxidative stress markers in amyotrophic lateral sclerosis: a systematic review and meta-analysis. *Oxid Med Cell Longev.* 2019;2019, 1712323.
36. Tan VX, Guillemin GJ. Kynurenine pathway metabolites as biomarkers for amyotrophic lateral sclerosis. *Front Neurosci.* 2019;13:1013.
37. Chen Y, Stankovic R, Cullen KM, et al. The Kynurenine pathway and inflammation in amyotropic lateral sclerosis. *Neurotox Res.* 2010;18:132–142.
38. Oxenkrug GF. Interferon-gamma-inducible kynurenines/pteridines inflammation cascade: implications for aging and aging-associated psychiatric and medical disorders NIH public access. *J Neural Transm.* 2011;118:75–85.
39. Maddison DC, Giorgini F. The kynurenine pathway and neurodegenerative disease. *Semin Cell Dev Biol.* 2015;40:134–141.

40. Ruddick JP, Evans AK, Nutt DJ, Lightman SL, Rook GAW, Lowry CA. Tryptophan metabolism in the central nervous system: medical implications. *Expert Rev Mol Med.* 2006;8:1–27.
41. Owe-Young R, Webster NL, Mukhtar M, et al. Kynurenine pathway metabolism in human blood-brain-barrier cells: implications for immune tolerance and neurotoxicity. *J Neurochem.* 2008;105:1346–1357.
42. Gao J, Xu K, Liu H, Liu G, Bai M, et al. Impact of the gut microbiota on intestinal immunity mediated by tryptophan metabolism. *Front Cell Infect Microbiol.* 2018;8:13.
43. Barber SC, Shaw PJ. Oxidative stress in ALS: key role in motor neuron injury and therapeutic target. *Free Radic Biol Med.* 2010;48:629–641.
44. Rao SD, Weiss JH. Excitotoxic and oxidative cross-talk between motor neurons and glia in ALS pathogenesis. *Trends Neurosci.* 2004;27:17–23.
45. Johri A, Beal MF. Antioxidants in Huntington's disease. *Biochim Biophys Acta.* 2012;1822:664–674.
46. Pfeiffer RF. Gastrointestinal dysfunction in Parkinson's disease. *Parkinsonism Relat Disord.* 2011;17:10–15.
47. Hsiao EY, McBride SW, Hsien S, et al. Microbiota modulate behavioral and physiological abnormalities associated with neurodevelopmental disorders. *Cell.* 2013;155:1451–1463.
48. Steyn FJ, Ioannides ZA, van Eijk RPA, et al. Hypermetabolism in ALS is associated with greater functional decline and shorter survival. *J Neurol Neurosurg Psychiatry.* 2018;89:1016–1023.
49. Fournier CN, Houser M, Tansey MG, Glass JD, Hertzberg VS. The gut microbiome and neuroinflammation in amyotrophic lateral sclerosis? Emerging clinical evidence. *Neurobiol Dis.* 2018;12:104300.
50. Zhai C-D, Zheng J-J, An B-C, Huang H-F, Tan Z-C. Intestinal microbiota composition in patients with amyotrophic lateral sclerosis: establishment of bacterial and archaeal communities analyses. *Chin Med J (Engl).* 2019;132:1815–1822.
51. Zhang R, Miller RG, Gascon R, et al. Circulating endotoxin and systemic immune activation in sporadic amyotrophic lateral sclerosis (sALS). *J Neuroimmunol.* 2009;206:121–124.
52. Rowin J, Xia Y, Jung B, Sun J. Gut inflammation and dysbiosis in human motor neuron disease. *Physiol Rep.* 2017;5:e13443.
53. Correia AS, Patel P, Dutta K, Julien JP. Inflammation induces TDP-43 mislocalization and aggregation. *PLoS One.* 2015;10:e0140248.
54. Wright ML, Fournier C, Houser MC, Tansey M, Glass J, Hertzberg VS. Potential role of the gut microbiome in ALS: a systematic review. *Biol Res Nurs.* 2018;20:513–521.
55. Roos RA. Huntington's disease: a clinical review. *Orphanet J Rare Dis.* 2010;5:40.
56. Suhr ST, Senut MC, Whitelegge JP, Faull KF, Cuizon DB, Gage FH. Identities of sequestered proteins in aggregates from cells with induced polyglutamine expression. *J Cell Biol.* 2001;153:283–294.
57. Saudou F, Humbert S. The biology of huntingtin. *Neuron.* 2016;89:910–926.
58. Kay C, Collins JA, Skotte NH, et al. Huntingtin haplotypes provide prioritized target panels for allele-specific silencing in Huntington disease patients of European ancestry. *Mol Ther.* 2015;23:1759–1771.
59. Ross CA, Aylward EH, Wild EJ, et al. Huntington disease: natural history, biomarkers and prospects for therapeutics. *Nat Rev Neurol.* 2014;10:204–216.
60. Waldvogel HJ, Kim EH, Thu DC, Tippett LJ, Faull RL. New perspectives on the neuropathology in Huntington's disease in the human brain and its relation to symptom variation. *J Huntingtons Dis.* 2012;1:143–153.
61. Gutekunst CA, Li SH, Yi H, et al. Nuclear and neuropil aggregates in Huntington's disease: relationship to neuropathology. *J Neurosci.* 1999;19:2522–2534.

62. Bañez-Coronel M, Ayhan F, Tarabochia AD, et al. RAN translation in Huntington disease. *Neuron*. 2015;88:667–677.
63. Burnett BG, Andrews J, Ranganathan S, Fischbeck KH, Di Prospero NA. Expression of expanded polyglutamine targets profilin for degradation and alters actin dynamics. *Neurobiol Dis*. 2008;30:365–374.
64. Crotti A, Glass CK. The choreography of neuroinflammation in Huntington's disease. *Trends Immunol*. 2015;36:364–373.
65. Estrada Sánchez AM, Mejía-Toiber J, Massieu L. Excitotoxic neuronal death and the pathogenesis of Huntington's disease. *Arch Med Res*. 2008;39:265–276.
66. Jodeiri Farshbaf M, Ghaedi K. Huntington's disease and mitochondria. *Neurotox Res*. 2017;32:518–529.
67. Ramdzan YM, Trubetskov MM, Ormsby AR, et al. Huntingtin inclusions trigger cellular quiescence, deactivate apoptosis, and lead to delayed necrosis. *Cell Rep*. 2017;19:919–927.
68. Thevandavakkam MA, Schwarcz R, Muchowski PJ, Giorgini F. Targeting kynurenine 3-monooxygenase (KMO): implications for therapy in Huntington's disease. *CNS Neurol Disord Drug Targets*. 2010;9:791–800.
69. Sun M, Ma N, He T, Johnston LJ, Ma X. Tryptophan (Trp) modulates gut homeostasis via aryl hydrocarbon receptor (AhR). *Crit Rev Food Sci Nutr*. 2020;60:1760–1768.
70. Rothhammer V, Borucki DM, Tjon EC, Takenaka MC, Chao CC, et al. Microglial control of astrocytes in response to microbial metabolites. *Nature*. 2018;557:724–728.
71. Kong G, Cao KL, Judd LM, Li S, Renoir T, Hannan AJ. Microbiome profiling reveals gut dysbiosis in a transgenic mouse model of Huntington's disease. *Neurobiol Dis*. 2018;5:104268.

CHAPTER 6

The gut microbiome in neurodegenerative disorders

Shaadi J. Abughazaleh[a], Lindsay A. Euers[a], and Eamonn M.M. Quigley[a,b]
[a]Department of Internal Medicine, Houston Methodist Hospital and Weill Cornell Medical College, Houston, TX, United States
[b]Lynda K and David M Underwood Center of Digestive Disorders, Houston Methodist Hospital and Weill Cornell Medical College, Houston, TX, United States

1 Introduction

The gut microbiome refers to the collective genomes of trillions of microorganisms including bacteria, viruses, protozoa, and fungi that form a complex ecosystem that resides in the gastrointestinal (GI) system of its host. These organisms and their interactions amongst both themselves and with their host environment result in complex relationships with effects extending well beyond the confines of the GI system. The relationship between the host and its microbiome is considered mutualistic as the human body provides the environment and sustenance necessary to maintain microbiota and the microbes provide a vast array of benefits to the host whose extent continues to be revealed.[1]

While these organisms may be microscopic, their sheer number is such that they comprise 1%–3% of our total body mass. Bacterial cells, in an average human body, number 10 times more than human cells and include a 1000-fold more genes than are present in the human genome.[2] Unlike the genome of its host, the gut microbiome is unique in the fact that it is constantly evolving and changing. As the composition of the gut microbiome changes, so do its effects, both beneficial and detrimental, on the health of the host.[3] As scientists continue to uncover information about this relationship, it has become apparent that there is a connection between the microbiome and both neuronal development and neurodegenerative diseases.

2 The microbiome: Development and function

Initially coined the *sterile wound paradigm*, it was once believed that the microbes that inhabit the intestinal tract of infants were obtained during

Gut Microbiota in Neurologic and Visceral Diseases
https://doi.org/10.1016/B978-0-12-821039-0.00006-X

both the initial birthing process (vertical transmission) and from other humans and their environments early during infancy (horizontal transmission). However, more recently, using modern sequencing technology, this once-accepted notion has been challenged by those proposing that neither the fetus, placenta, nor the amniotic fluid are sterile and are, in fact, the host of what is being termed the placental microbiome.[4–6] This "in utero colonization hypothesis" states that the fetus is exposed to these microorganisms in utero where the colonization of the GI tract begins.[7] Although there is still no firm consensus on the prenatal microbiome, both hypotheses emphasize the importance of the first 1–3 years of life in the development of what will become the mature gut microbiome. Factors that most heavily influence its composition include method of delivery (vaginal vs cesarean section), source of nutrition (breast milk vs formula), maternal weight, prenatal diet, location of birth, and exposure to antibiotics.[8–16]

Over time, the composition of gut microbiota becomes more stable, with multiple members of Bacteroidetes, including those with butyrate-producing capacity, establishing a presence. It is well documented that the initial gut microbiome consists of bacteria that are able to metabolize lactose from either breast milk or cow's milk. As solid food is introduced into the diet, the microbiome evolves to one that can metabolize carbohydrates, proteins, and fats and synthesize vitamins.[17] By preadolescence (7–12 years of age), the number of bacterial taxa and functional genes present in the gut microbiome has matured to what will persist throughout most of adulthood.[18] The adult microbiome is largely dominated by two phyla, Firmicutes and Bacteroidetes, which some have subclassified into so-called enterotypes based on the prominence of one of the following genera: *Prevotella*, *Bacteroides*, or *Ruminococcus*.[19] Later in life the gut microbiome appears to undergo some age-related changes characterized by a proliferation of opportunistic *Proteobacteria* at the expense of symbiont Firmicutes and Bacteroidetes; phyla that include species with recognized antiinflammatory properties.[20] It is fair to state that the precise nature and clinical significance of aging-related changes in gut microbiota remain to be defined; given that many neurodegenerative diseases occur in the elderly, defining a "normal" older persońs microbiota is of critical importance before we can ascribe any observed changes to disease.[21] Moreover, diet may well be a critical determinant of age-related changes in microbiota of the elderly, and, if inadequate or poor in diversity and quality, may lead to a reduction in species richness and a distorted gut ecology.[22,23]

What exactly does our microbiome do? These microorganisms are involved in a whole host of functions including the development and maturation of the mucosal immune system,[24–26] maintaining the integrity of the gut barrier,[27] modulating gut neuromuscular function,[28–30] and performing a number of key metabolic functions.[31,32] These functions include both the metabolism of dietary fibers and starches into short-chain fatty acids (SCFAs), which, in turn, support the proliferation of microbial species and are beneficial to host health,[33] the metabolism of bile acids, and the biosynthesis of neurotransmitters and other potentially neuro-active molecules.[18,34] Diet has been found to substantially influence microbiota throughout all phases of life and is an important confounder in studies of gut microbiota in disease. The features (i.e., total calories, consumption of highly processed foods vs a vegetable- and fruit-based diet)[22,35–37] as well as the individual components of our diet such as carbohydrates,[38,39] proteins,[40] fats,[41] fiber,[42–44] and vitamins[45] all influence the composition of microbiota. While dietary changes generally take effect over long periods of time, they can, if sufficiently drastic, evoke alterations in microbiota composition in the short term.[19,21,36,37,46–49] Many other factors play a role in microbiome composition and diversity, including genetic predisposition,[50] environmental stressors,[51,52] and medications such as antibiotics, proton pump inhibitors, and metformin.[53–57] Any or all of these factors may confound the interpretation of microbiome studies in neurodegenerative disease.

3 Overview of the gut-brain axis

The gut-brain axis is a complex, bi-directional, communication system that is not only fundamental to the function of the GI tract, but also influences affect, motivation, and higher cognitive functions. The gut-brain axis communicates bi-directionally using neuro-immuno-endocrine mediators to monitor and influence gut and central nervous system function. This includes linking emotional and cognitive areas of the brain with the enteric nervous system, which, in turn, regulates motility, secretion, immune responses, entero-endocrine signaling, gut permeability, and local reflexes. This bi-directional signaling network encompasses all components of the nervous system including the central (CNS), autonomic (ANS), and enteric (ENS) nervous systems as well as the hypothalamic-pituitary-adrenal axis (HPA), which in conjunction with hormonal signaling directly modulates the activities of immune, epithelial, smooth muscle, and enterochromaffin cells.[52] Its influence can be readily appreciated in the context of GI dysfunction and symptomatology induced by stress.

This gut brain-axis has been linked to a growing number of diseases and syndromes. In autism spectrum disorder (ASD), suspicion of its relevance first arose when it was noted that up to 70% of patients reported abdominal symptoms.[58] A subsequent report of significant decreases in the Bacteroidetes/Firmicutes ratio and elevations in *Lactobacillus* spp. in children diagnosed with ASD and of the normalization of these microbial alterations following initiation of a probiotic-supplemented diet suggested a role for the gut microbiome.[59] It should be stressed, however, that a recent meta-analysis incorporating 26 studies failed to identify a consistent gut microbiome signal in ASD.[60]

Another prototypical disorder exemplifying the clinical manifestations of the gut-brain axis is irritable bowel syndrome (IBS). Stress has been considered a key factor in not only the initiation but the ongoing recurrence and severity of symptoms in IBS; stress has also been found to impact nearly all aspects of gut physiology including motility, visceral perception, GI permeability, and GI secretions.[61] Mounting evidence has shown that an abnormal and exaggerated stress response may be mediated via gut microbiota.[62]

4 Microbiota-gut-brain axis

In animal models, the gut-brain axis has been found to play a key role in brain development through the upregulation of genes associated with plasticity and metabolism including the expression of brain-derived neurotrophic factor (BDNF) in the cortex, hippocampus, and amygdala.[63]

Aside from its key role in brain development, some of the most compelling studies documenting the influence gut microbiota have on the brain come from germ-free (GF) animals. These animals, which lack a gut microbiome, have been found to have impaired adult hippocampal neurogenesis,[64] altered neural morphology in the amygdala,[65] and perturbation in the development and maturation of microglia in the brain.[66,67] It has also been shown that the absence of the microbiome is associated with altered expression and metabolism of neurotransmitters in both the CNS and peripheral nervous systems (PNS); consequences for germ-free (GF) animals include abnormalities in gut sensory-motor functions (including dysmotility),[68,69] modulation of the set point for HPA activation (exaggerated stress response),[62,70] and reduced expression of BDNF (memory dysfunction).[71] Many of these proved to be reversible following reconstitution of the gut microbiome.

Reflecting the bi-directional nature of the microbiota-gut-brain axis, not only does the microbiome influence the CNS, but the brain can modulate the composition and total biomass of GI microbiota. Studies have shown that periods of stress as short as 2 hours can result in significant changes to the makeup of the microbiome.[72] It is believed that the ANS and HPA act directly both on neurotransmitter and neuroendocrine receptors on organisms within microbiota to ultimately influence the function of components of microbiota. This, in turn, can alter the host's predisposition to inflammatory and infectious stimuli.[73] Aside from the braińs interactions directly on microbiota, its modulation of such gut functions as intestinal permeability, motility, mucosal immune responses, and secretion of fluid, electrolytes, and mucus may influence microbiota composition and function.[74]

5 Microbiota and neurodegenerative diseases

A link between the microbiome and disease has been described in both CNS and psychiatric disorders. A number of hypotheses have been advanced to explain how microbiota-gut-brain interactions impact the maturation and development of the CNS. These include the influence of microbiota on the regulation of the immune system and inflammatory responses and the effects of metabolic byproducts of the microbiome.[75] It has been hypothesized that these microorganisms can control the upregulation or downregulation of various proteins, signaling factors, and receptors in a manner that ultimately manifests as a neuropsychological disorder.[76] Here, we will focus on the impact the microbiota has on neurodegenerative diseases.

Neurodegenerative diseases can be defined as a group of disorders that result in progressive degeneration in the structure and function of either the CNS, PNS, or both. This group of diseases includes Alzheimer's disease (AD), Parkinsońs disease (PD), Huntingtońs disease (HD), amyotrophic lateral sclerosis (ALS), and multiple sclerosis (MS). These diseases share multiple neuropathological traits and are characterized by the accumulation of abnormally folded proteins, neuronal loss in characteristic brain areas, and inflammatory components. Despite efforts to better understand their pathophysiology and uncover targets for treatment, success in limiting disease progression has been limited. With growing evidence implicating an impact of microbiota on brain metabolism and function, it is now believed that alterations in the microbiome may play a role in the etiopathology of some, if not all, of these neurodegenerative diseases.[77]

5.1 Microbiota and Alzheimer's disease

AD is a neurodegenerative disease characterized by the accumulation of amyloid-β and tau protein in the brain forming senile plaques, and typically manifests as progressive cognitive impairment along with memory and language deficits. The microbiome has been directly linked to the development and progression of AD. It has been hypothesized that the microbiome's effect stems from the immense amounts of lipopolysaccharide (LPS) and amyloid that it can produce. Additionally, during the aging process, both the gut barrier and the blood-brain barrier become increasing permeable.[78] Accordingly, it is hypothesized that LPS and amyloid pass through permeable gut and blood-brain barriers and then, through their activation of cytokines and other small proinflammatory molecules, lead to potentially harmful effects on the CNS.[79,80]

LPS, elaborated by Gram-negative bacteria, has been shown to exert pro-inflammatory effects in a variety of experimental animal models. Chronic infusion of LPS into the ventricles of rats has been shown to produce many of the same inflammatory and pathological features that characterize AD.[81] Along with this observation, studies have linked LPS to the potentiation of the fibrillogenesis that is fundamental to the development of amyloid-β.[82] In one study it was shown that lipopolysaccharaide from *Bacteroides fragilis* (BF-LPS) was an exceedingly potent inducer of the nuclear factor kappa-light-chain-enhancer of activated B cells (NF-κB) (p50/p65) complex, which is involved in the inflammatory neurodegeneration seen in AD.[83]

Amyloid, another known contributor in the development of AD, is also generated in significant amounts by bacterial strains that are common members of human gut microbiota including *Escherichia coli*, *Bacillus subtilis*, *Salmonella typhimurium*, *Salmonella enterica*, *Mycobacterium tuberculosis*, and *Staphylococcus aureus*.[84,85] Many believe that this bacteria-produced amyloid may be the source of misfolding of neuronal proteins via a method termed cross-seeding, which then initiates neuroinflammation.[86,87] One specific amyloid fiber that has been studied in great detail is the curli protein. Experiments have demonstrated that these proteins play a major role in promoting immune tolerance to commensal gut and oral bacteria and assist in maintaining the integrity of the intestinal barrier.[88,89] These curli proteins play a protective role when confined to the GI system; however, outside of the GI tract, these fibers likely act as antigens that trigger a defensive innate response, which, in turn, results in β-amyloid synthesis and CNS deposition.[90]

There have been several approaches to link gut microbiota to AD; however, some of the most compelling evidence stems from GF animal studies. Here, studies have demonstrated that the accumulation of amyloid-β increases significantly in wild-type or GF mice on receipt of microbiota transplants from AD mouse lines. Others have shown that compared to conventionally reared mice, GF mice have been found to have reduced BDNF in the hippocampus, a characteristic found in AD patients that correlates with amyloid-β burden.[91–93] Decreased expression of *N*-methyl-D-aspartate (NMDA) receptor 2A (NR2A) mRNA in the cortex and hippocampus of GF mice has been documented when compared to their normal counterparts.[70] This NMDA receptor is a vital component of amyloid-β-dependent synaptic dysfunction, which is characteristic in AD.[94] These findings have led researchers to believe that, not only does a perturbed microbiota lead to undesirable effects, but that the lack of a healthy microbiome results in the loss of its potentially neuroprotective role.

Ho et al. found that valeric acid, butyric acid, and other SCFAs produced by healthy gut microbiota are strong inhibitors of the aggregation of amyloid-β.[95] Studies completed on human participants comparing the microbiome of cognitively impaired patients known to have amyloid brain deposition showed suppression of neuroprotective antiinflammatory taxa such as *Eubacterium rectale* and increased amounts of *Escherichia* and *Shigella*, which are known to be pro-inflammatory.[96]

5.2 Parkinsońs disease

PD is an incurable neurodegenerative disorder hallmarked by damage to dopaminergic neurons in the substantia nigra (SN) with findings of αSynuclein (αSyn) containing inclusion bodies termed Lewy pathology (LP) in the remnant SN. This disease typically manifests with tremors, bradykinesia, limb rigidity, and gait and balance impairment. This was one of the first diseases thought to demonstrate a potential connection between the microbiome and neurodegenerative disease with the discovery of LP in the enteric nervous system.[97,98] Since that discovery, many have found further evidence linking dysfunction in the gut-brain axis to PD. One of the most widespread concepts describes that PD may be triggered in the gut and olfactory system by some unknown neurotropic agent that propagates along the vagus nerve to its final destination in the SN through the dorsal motor nucleus of the vagus (DMV); this has been termed Braak's hypothesis.[99]

PD patients have been found to have increased expression of pro-inflammatory cytokines including tumor necrosis factor, interferon gamma, interleukin-6, and interleukin-1 beta in the colon. Levels of expression correlated with disease duration but not with symptom severity, GI symptomatology, or lifetime cumulative dose of L-dopa.[100] However, increased inflammation in the colon has been linked to increased permeability of the intestinal barrier, which, in turn, has been associated with increased numbers of bacteria, oxidative stress, and αSyn.[101] This pro-inflammatory state and associated increased permeability has been hypothesized to allow mediators of oxidative stress to enter the ENS where they result in αSyn misfolding and aggregation that may spread to the CNS leading to the characteristic motor findings seen in PD.[102]

One study by Gorecki et al. demonstrated that although the diversity of gut microbiota did not significantly differ between healthy controls and PD patients overall, diversity among those with severe PD was significantly reduced compared to those with mild PD. It was also noted that while the phyla Firmicutes, Bacteroidetes, Proteobacteria, and Actinobacteria comprised 98.4% of microbiota in control subjects they collectively accounted for only 88.1% in those with PD; concomitant increases in Gammaproteobacteria andVerrucomicrobiae were observed in PD.[103] Within PD subjects, not only were there greater amounts of Enterobacteriaceae, but one study found there was a positive correlation between abundance of Enterobacteriaceae in the feces of PD patients and severity of postural and gait symptoms and an inverse correlation to the severity of tremors.[104] PD has also been linked to common GI disorders including small intestinal bacterial overgrowth and *Helicobacter pylori* infection.[105–107] One study found that patients with PD and *H. pylori* experienced significant improvements in their levodopa onset time, on duration (in patients who suffer from on-off phenomena), motor severity, and quality of life parameters after eradication of *H. pylori* infection.[108]

5.3 Huntingtońs disease

HD is a progressive neurodegenerative disorder caused by a CAG trinucleotide repeat expansion in the *huntingtin* (HTT) gene leading to an abnormally long huntingtin protein. This results in neuronal abnormalities that manifest as uncontrolled movement, emotional problems, and loss of cognition. Although this aberrant HTT gene is expressed from embryogenesis, significant resources have been dedicated to better understanding the factors that lead to symptom manifestations decades later.

Although there are limited studies documenting gut microbiota changes in HD, there have been studies that investigated the microbiome in the brain. Alonso et al. described a variety of fungal structures in the striatum and frontal cortex of several HD patients; some, such as *Ramularia*, seemed to be unique to HD.[109] Using next-generation sequencing, they also identified prevalent bacterial genera including *Pseudomonas*, *Acinetobacter*, and *Burkholderia* in HD; findings not previously reported in other neurodegenerative diseases. It is believed that microbial colonization in the brain may be a risk factor that could explain why the onset of disease occurs after decades of life.[109]

Given the known finding of increased gut permeability associated with microbiota dysbiosis and inflammation, it is not unreasonable to believe that there may be ties between bacterial populations in the brain and gut microbiome. Indeed, one study, using a transgenic mouse model of HD, revealed gut dysbiosis relative to wild-type controls as early as 12 weeks of age. They noted an increase in Bacteroidetes and a proportional decrease in Firmicutes in the transgenic HD mouse model gut microbiome compared to their wild-type control counterparts. This microbial pattern coincided with impaired body weight gain along with motor defects.[110]

5.4 Amyotrophic lateral sclerosis

ALS, commonly known as Lou Gehrig's disease, is a progressive neurodegenerative disorder involving motor neurons in both the brain and spinal cord that ultimately leads to weakness, muscle wasting, and death. Although the exact pathogenesis is unclear, evidence supporting roles for inflammatory and immune-mediated processes continues to grow. As with many of the other neurodegenerative conditions, GI manifestations are common. Constipation has been shown to be a very frequent complaint in ALS patients.[111] While changes in gut microbial populations have been well described in relation to constipation per se, there have been no studies investigating relationships between the microbiome and constipation in ALS.[112]

Alterations in gut permeability may be the avenue through which molecules, including LPS, leave the gut and are found in elevated concentrations in ALS patients.[113] LPS levels in plasma has been significantly and positively associated with blood monocyte and macrophage activation and ALS disease severity.[113] These findings support the hypothesis that the microbiome may promote systemic inflammation in ALS. Another link between ALS and the microbiome stems from the role that the microbiome plays in immune system programming. Microbial metabolites are diverse and are formed from both anaerobic fermentation and through endogenous compounds that are generated by the microbiome and the host.[114] These metabolites, including SCFAs, contribute to the activation and maintenance of microglia,[115] fortify intestinal epithelial cell barrier function,[114] bind to G protein-coupled receptors (GPCRs) (where they act as signaling molecules leading to expansion of hematopoietic and nonhematopoietic cell lineages),[114] and metabolize and/or inhibit histone deacetylase (HDAC), which in turn promotes a tolerogenic antiinflammatory cell phenotype necessary for immune homeostasis.[114] A disturbed microbiome could contribute to the deficiency of antiinflammatory CD25, CD4, and FOXP3 regulatory T cells, which is commonly seen in ALS patients and has been positively associated with ALS severity.[116] However, even with these associations, it is difficult to pinpoint the exact role the microbiome has in ALS patients without further studies.

5.5 Multiple sclerosis

MS is an auto-immune disease of the CNS characterized by demyelination, axonal damage, and progressive neuronal dysfunction resulting in a wide range of symptoms including weakness, sensory dysfunction, loss of coordination and, at times, psychiatric manifestations.[117] It is believed that genetic factors account for approximately 30% of disease risk and the remaining 70% is accounted for by nongenetic components including environmental factors.[118] As we learn more on the role of the microbiome in inflammation in auto-immunity, the idea that gut dysbiosis may be a significant contributor to this environmental component of MS pathogenesis has gained traction.

It is hypothesized that a disturbed microbiome leads to alterations in host immune responses through interactions with lymphocytes and dendritic cells located in the lamina propria of the GI tract, most notably in the cecum and ileum.[119] Initially, T cells were believed to play a key role in the pathogenesis of MS; however, more recent studies involving B-cell-depleting therapies have shed light on the importance of B cells in MS.[120] Several studies highlighted the ways that gut microbiota and their microbial products directly regulate B-cell development, activation, and differentiation.[121] It was previously believed that B cells developed in the bone marrow; however, there is mounting evidence that B-cell development also occurs in the gut where it is stimulated by gut commensals.[122] More recent studies have shown the presence of antibodies in the serum of MS patients, and not in controls, toward bacterial peptides from *Acinetobacter* species and *Pseudomonas aeruginosa*, which have been known to share sequence homology to myelin basic protein and myelin oligodendrocyte glycoprotein.[123]

This has led scientists to hypothesize a contribution of molecular mimicry to the pathogenesis of MS.

A number of studies have shown variations in the composition of the gut microbiota in MS patients. There have been documented decreases in several bacteria including *Prevotella* and *Parabacteroides* in patients with relapsing and remitting multiple sclerosis (RRMS) compared to healthy controls.[124] It was also noted that after treatment with disease-modifying therapies, there was a subsequent increase in abundance of *Prevotella*.[125] Other studies have shown increases in the abundance of *Dorea*, *Mycoplana*, *Clostridium*, and *Pseudomonas* in MS patients, which have also been linked with other inflammatory diseases, further pointing toward the inflammatory role that gut microbiota may play in MS.[125–128]

6 Role of diet in the microbiome and neurodegenerative disease

Given the known influence that both the composition of the microbiome and the gut environment play in the development and progression of disease, many have turned to evaluating dietary intake as an area ripe for intervention. Long-term dietary patterns have been shown to have a significant and reproducible effect on intestinal microbial communities.[129,130]

The Mediterranean Diet (MeDi) has been one of the most widely recommended diets in patients with neurodegenerative diseases due to mounting evidence supporting its protective role against cognitive decline and dementia.[131] This diet is characterized by the consumption of significant amounts of fruits, vegetables, and legumes with the addition of olive oil as the main lipid component combined with moderate consumption of alcohol, fish, and dairy products, and more limited amounts of red meat. A myriad of studies has shown that close adherence to the MeDi was associated with less decline in Mini-Mental State Examinations, better performance in executive function, and decreased decline in the visuospatial cognitive domain in healthy elderly individuals.[132–134] It is believed that this diet, which is rich in unsaturated fats and fiber, leads to reduced metabolic endotoxemia; it has also been linked to increases in fecal SCFAs along with increased abundance of *Prevotella*, Firmicutes, Christensenellaceae, and *Bacteriodes*.[135–137]

Another well-studied diet is the Dietary Approach to Stop Hypertension (DASH) diet. This diet is characterized by the consumption of fruits, vegetables, whole grains, and low-fat dairy foods; it includes fish, lean meats, poultry, nuts, and beans, and emphasizes limitations on sugar-sweetened foods/beverages along with red meat and added fats. The Memory and Aging Project (MAP) followed older adults over 4 years and found that the DASH diet was associated with slower rates of cognitive decline and every one-unit increase in the DASH score (a marker of adherence to DASH diet) was equivalent to being at least 4.4 years younger in terms of expected cognitive function.[138] This same study showed that increased adherence to the DASH diet was associated with a greater reduction in the incidence of AD even when compared to improved adherence to MeDi (54% vs 39%, respectively).[139] It is believed that this diet's proven benefit in reducing blood pressure, low-density lipoprotein, and oxidative stress/inflammation resulted in these protective effects on cognition. Unfortunately, there have been no studies documenting the DASH diet's effect on the gut microbiome.

The Western diet is characteristically high in saturated fatty acids, sugar, and protein; this diet has been linked to numerous adverse health conditions including hypertension, type 2 diabetes, and dementia.[140] Consumption of a high-fat diet has been linked to changes in intestinal microbial composition and decreases in gut microbial diversity.[41,141,142] After each dietary perturbation, there is typically a 3.5-day response time for diet-responsive bacterial groups to reach a new steady state.[143] This high animal-protein diet has been shown to increase levels of bile-tolerant bacteria like *Bacteroides*, *Alistipes*, and *Bilophila* and decrease amounts of fecal SCFAs; changes that have been linked to a variety of both inflammatory states and neurodegenerative disorders.[49,144]

Lastly, the Mediterranean-DASH Intervention of Neurodegenerative Delay (MIND) diet, which combines the aforementioned MeDi and DASH diets, not only emphasizes fresh fruit, vegetables, and legumes, but also includes recommendations for specific foods shown to slow cognitive decline including leafy greens and berries.[139] A MIND diet score was not only associated with slower decline in cognitive abilities, but also showed that those in the highest quartile of diet-compliance scores showed a level of cognitive difference that was equivalent to being 7.5 years younger compared to those in the lowest compliance quartile.[139]

7 Potential therapies

New therapies based on altering and/or diversifying microbial composition are being explored. For millennia preceding germ theory, humans have manipulated fermented foods for their nutritional and therapeutic properties.

Elie Metchnikoff, a Russian scientist, was the first to associate enhanced longevity with regular consumption of fermented dairy products and is credited with first describing the "probiotic concept."[145] Probiotics are live microorganisms that when administered in adequate amounts confer a health benefit on the host and have been used in the treatment of diarrheal conditions, colitis, irritable bowel disease, and liver disease.[146] Current evidence suggests that probiotic effects are strain specific; efforts continue to identify microbial signatures that can predict responses to specific probiotic prescriptions.[147] Prebiotics are nondigestible food ingredients that beneficially affect the host by selectively stimulating the growth and/or activity of one or a limited number of bacteria in the colon[1]; synbiotics refer to the combination of a pro and prebiotic.

Studies evaluating the use of probiotics for neurodegenerative disorders are few; however, two studies found benefits from supplementation with a strain of *Bifidobacteria* in AD mice models.[148,149] Another study compared normal mice and mice injected with LPS to normal mice on probiotics and LPS-injected mice on probiotics and found decreased levels of pro-inflammatory cytokines in both normal and LPS-injected mice following probiotic treatment with *Lactobacillus* and *Bifidobacterium*.[150] While the use of probiotics by both the general public and in clinical practice has become increasingly common, the evidence supporting their use is difficult to interpret given the heterogeneity of the many probiotic products studied and outcomes assessed.[146]

Given the known connections between the microbiome and neurodegenerative diseases, some have postulated that using antibiotics to regulate the gut microbiota may be a future treatment option. However, a recent study showed that in AD APP/PS1 transgenic mice the use of a cocktail of antibiotics resulted in an increase in both the neuroinflammatory state and cytokine levels, indicating an increase in the disease state.[151] Other studies have shown that antibiotics such as streptozotocin have been used to induce sporadic AD forms in animal models with significant alterations in learning and memory performance.[152] On the other hand, a number of studies have shown that some antibiotics may be beneficial. The use of rifampicin, minocycline, rapamycin, and D-cycloserin in rodent AD models was found to have an antitau, antiamyloid, and cholinergic effect.[153–157] However, in human studies, although initial trials showed promise, subsequent studies showed no significant effect on cognition after the use of doxycycline, rifampin, and D-cycloserin.[157–159] At this time, there is a lack of evidence in regard to the use of antibiotics as a therapeutic option for patients with neurodegenerative diseases.

Another potential avenue for intervention includes fecal microbiota transplantation (FMT). FMT involves the transfer of stool containing both microbes and bioactive molecules from a healthy donor to the patient. Although many believe that FMT may be a potential therapeutic option for patients who suffer from neurodegenerative diseases, presently there is insufficient data to draw any firm conclusions.[160] There are currently multiple randomized control trials in progress that may shed light on this potential treatment modality.

8 Conclusion

With increasing understanding of the microbiota-gut-brain axis, it is becoming apparent that there is a significant link between the gut microbiome and the development of the central nervous system and the onset and progression of neurodegenerative diseases. Although supported by work in animal models and limited human studies, there are still large gaps in knowledge that must be filled in order to finally define the precise role of the gut microbiome in disease development, expression, and/or progression. As resources continue to be directed toward advancing this field, there is a great deal of optimism that, not only will there be further developments in disease understanding, but that new avenues for therapy may be revealed.

References

1. Gibson GR, Probert HM, Loo JV, Rastall RA, Roberfroid MB. Dietary modulation of the human colonic microbiota: updating the concept of prebiotics. *Nutr Res Rev.* 2004;17:259–275.
2. About the Human Microbiome. *NIH Human Microbiome Project—About the Human Microbiome.* https://hmpdacc.org/hmp/overview/. Accessed 17 April 2019.
3. Amon P, Sanderson I. What is the microbiome? *Arch Dis Child Educ Pract Ed.* 2017;102:258–261.
4. Jiménez E, Marín ML, Martín R, et al. Is meconium from healthy newborns actually sterile? *Res Microbiol.* 2008;159:187–193.
5. Aagaard K, Ma J, Antony KM, Ganu R, Petrosino J, Versalovic J. The placenta harbors a unique microbiome. *Sci Transl Med.* 2014;6:237ra65.
6. Collado MC, Rautava S, Aakko J, Isolauri E, Salminen S. Human gut colonization may be initiated in utero by distinct microbial communities in the placenta and amniotic fluid. *Sci Rep.* 2016;6:23129.
7. Perez-Munoz ME, Arrieta MC, Ramer-Tait AE, Walter J. A critical assessment of the "sterile womb" and "in utero colonization" hypotheses: implications for research on the pioneer infant microbiome. *Microbiome.* 2017;5:48.
8. Neu J. The microbiome during pregnancy and early postnatal life. *Semin Fetal Neonatal Med.* 2016;21:373–379.

9. Jakobsson HE, Abrahamsson TR, Jenmalm MC, et al. Decreased gut microbiota diversity, delayed Bacteroidetes colonisation and reduced Th1 responses in infants delivered by caesarean section. *Gut.* 2014;63:559–566.
10. Dogra S, Sakwinska O, Soh SE, et al. Dynamics of infant gut microbiota are influenced by delivery mode and gestational duration and are associated with subsequent adiposity. *MBio.* 2015;6. e02419-14.
11. Cong X, Xu W, Janton S, et al. Gut microbiome developmental patterns in early life of preterm infants: impacts of feeding and gender. *PLoS One.* 2016;11:e0152751.
12. Bäckhed F, Roswall J, Peng Y, et al. Dynamics and stabilization of the human gut microbiome during the first year of life. *Cell Host Microbe.* 2015;17:690–703.
13. Yatsunenko T, Rey FE, Manary MJ, et al. Human gut microbiome viewed across age and geography. *Nature.* 2012;486:222–227.
14. Pannaraj PS, Li F, Cerini C, et al. Association between breast milk bacterial communities and establishment and development of the infant gut microbiome. *JAMA Pediatr.* 2017;171:647–654.
15. Tun HM, Bridgman SL, Chari R, et al. Roles of birth mode and infant gut microbiota in intergenerational transmission of overweight and obesity from mother to offspring. *JAMA Pediatr.* 2018;172(4):368–377.
16. Lundgren SN, Madan JC, Emond JA, et al. Maternal diet during pregnancy is related with the infant stool microbiome in a delivery mode-dependent manner. *Microbiome.* 2018;6(1):109.
17. Koenig JE, Spor A, Scalfone N, et al. Succession of microbial consortia in the developing infant gut microbiome. *Proc Natl Acad Sci U S A.* 2011;108(Suppl. 1):4578–4585.
18. Lynch SV, Pedersen O. The human intestinal microbiome in health and disease. *N Engl J Med.* 2016;375:2369–2379.
19. Arumugam M, Raes J, Pelletier E, et al. Enterotypes of the human gut microbiome. *Nature.* 2011;473:174–180.
20. Santoro A, Ostan R, Candela M, et al. Gut microbiota changes in the extreme decades of human life: a focus on centenarians. *Cell Mol Life Sci.* 2018;75:129–148.
21. Kumar M, Babaei P, Ji B, Nielsen J. Human gut microbiota and healthy aging: recent developments and future prospective. *Nutr Healthy Aging.* 2016;4:3–16.
22. Claesson MJ, Jeffery IB, Conde S, et al. Gut microbiota composition correlates with diet and health in the elderly. *Nature.* 2012;488:178–184.
23. Odamaki T, Kato K, Sugahara H, et al. Age-related changes in gut microbiota composition from newborn to centenarian: a cross-sectional study. *BMC Microbiol.* 2016;16:90.
24. Fung TC, Olson CA, Hsiao EY. Interactions between the microbiota, immune and nervous systems in health and disease. *Nat Neurosci.* 2017;20:145–155.
25. Kau AL, Ahern PP, Griffin NW, Goodman AL, Gordon JI. Human nutrition, the gut microbiome and the immune system. *Nature.* 2011;474:327–336.
26. Surana N, Kasper DL. Deciphering the tête-à-tête between the microbiota and the immune system. *J Clin Invest.* 2014;124:4197–4203.
27. Wells JM, Brummer RJ, Derrien M, et al. Homeostasis of the gut barrier and potential biomarkers. *Am J Physiol Gastrointest Liver Physiol.* 2017;312:G171–G193.
28. Dey N, Wagner VE, Blanton LV, et al. Regulators of gut motility revealed by a gnotobiotic model of diet-microbiome interactions related to travel. *Cell.* 2015;163:95–107.
29. Kabouridis PS, Lasrado R, McCallum S, et al. The gut microbiota keeps enteric glial cells on the move; prospective roles of the gut epithelium and immune system. *Gut Microbes.* 2015;6:398–403.
30. Savidge TC. Epigenetic regulation of enteric neurotransmission by gut bacteria. *Front Cell Neurosci.* 2016;9:503.
31. Turnbaugh PJ, Gordon JI. The core gut microbiome, energy balance and obesity. *J Physiol.* 2009;587:4153–4158.

32. Carmody RN, Turnbaugh PJ. Host-microbial interactions in the metabolism of therapeutic and diet-derived xenobiotics. *J Clin Invest.* 2014;124:4173–4181.
33. Vonk RJ, Reckman G. Progress in the biology and analysis of short chain fatty acids. *J Physiol.* 2017;595:419–420.
34. Yano JM, Yu K, Donaldson GP, et al. Indigenous bacteria from the gut microbiota regulate host serotonin biosynthesis. *Cell.* 2015;161:264–276.
35. Shanahan F, van Sinderen D, O'Toole PW, Stanton C. Feeding the microbiota: transducer of nutrient signals for the host. *Gut.* 2017;66:1709–1717.
36. Smith MI, Yatsunenko T, Manary MJ, et al. Gut microbiomes of Malawian twin pairs discordant for kwashiorkor. *Science.* 2013;339:548–554.
37. Subramanian S, Huq S, Yatsunenko T, et al. Persistent gut microbiota immaturity in malnourished Bangladeshi children. *Nature.* 2014;510:417–421.
38. Sonnenburg ED, Sonnenburg JL. Starving our microbial self: the deleterious consequences of a diet deficient in microbiota-accessible carbohydrates. *Cell Metab.* 2014;20:779–786.
39. McIntosh K, Reed DE, Schneider T, et al. FODMAPs alter symptoms and the metabolome of patients with IBS: a randomised controlled trial. *Gut.* 2017;66:1241–1251. https://doi.org/10.1136/gutjnl-2015-311339.
40. Clarke SF, Murphy EF, O'Sullivan O, et al. Exercise and associated dietary extremes impact on gut microbial diversity. *Gut.* 2014;63:1913–1920.
41. Hildebrandt MA, Hoffmann C, Sherrill-Mix SA, et al. High-fat diet determines the composition of the murine gut microbiome independently of obesity. *Gastroenterology.* 2009;137:1716–1724.
42. Heinritz SN, Weiss E, Eklund M, et al. Intestinal microbiota and microbial metabolites are changed in a pig model fed a high-fat/low-fiber or a low-fat/high-fiber diet. *PLoS One.* 2016;11:e0154329.
43. Kovatcheva-Datchary P, Nilsson A, Akrami R, et al. Dietary fiber-induced improvement in glucose metabolism is associated with increased abundance of Prevotella. *Cell Metab.* 2015;22:971–982.
44. Sonnenburg ED, Smits SA, Tikhonov M, Higginbottom SK, Wingreen NS, Sonnenburg JL. Diet-induced extinctions in the gut microbiota compound over generations. *Nature.* 2016;529:212–215.
45. Degnan PH, Taga ME, Goodman AL. Vitamin B12 as a modulator of gut microbial ecology. *Cell Metab.* 2014;20:769–778.
46. Wu GD, Chen J, Hoffmann C, et al. Linking long-term dietary patterns with gut microbial enterotypes. *Science.* 2011;334:105–108.
47. Halmos EP, Christophersen CT, Bird AR, Shepherd SJ, Gibson PR, Muir JG. Diets that differ in their FODMAP content alter the colonic luminal microenvironment. *Gut.* 2015;64:93–100.
48. Bonder MJ, Tigchelaar EF, Cai X, et al. The influence of a short-term gluten-free diet on the human gut microbiome. *Genome Med.* 2016;8:45.
49. David LA, Maurice CF, Carmody RN, et al. Diet rapidly and reproducibly alters the human gut microbiome. *Nature.* 2014;505:559–563.
50. Luca F, Kupfer SS, Knights D, Khoruts A, Blekhman R. Functional genomics of host-microbiome interactions in humans. *Trends Genet.* 2017;34:30–40.
51. Dong TS, Gupta A. Influence of early life, diet, and the environment on the microbiome. *Clin Gastroenterol Hepatol.* 2019;17:231–242.
52. Mayer EA. Gut feelings: the emerging biology of gut-brain communication. *Nat Rev Neurosci.* 2011;12:453–466.
53. Modi SR, Collins JJ, Relman DA. Antibiotics and the gut microbiota. *J Clin Invest.* 2014;124:4212–4218.

54. Blaser MJ. Antibiotic use and its consequences for the normal microbiome. *Science.* 2016;352:544–545.
55. Freedberg DE, Toussaint NC, Chen SP, et al. Proton pump inhibitors alter specific taxa in the human gastrointestinal microbiome: a crossover trial. *Gastroenterology.* 2015;149:883–885.
56. Jackson MA, Goodrich JK, Maxan ME, et al. Proton pump inhibitors alter the composition of the gut microbiota. *Gut.* 2016;65:749–756.
57. Forslund K, Hildebrand F, Nielsen T, et al. Disentangling type 2 diabetes and metformin treatment signatures in the human gut microbiota. *Nature.* 2015;528:262–266.
58. Wasilewska J, Klukowski M. Gastrointestinal symptoms and autism spectrum disorder: links and risks—a possible new overlap syndrome. *Pediatr Health Med Ther.* 2015;6:153–166.
59. Tomova A, Husarova V, Lakatosova S, et al. Gastrointestinal microbiota in children with autism in Slovakia. *Physiol Behav.* 2015;138:179–187.
60. Ho LKH, Tong VJW, Syn N, et al. Gut microbiota changes in children with autism spectrum disorder: a systematic review. *Gut Pathog.* 2020;12:6.
61. O'Mahony SM, Marchesi JR, Scully P, et al. Early life stress alters behavior, immunity, and microbiota in rats: implications for irritable bowel syndrome and psychiatric illnesses. *Biol Psychiatry.* 2009;65:263–267.
62. Sudo N, Chida Y, Aiba Y, et al. Postnatal microbial colonization programs the hypothalamic-pituitary-adrenal system for stress response in mice. *J Physiol.* 2004;558:263–275.
63. Diaz Heijtz R, Wang S, Anuar F, et al. Normal gut microbiota modulates brain development and behavior. *Proc Natl Acad Sci U S A.* 2011;108:3047–3052.
64. Ogbonnaya ES, Clarke G, Shanahan F, Dinan TG, Cryan JF, O'Leary OF. Adult hippocampal neurogenesis is regulated by the microbiome. *Biol Psychiatry.* 2015;78:e7–e9.
65. Luczynski P, Whelan SO, O'Sullivan C, et al. Adult microbiota-deficient mice have distinct dendritic morphological changes: differential effects in the amygdala and hippocampus. *Eur J Neurol.* 2016;44:2654–2666.
66. Erny D, Hrabě de Angelis AL, et al. Host microbiota constantly control maturation and function of microglia in the CNS. *Nat Neurosci.* 2015;18:965–977.
67. Thion MS, Low D, Silvin A, et al. Microbiome influences prenatal and adult microglia in a sex-specific manner. *Cell.* 2018;172:500–516. e16.
68. Barbara G, Stanghellini V, Brandi G, et al. Interactions between commensal bacteria and gut sensorimotor function in health and disease. *Am J Gastroenterol.* 2005;100:2560–2568.
69. Clarke G, Grenham S, Scully P, et al. The microbiome-gut-brain axis during early life regulates the hippocampal serotonergic system in a sex-dependent manner. *Mol Psychiatry.* 2013;18:666–673.
70. Neufeld KM, Kang N, Bienenstock J, Foster JA. Reduced anxiety-like behavior and central neurochemical change in germ-free mice. *Neurogastroenterol Motil.* 2011;23:255–264.
71. Gareau MG, Wine E, Rodrigues DM, et al. Bacterial infection causes stress-induced memory dysfunction in mice. *Gut.* 2011;60:307–317.
72. Galley JD, Nelson MC, Yu Z, et al. Exposure to a social stressor disrupts the community structure of the colonic mucosa-associated microbiota. *BMC Microbiol.* 2014;14:189.
73. Carabotti M, Scirocco A, Maselli MA, Severi C. The gut-brain axis: interactions between enteric microbiota, central and enteric nervous systems. *Ann Gastroenterol.* 2015;28:203–209.
74. Bischoff SC, Barbara G, Buurman W, et al. Intestinal permeability—a new target for disease prevention and therapy. *BMC Gastroenterol.* 2014;14:189.
75. Ma Q, Xing C, Long W, et al. Impact of microbiota on central nervous system and neurological diseases: the gut-brain axis. *J Neuroinflammation.* 2019;16:53.

76. Zhu X, Han Y, Du J, Liu R, Jin K, Yi W. Microbiota-gut-brain axis and the central nervous system. *Oncotarget*. 2017;8:53829–53838.
77. Franceschi C, Bonafè M, Valensin S, et al. Inflamm-aging. An evolutionary perspective on immunosenescence. *Ann NY Acad Sci*. 2000;908:244–254.
78. Kohler CA, Maes M, Slyepchenko A, et al. The gut-brain axis, including the microbiome, leaky gut and bacterial translocation: mechanisms and pathophysiological role in Alzheimer's disease. *Curr Pharm Des*. 2016;22:6152–6166.
79. Zhao Y, Lukiw WJ. Microbiome-generated amyloid and potential impact on amyloidogenesis in Alzheimer's disease (AD). *J Nat Sci*. 2015;1:e138.
80. Zhao Y, Dua P, Lukiw WJ. Microbial sources of amyloid and relevance to amyloidogenesis and Alzheimer's disease (AD). *J Alzheimers Dis Parkinsonism*. 2015;5:177.
81. Hauss-Wegrzyniak B, Vraniak PD, Wenk GL. LPS induced neuroinflammatory effects do not recover with time. *Neuroreport*. 2000;11:1759–1763.
82. Asti A, Gioglio L. Can a bacterial endotoxin be a key factor in the kinetics of amyloid fibril formation? *J Alzheimers Dis*. 2014;39:169–179.
83. Lukiw WJ. *Bacteroides fragilis* lipopolysaccharide and inflammatory signaling in Alzheimer's disease. *Front Microbiol*. 2016;7:1544.
84. Schwartz K, Boles BR. Microbial amyloids–functions and interactions within the host. *Curr Opin Microbiol*. 2013;16:93–99.
85. Hufnagel DA, Tukel C, Chapman MR. Disease to dirt: the biology of microbial amyloids. *PLoS Pathog*. 2013;9:e1003740.
86. Friedland RP. Mechanisms of molecular mimicry involving the microbiota in neurodegeneration. *J Alzheimers Dis*. 2015;45:349–362.
87. Chen S, Stribinskis V, Rane M, et al. Exposure to the functional bacterial amyloid protein curli enhances alpha-synuclein aggregation in aged Fischer 344 rats and *Caenorhabditis elegans*. *Sci Rep*. 2016;6:34477.
88. Barnhart MM, Chapman MR. Curli biogenesis and function. *Annu Rev Microbiol*. 2006;60:131–147.
89. Tursi SA, Tükel Ç. Curli-containing enteric biofilms inside and out: matrix composition, immune recognition, and disease implications. *Microbiol Mol Biol Rev*. 2018;82. e00028–e00018.
90. Friedland RP, Chapman MR. The role of microbial amyloid in neurodegeneration. *PLoS Pathog*. 2017;13:e1006654.
91. Lu B, Nagappan G, Guan X, Nathan PJ, Wren P. BDNF-based synaptic repair as a disease-modifying strategy for neurodegenerative diseases. *Nat Rev Neurosci*. 2013;14:401–416.
92. Michalski B, Corrada MM, Kawas CH, Fahnestock M. Brain-derived neurotrophic factor and TrkB expression in the "oldest-old," the 90+ study: correlation with cognitive status and levels of soluble amyloid-beta. *Neurobiol Aging*. 2015;36:3130–3139.
93. Carlino D, De Vanna M, Tongiorgi E. Is altered BDNF biosynthesis a general feature in patients with cognitive dysfunctions? *Neuroscientist*. 2013;19:345–353.
94. Tu S, Okamoto S, Lipton SA, Xu H. Oligomeric Abeta-induced synaptic dysfunction in Alzheimer's disease. *Mol Neurodegener*. 2014;9:48.
95. Ho L, Ono K, Tsuji M, Mazzola P, Singh R, Pasinetti GM. Protective roles of intestinal microbiota derived short chain fatty acids in Alzheimer's disease-type beta-amyloid neuropathological mechanisms. *Expert Rev Neurother*. 2018;18:83–90.
96. Cattaneo A, Cattane N, Galluzzi S, et al. Association of brain amyloidosis with pro-inflammatory gut bacterial taxa and peripheral inflammation markers in cognitively impaired elderly. *Neurobiol Aging*. 2017;49:60–68.
97. Qualman SJ, Haupt HM, Yang P, Hamilton SR. Esophageal Lewy bodies associated with ganglion cell loss in achalasia. Similarity to Parkinsońs disease. *Gastroenterology*. 1984;87:848–856.

98. Wakabayashi K, Takahashi H, Ohama E, Ikuta F. Parkinsonś disease: an immunohistochemical study of Lewy body-containing neurons in the enteric nervous system. *Acta Neuropathol.* 1990;79:581–583.
99. Braak H, de Vos RAI, Bohl J, Del Tredici K. Gastric alpha-synuclein immunoreactive inclusions in Meissner's and Auerbach's plexuses in cases staged for Parkinsonś disease-related brain pathology. *Neurosci Lett.* 2006;396:67–72.
100. Devos D, Lebouvier T, Lardeux B, et al. Colonic inflammation in Parkinsonś disease. *Neurobiol Dis.* 2013;50:42–48.
101. Forsyth CB, Shannon KM, Kordower JH, et al. Increased intestinal permeability correlates with sigmoid mucosa alpha-synuclein staining and endotoxin exposure markers in early Parkinsonś disease. *PLoS One.* 2011;6. e28032.10.
102. Rietdijk C, Perez-Pardo P, Garssen J, van Wezel R, Kraneveld A. Exploring Braak's hypothesis of Parkinsonś diseases. *Front Neurol.* 2017;8:37.
103. Gorecki A, Preskey L, Bakeberg M, et al. Altered gut microbiome in Parkinsonś disease and the influence of lipopolysaccharide in a human α-synuclein over-expressing mouse model. *Front Neurosci.* 2019;13:839.
104. Scheperjans F, Aho V, Pereira PA, et al. Gut microbiota are related to Parkinson's disease and clinical phenotype. *Mov Disord.* 2015;30:350–358.
105. Fasano A, Bove F, Gabrielli M, et al. The role of small intestinal bacterial overgrowth in Parkinson's disease. *Mov Disord.* 2013;28:1241–1249.
106. Cassani E, Barichella M, Cancello R, et al. Increased urinary indoxyl sulfate (indican): new insights into gut dysbiosis in Parkinson's disease. *Parkinsonism Relat Disord.* 2015;21:389–393.
107. Shen X, Yang H, Wu Y, Zhang D, Jiang H. Association of *Helicobacter pylori* infection with Parkinson's diseases: a meta-analysis. *Helicobacter.* 2017;22:e12398.
108. Hashim H, Azmin S, Razlan H, et al. Eradication of *Helicobacter pylori* infection improves levodopa action, clinical symptoms and quality of life in patients with Parkinson's disease. *PLoS One.* 2014;9:e112330.
109. Alonso R, Pisa D, Carrasco L. Brain microbiota in Huntingtonś disease patients. *Front Microbiol.* 2019;10:2622.
110. Kong G, Cao K-A, Judd L, Li S, Renoid T, Hannan A. Microbiome profiling reveals gut dysbiosis in a transgenic mouse model of Huntingtonś disease. *Neurobiol Dis.* 2018;135:104268.
111. Nübling GS, Mie E, Bauer RM, et al. Increased prevalence of bladder and intestinal dysfunction in amyotrophic lateral sclerosis. *Amyotroph Lateral Scler Frontotemporal Degener.* 2014;15:174–179.
112. Wright M, Fournier C, Houser MC, Tansey M, Glass J, Hertzberg VS. Potential role of the gut microbiome in ALS: a systematic review. *Biol Res Nurs.* 2018;20:513–521.
113. Zhang R, Miller R, Gascon R. Circulating endotoxin and systemic immune activation in sporadic Amyotrophic Lateral Sclerosis (sALS). *J Neuroimmunol.* 2019;206:121–124.
114. Rooks MG, Garrett WS. Gut microbiota, metabolites and host immunity. *Nat Rev Immunol.* 2016;16:341–352.
115. Abdel-Haq R, Schlachetzki J, Glass CK, Mazmanian SK. Microbiome-microglia connections via the gut-brain axis. *J Exp Med.* 2019;216:41–59.
116. He F, Balling R. The role of regulatory T cells in neurodegenerative diseases. *Wiley Interdiscip Rev Syst Biol Med.* 2013;5:153–180.
117. Pröbstel A-K, Baranzini S. The role of the gut microbiome in multiple sclerosis risk and progression: towards characterization of the "MS microbiome". *Neurotherapeutics.* 2018;15:126–134.
118. Yadav S, Mindur J, Ito K, Dhib-Jalbut S. Advances in the immunopathogenesis of multiple sclerosis. *Curr Opin Neurol.* 2015;28:206–219.

119. Honda K, Littman DR. The microbiota in adaptive immune homeostasis and disease. *Nature.* 2016;535:75–84.
120. Hauser SL. The Charcot lecture | beating MS: a story of B cells, with twists and turns. *Mult Scler.* 2015;21:8–21.
121. Kim M, Kim CH. Regulation of humoral immunity by gut microbial products. *Gut Microbes.* 2017;4:1–8.
122. Rosser EC, Oleinika K, Tonon S, et al. Regulatory B cells are induced by gut microbiota-driven interleukin-1beta and interleukin-6 production. *Nat Med.* 2014;20:1334–1339.
123. Hughes LE, Smith PA, Bonell S, et al. Cross-reactivity between related sequences found in Acinetobacter sp., *Pseudomonas aeruginosa*, myelin basic protein and myelin oligodendrocyte glycoprotein in multiple sclerosis. *J Neuroimmunol.* 2003;144:105–115.
124. Shahi S, Freedman S, Mangalam A. Gut microbiome in multiple sclerosis: the players involved and the roles they play. *Gut Microbes.* 2017;8:1–9.
125. Jangi S, Gandhi R, Cox LM, et al. Alterations of the human gut microbiome in multiple sclerosis. *Nat Commun.* 2016;7:12015.
126. Chen J, Chia N, Kalari KR, et al. Multiple sclerosis patients have a distinct gut microbiota compared to healthy controls. *Sci Rep.* 2016;6:28484.
127. Rajilic-Stojanovic M, Biagi E, Heilig HG, et al. Global and deep molecular analysis of microbiota signatures in fecal samples from patients with irritable bowel syndrome. *Gastroenterology.* 2011;141:1792–1801.
128. Mukhopadhya I, Hansen R, El-Omar EM, Hold GL. IBD what role do proteobacteria play? *Nat Rev Gastroenterol Hepatol.* 2012;9:219–230.
129. Herpertz-Dahlmann B, Seitz J, Baines J. Food matters: how the microbiome and gut–brain interaction might impact the development and course of anorexia nervosa. *Eur Child Adolesc Psychiatry.* 2017;26:1031–1041.
130. Ussar S, Griffin NW, Bezy O, et al. Interactions between gut microbiota, host genetics and diet modulate the predisposition to obesity and metabolic syndrome. *Cell Metab.* 2015;22:516–530.
131. Solfrizzi V, Custodero C, Lozupone M, et al. Relationships of dietary patterns, foods, and micro- and macronutrients with Alzheimer's disease and late-life cognitive disorders: a systematic review. *J Alzheimers Dis.* 2017;59:815–849.
132. Trichopoulou A, Kyrozis A, Rossi M, et al. Mediterranean diet and cognitive decline over time in an elderly Mediterranean population. *Eur J Nutr.* 2013;54:1311–1321.
133. Qin B, Adair LS, Plassman BL, et al. Dietary patterns and cognitive decline among Chinese older adults. *Epidemiology.* 2015;26:758–768.
134. Gardener SL, Rainey-Smith SR, Barnes MB, et al. Dietary patterns and cognitive decline in an Australian study of ageing. *Mol Psychiatry.* 2015;20:860–866.
135. De Filippis F, Pellegrini N, Vannini L, et al. High-level adherence to a Mediterranean diet beneficially impacts the gut microbiota and associated metabolome. *Gut.* 2016;65:1812–1821.
136. Garcia-Mantrana I, Selma-Royo M, Alcantara C, Collado MC. Shifts on gut microbiota associated to Mediterranean diet adherence and specific dietary intakes on general adult population. *Front Microbiol.* 2018;9:890.
137. Mitsou EK, Kakali A, Antonopoulou S, et al. Adherence to the Mediterranean diet is associated with the gut microbiota pattern and gastrointestinal characteristics in an adult population. *Br J Nutr.* 2017;117:1645–1655.
138. Tangney CC, Li H, Wang Y, et al. Relation of DASH- and Mediterranean-like dietary patterns to cognitive decline in older persons. *Neurology.* 2014;83:1410–1416.
139. Morris MC, Tangney CC, Wang Y, Sacks FM, Bennett DA, Aggarwal NT. MIND diet associated with reduced incidence of Alzheimer's disease. *Alzheimers Dement.* 2015;11:1007–1014.
140. Whitmer RA, Gunderson EP, Barrett-Connor E, et al. Obesity in middle age and future risk of dementia: a 27-year longitudinal population-based study. *BMJ.* 2005;330:1360.

141. Beilharz JE, Kaakoush NO, Maniam J, et al. Cafeteria diet and probiotic therapy: cross talk among memory, neuroplasticity, serotonin receptors and gut microbiota in the rat. *Mol Psychiatry.* 2018;23:351–361.
142. Agus A, Denizot J, Thevenot J, et al. Western diet induces a shift in microbiota composition enhancing susceptibility to Adherent-Invasive *E. coli* infection and intestinal inflammation. *Sci Rep.* 2016;6:19032.
143. Carmody RN, Gerber GK, Luevano Jr JM, et al. Diet dominates host genotype in shaping the murine gut microbiota. *Cell Host Microbe.* 2015;17:72–84.
144. Singh RK, Chang HW, Yan D, et al. Influence of diet on the gut microbiome and implications for human health. *J Transl Med.* 2017;15:73.
145. Gasbarrini G, Bonvicini F, Gramenzi A. Probiotics history. *J Clin Gastroenterol.* 2016;50:S116–S119.
146. Parker EA, Roy T, D'Adamo CR, Wieland LS. Probiotics and gastrointestinal conditions: an overview of evidence from the Cochrane collaboration. *Nutrition.* 2017;45:125–134. e11.
147. FAO/WHO. *Guidelines for the evaluation of probiotics in food.* Report of a Joint FAO/WHO Working Group on Drafting Guidelines for the Evaluation of Probiotics in Food, London: World Health Organization; 2002.
148. Bonfili L, Cecarini V, Berardi S, et al. Microbiota modulation counteracts Alzheimer's disease progression influencing neuronal proteolysis and gut hormones plasma levels. *Sci Rep.* 2017;7:2426.
149. Kobayashi Y, Sugahara H, Shimada K, et al. Therapeutic potential of *Bifidobacterium breve* strain A1 for preventing cognitive impairment in Alzheimer's disease. *Sci Rep.* 2017;7:13510.
150. Mohammadi G, Dargahi L, Peymani A, et al. The effects of probiotic formulation pretreatment (*Lactobacillus helveticus* R0052 and *Bifidobacterium longum* R0175) on a lipopolysaccharide rat model. *J Am Coll Nutr.* 2019;38:209–217.
151. Minter MR, Zhang C, Leone V, et al. Antibiotic-induced perturbations in gut microbial diversity influences neuro-inflammation and amyloidosis in a murine model of Alzheimer's disease. *Sci Rep.* 2016;6:30028.
152. Ravelli KG, Rosário B dos A, Camarini R, Hernandes MS, Britto LR. Intracerebroventricular streptozotocin as a model of Alzheimer's disease: neurochemical and behavioral characterization in mice. *Neurotox Res.* 2016;31:327–333.
153. Yulug B, Hanoglu L, Ozansoy M, et al. Therapeutic role of rifampicin in Alzheimer's disease. *Psychiatry Clin Neurosci.* 2018;72:152–159.
154. Budni J, Garcez ML, de Medeiros J, et al. The anti-inflammatory role of minocycline in Alzheimer's disease. *Curr Alzheimer Res.* 2016;13:1319–1329.
155. Wang C, Yu J-T, Miao D, Wu Z-C, Tan M-S, Tan L. Targeting the mTOR signaling network for Alzheimer's disease therapy. *Mol Neurobiol.* 2013;49:120–135.
156. Portero-Tresserra M, Martí-Nicolovius M, Tarrés-Gatius M, Candalija A, Guillazo-Blanch G, Vale-Martínez A. Intra-hippocampal d-cycloserine rescues decreased social memory, spatial learning reversal, and synaptophysin levels in aged rats. *Psychopharmacology.* 2018;235:1463–1477.
157. Tsai GE, Falk WE, Gunther J, Coyle JT. Improved cognition in Alzheimer's disease with short-term D-cycloserine treatment. *Am J Psychiatry.* 1999;156:467–469.
158. Loeb MB, Molloy DW, Smieja M, et al. A randomized, controlled trial of doxycycline and rifampin for patients with Alzheimer's disease. *J Am Geriatr Soc.* 2004;52:381–387.
159. Molloy DW, Standish TI, Zhou Q, Guyatt G. A multicenter, blinded, randomized, factorial controlled trial of doxycycline and rifampin for treatment of Alzheimer's disease: the DARAD trial. *Int J Geriatr Psychiatry.* 2012;28:463–470.
160. Vendrik KEW, Ooijevaar RE, de Jong PRC, et al. Fecal microbiota transplantation in neurological disorders. *Front Cell Infect Microbiol.* 2020;10:98.

CHAPTER 7

The contribution of gut microbiota in the pathogenesis of Parkinson's disease

Tahira Farooqui
Department of Entomology, The Ohio State University, Columbus, OH, United States

1 Introduction

Parkinson's disease is a multifactorial disorder of the nervous system in which there is a progressive loss of dopaminergic neurons. Parkinson's disease is the second most common neurodegenerative disease (after Alzheimer's disease) of aging,[1] with median age-standardized annual incidence rates in high-income countries of 14 per 100, 000 people in the total population, and 160 per 100, 000 people aged 65 years or older.[2] Parkinson's disease is expected to impose an increasing social and economic burden on societies as populations age.

Parkinson's disease is caused by the interaction between environmental and genetic risk factors. Parkinson's disease is hallmarked by a distinctive degeneration of dopaminergic neurons in the substantia nigra pars compacta (SNc) with consequent dopamine deficiency within the striatum and by the manifestation of intracellular eosinophilic inclusions, the so-called Lewy bodies and Lewy neurites in the remaining neurons. This neuropathology is characterized by motor and nonmotor symptoms, which ultimately have a negative impact on quality of life. As a debilitating neuromotor disorder, Parkinson's affects the nigrostriatal pathway in the midbrain, resulting in an array of classical motor symptoms ranging from resting tremors (typically starting in one hand, foot, or leg, and then eventually affecting both sides of the body), rigidity (stiff or inflexible muscles), bradykinesia (slowness of movement), and postural instability (impaired balance) (Fig. 1).[3] Parkinson's disease patients also show additional motor deficits, such as gait disturbance (walking abnormality), akinesia (losing ability to move muscles voluntarily), micrographia (impaired hand writing), weaker grip force, and speech deficits (such as slur words, mumble or trail off at the end of a sentence), among others.[3] However, this disease is highly variable and not every patient experiences all the symptoms.

Gut Microbiota in Neurologic and Visceral Diseases
https://doi.org/10.1016/B978-0-12-821039-0.00008-3

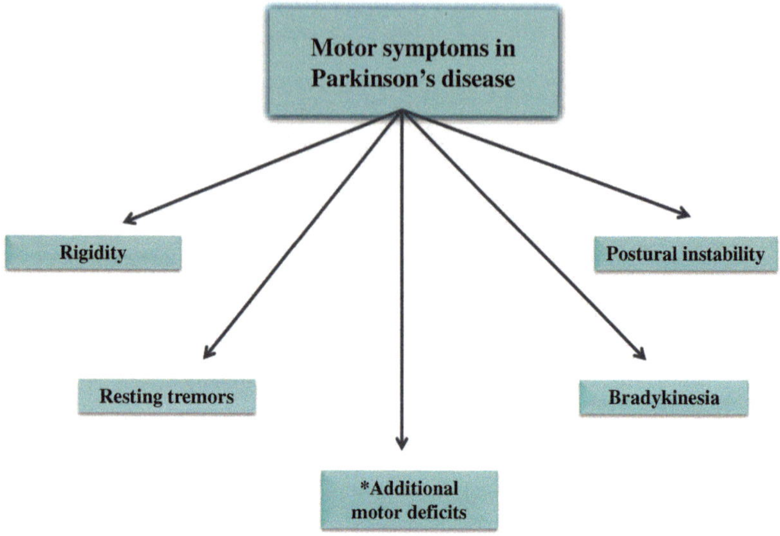

Fig. 1 Motor symptoms of Parkinson's disease.
*Include gait disturbance, akinesia, micrographia, grip force abnormality, and speech deficits.

In addition to motor symptoms, Parkinson's disease is also associated with a broad spectrum of nonmotor symptoms, including autonomic dysfunction, cognitive impairment, and neuropsychiatric symptoms (Fig. 2).[4–8] Autonomic dysfunction has been considered as an important nonmotor phenotype of Parkinson's disease, and can help in the prediction and early diagnosis of this disease. Autonomic dysfunction includes cardiovascular dysregulation (orthostatic hypotension, supine hypertension, and increased resting heart rate), gastrointestinal malfunction (constipation/diarrhea/gastroparesis), sensory issues (loss of sense of smell), urinary dysfunction (urinary urgency, frequency, and inconsistence), sexual

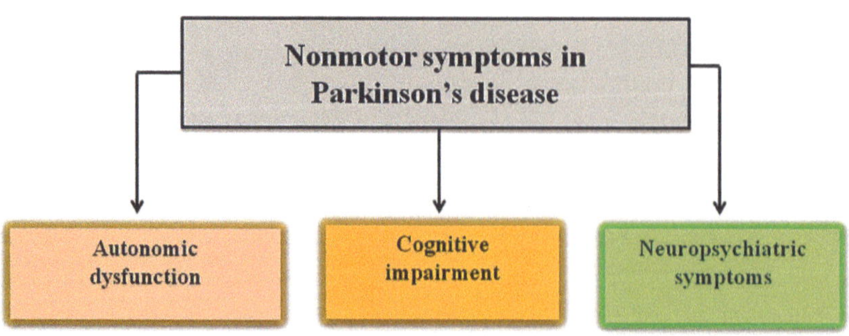

Fig. 2 Nonmotor symptoms in Parkinson's disease.

dysfunction (erectile dysfunction/ejaculation abnormality), thermoregulatory aberrance (heat/cold intolerance), and pupillary irregularities (pupil movement, focusing, lacrimation). Cognitive impairment is another common nonmotor complication of Parkinson disease, which is characterized by predominant executive function deficits, attention difficulties, problems with visuospatial and visuoconstructive abilities, speech deficits (difficulty in verbal fluency), memory deficits (trouble in recall, working memory, olfactory memory), and sleep disorders (sleep fragmentation, insomnia, rapid eye movement (REM) sleep behavior disorder, excessive daytime sleepiness (EDS), periodic limb movements of sleep (PLMS), and circadian rhythm (CR) dysregulation. Neuropsychiatric symptoms include depression, anxiety, psychosis, apathy, and impulse control disorders, which have been reported to occur in the majority of Parkinson's disease patients (Fig. 2).[4–8] Neuropsychiatric symptoms are associated with impaired quality of life for Parkinson's disease patients, which may result in additional deterioration of function thereby may require increased use of health resources.[8]

Some nonmotor symptoms such as loss of smell, constipation, depression, and REM disorder can be more troublesome and disabling, and can occur years before classic motor symptoms of Parkinson's disease. Several studies have observed that Parkinson's disease patients suffer from gastrointestinal disorders and gut microbiota dysbiosis prior to displaying motor symptoms, but the specific link between the gut microbiota and Parkinson's disease is still not clear.[8] Therefore, it has been hypothesized that nonmotor symptoms may occur due to the disruption in the interactions between enteric microbiota and the central (CNS) and enteric nervous systems (ENS), otherwise known as the gut-brain axis (GBA).

Incidence of Parkinson's disease has grown substantially, triggered by the discovery of several causative monogenetic mutations. These mutations likely only explain a small proportion (5%–10%) of all Parkinson's disease cases; however, ~90% of cases are apparently sporadic.[4] In sporadic Parkinson's disease, mitochondrial dysfunction, reactive oxygen species (ROS)-mediated oxidative stress, cytokine-induced toxicity associated with neuroinflammation, and protein mishandling due to an impaired ubiquitin-proteasome system (UPS) may play a significant role in the associated neuronal damage.[9] In addition, Parkinson's disease is associated with an accumulation of Lewy bodies, which are intra-cytoplasmic eosinophilic deposits composed of a misfolded protein (α-synuclein) in the basal ganglia neurons, especially in the caudate nucleus and the putamen. α-Synuclein is a small protein of the synuclein family that is encoded by the *SCNA* gene, which is highly expressed in the human brain, primarily at presynaptic terminals.[10] The toxic phenotype observed in neurons seems to derive from the noncell autonomous interaction between neurons and glia that is carried out by α-synuclein, which may lead to chronic inflammation. Both Parkinson's disease patients as well as animal models show signs of chronic inflammation.[11] The α-synuclein plays a role in exocytosis, recycling of the synaptic vesicles, and α-synuclein accumulation as a result of constant internalization, which could also alter the physiological state of membrane trafficking through vesicular transport and synaptic transmission.[11]

Several factors such as oxidative stress, proteolysis, fatty acid concentration, phospholipids, and metal ions can modulate α-synuclein structure, leading to alternative formations of the protein, which include oligomers and fibrils, the latter of which can develop into cytoplasmic inclusions.[10] Additionally, posttranslational modifications such as phosphorylation, ubiquitination, nitration, and truncation can alter protein size, structure, and charge. Because ~ 90% of insoluble α-synuclein in Lewy bodies is phosphorylated, phosphorylation has been the most widely studied posttranslational modification.[12,13] Although phosphorylation at Ser-129 (a major phosphorylation site) is characteristic of Parkinson's disease and related synucleinopathies, whether phosphorylation affects fibril formation, a necessary step in the formation of cytoplasmic inclusions and subsequently Lewy bodies, is still unclear.[14] The brains of Parkinson's disease patients contain abnormal clumps of proteins (Lewy bodies) that are largely made up of α-synuclein, which plays a role in crosstalk between brain cells. Lewy bodies interrupt this communication and are thought to be toxic to certain neurons in the brain. The presence of α-synuclein deposits in the postmortem brain biopsies of Parkinson's disease patients has further substantiated the role of inflammation.[15]

Moreover, it has been suggested that α-synuclein misfolding might begin in the gut and spread "prion like" via the vagus nerve into the lower brainstem and ultimately to the midbrain; this is known as Braak's hypothesis.[16] By using immunocytochemistry, this group investigated the gastric myenteric and submucosal plexuses on paraffin sections from autopsy individuals, whose brains were staged for Parkinson-associated synucleinopathy. These investigators successfully reported the synuclein immunoreactive inclusions in neurons of the submucosal Meissner plexus, with axons projecting into the gastric mucosa and terminating in direct proximity to fundic glands. These elements could provide the first link in an uninterrupted series of susceptible neurons that extend from the ENS to the CNS. The existence of such a neuronal chain supports Braak's hypothesis that a putative environmental pathogen capable of passing the gastric epithelial lining may induce α-synuclein misfolding and aggregation in specific cell types of the submucosal plexus and reach the brain via a consecutive series of projection neurons.[16]

Brundin and Melki summarized the evidence in favor of the hypothesis that prion-like spreading of α-synuclein fibrils contributes to pathogenesis in Parkinson's disease.[17] Although multiple molecular mechanisms are involved in triggering the disease process, neuron-to-neuron transfer is a crucial event that is essential for Lewy pathology to spread from one brain region to another. Whether the initial cellular α-synuclein misfolding actually leads to a cascade of Lewy pathology spreading may depend on several factors, including the age and genetic background of the individual as well as environmental factors. The cellular mechanisms (mitochondrial dysfunction, oxidative stress, impaired lysosomal autophagy and ubiquitin proteasome systems, and neuroinflammation) may also play crucial roles in enabling the initial misfolding of α-synuclein or in facilitating the cell-to-cell transfer of pathogenic α-synuclein assemblies (Fig. 3). Despite insights derived from genetic research, the exact pathogenetic mechanisms underlying selective dopaminergic cell loss in Parkinson's disease are still not understood.

Lewy body pathology is also observed in the peripheral autonomic nervous system (ANS) and ENS in the gastrointestinal tract, the olfactory system, and the cardiac sympathetic system, implicating that α-synuclein pathology in Parkinson's disease may start in these tissues.[18] Parkinson's disease patients frequently exhibit nonmotor symptoms, including signs and symptoms of gastrointestinal dysmotility (e.g., delayed gastric emptying and constipation).[19,20]

Similarly, in Parkinson's disease, accumulation of α-synuclein in the ENS in the gastrointestinal tract could commence 20 years before the onset of degenerative changes in the brain and before the associated typical motor features of the disease appear.[21]

Marked olfactory dysfunction is observed as a frequent and early abnormality in Parkinson's disease.[22] There is considerable evidence that autonomic dysfunction is an early feature of Parkinson's disease. Therefore, a scintigraphical qualitative analysis of peripheral vascular sympathetic function with meta-[123I] iodobenzylguanidine is also useful in detecting peripheral adrenergic dysfunction in neurological patients with autonomic failure.[23] Moreover, constipation is another most common nonmotor symptom of

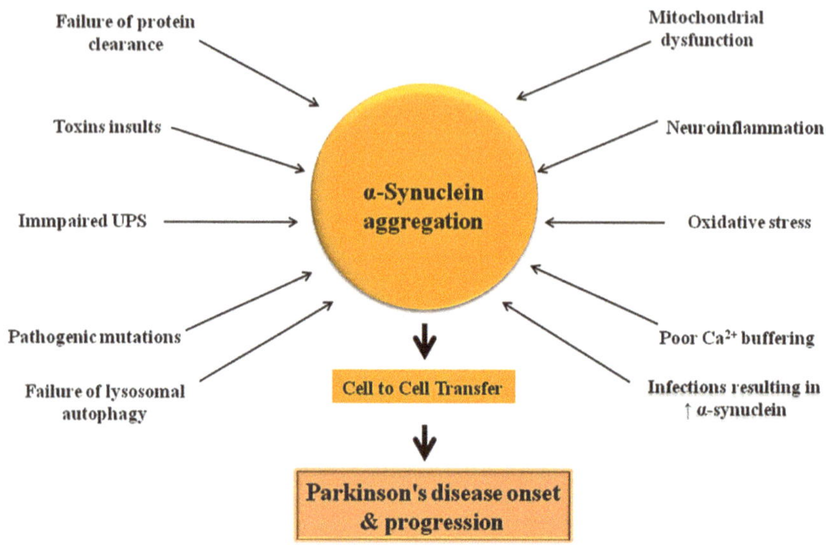

Fig. 3 Factors responsible for α-synuclein aggregation, resulting in the onset and progression of the disease. Ubiquitin-proteasome system (UPS).

Parkinson's disease.[24] The severity of constipation is associated with the risk of future Parkinson's disease diagnosis in a dose-dependent manner.[25] Therefore, olfactory dysfunction, autonomic failure, and constipation may be integrated with other nonmotor symptoms and biomarkers to identify individuals with an increased risk of developing Parkinson's disease.

Clinically, gastrointestinal symptoms (such as constipation) often appear in Parkinson's disease patients, which is often associated with α-synuclein accumulation in patients before other neurological signs, and aggregates of α-synuclein have been found in enteric nerves of Parkinson's disease patients.[26] Therefore it is possible that Lewy body pathology in the ENS (known as a lesion in the gut) occurring at early stages of the disease may be one of the triggering factors that contributes to the pathogenesis of Parkinson's disease.

2 Gut microbiota dysbiosis in Parkinson's disease

The gut microbiota is an important component of the intestinal environment that is involved in the regulation of immune function and development,[27] metabolism,[28] and GBA communication.[29,30] The interaction between microbiota and the GBA appears to be bidirectional, namely, through signaling from gut microbiota to brain and from brain to gut microbiota by means of neural, endocrine, immune, and humoral links.

Gut microbiota dysbiosis is the alteration in structural and/or functional configuration of the gut microbiota causing disruption of gut homeostasis, which can be disturbed by several factors, including dietary alterations, alcohol, smoking, antibiotic exposure, infections, disease status, aging, and genetics. Gut microbiota dysbiosis has been reported in several neurological disorders, including Alzheimer's disease,[31,32] multiple sclerosis,[33,34] schizophrenia,[35,36] autism,[37,38] depression,[39–41] and Parkinson's disease.[42–44]

Gut microbial dysbiosis and alterations in microbial metabolites in Parkinson's disease have been increasingly reported in literature.[42–47] The first study in 2015 by Scheperjans explored the relation between gut dysbiosis and Parkinson's disease and reported reduced abundance of Prevotellaceae and increased abundance of Enterobacteriaceae in Parkinson's disease patients, which were associated with postural instability and gait disorder.[47] In 2016, Sampson reported that gut microbiota dysbiosis leads to an altered ratio of short-chain fatty acids (SCFAs), especially butyrate, and altered microglial signaling in the brain leading to disease development and the appearance of Parkinson's disease-associated motor symptoms.[48] This study provided evidence for gut dysbiosis in Parkinson's disease pathogenesis based on the fact that microbial species harvested from Parkinson's disease patients showed decreased levels of *Prevotella*, *Lactobacillus*, *Peptostreptococcus*, and *Butyricicoccus* spp., but increased levels of *Proteus* and *Enterobacter* spp. compared to healthy controls.[48] Minato's group in 2017 examined changes in gut microbiota and demographic features over 2 years in 36 Parkinson's patients, suggesting the role of certain bacteria in the worsening of Parkinson's disease and associated symptoms.[49] *Bifidobacterium* is dominant in Asian people compared to Western people due to their respective cultures and dietary habits. According to metaanalysis, prevalence of Parkinson's disease in persons aged 70–79 years in Asia is significantly less compared to other regions such as North America, Europe, and Australia.[50] Minato's group's findings implicated that *Bifidobacterium* may be protective against progression of Parkinson's disease.[48] However, levels of *Bifidobacterium* did not differ much between Parkinson's disease patients and controls in Japan and the United States.[45,51] Therefore, more studies are needed with larger cohorts for longer times in multiple countries using the identical method of intestinal microbiota analysis to elucidate the causal association of *Bifidobacterium* and other bacteria in the development and progression of Parkinson's disease. A recent microbiome-wide association study by Wallen et al. with two large datasets (N = 333 and 507) characterized gut microbial dysbiosis in Parkinson's disease.[52] This group uncovered robust and reproducible signals, reaffirming (SCFA and probiotics) and generating leads (opportunistic pathogens) for experimentation into cause and effect, disease progression, and therapeutic targets.[52] Three clusters of co-occurring microorganisms were reported: cluster 1 of opportunistic pathogens (*Porphyromonas*, *Prevotella*, *Corynebacterium_1*, *Bifidobacterium*, and *Lactobacillus*) was elevated in Parkinson's disease compared to controls; cluster 2 of SCFA-producing bacteria (*Faecalibacterium*, *Agathobacter*, *Blautia*, *Roseburia*, *Fusicatenibacter*, *Lachnospira*, *Butyricicoccus*, *Lachnospiraceae_ND3007_*, *Lachnospiraceae_UCG-004*, and *Oscillospira*) was reduced in Parkinson's disease compared to controls; and cluster 3 of carbohydrate-metabolizing probiotics (*Lactobacillus* and *Bifidobacteria*) was elevated in Parkinson's disease compared to controls.[52] The overabundance of opportunistic pathogens in the gut microbiomes of Parkinson's patients is potentially the most exciting finding of this study because it supports Braak's hypothesis of involvement of a pathogen passing through the mucosal barrier of the gastrointestinal tract and spread to the brain via ENS in Parkinson's disease.[16,53] The findings regarding cluster 2 of 10 genera belonging to the Lachnospiraceae and Ruminococcaceae families (anti-inflammatory SCFA-producing bacteria) being diminished in Parkinson's disease could be due to medication or advancement of the disease. Lastly, findings regarding elevated levels of cluster 3 bacteria (*Bifidobacterium* and *Lactobacillus*) in Parkinson's disease are confirmatory.[52]

Bifidobacterium and *Lactobacillus* are generally considered as probiotics because they can metabolize carbohydrates from plants and dairy. When *Bifidobacterium* and *Lactobacillus* are administered as probiotics, the outcome shows reduction in negative mood in healthy controls. These probiotics considered to confer mental health benefits through interaction with other commensal gut microbiota are known as psychobiotics.[54] As Parkinson's disease advances clinically, there is dysfunction of ENS and dysautonomia (autonomic dysfunction), which leads to greater slowing of gastrointestinal motility, which in turn predisposes Parkinson's disease patients to small intestinal bacterial overgrowth,[55] highly observed in Parkinson's disease patients compared to controls.[56] These findings support the role of gut microbiota dysbiosis in Parkinson's disease.

3 Microbiota gut-brain axis role in Parkinson's disease

The GBA is a bidirectional signaling pathway between the gut and the brain.[30] In broadly defined terms, the GBA includes the CNS; neuroendocrine and neuroimmune systems, including the hypothalamic-pituitary-adrenal axis (HPA axis); and sympathetic and parasympathetic arms of the ANS, including the ENS, vagus nerve, and the gut microbiota, to regulate many physiological processes (Fig. 4). The extrinsic nerves of the gastrointestinal tract connect the gut to the brain through vagal and spinal afferent fibers, while the brain sends efferent sympathetic and parasympathetic fibers to the gastrointestinal tract.[57–59] The HPA axis is the main modulator of the physiological stress response, but it also modulates alimentary function during digestion to facilitate gluconeogenesis. The hypothalamus releases corticotrophin-releasing factor (CRF) and different proteins within this family that affect several gastrointestinal tract functions, including intestinal motility, permeability, and inflammation.[60] Rats experiencing chronic psychological stress showed a decreased level of intestinal epithelial tight junction proteins concurrent with increased intestinal permeability in the colon, implicating a novel, region-specific role for a corticosteroid as a mediator of chronic psychological stress-induced increased permeability to macromolecules across the colon epithelium.[61] In another study, investigators examined the HPA axis in patients with irritable bowel syndrome (IBS) to relate such response to plasma cytokine profiles.[62] Patients with IBS showed elevated levels of proinflammatory cytokines, including interleukin (IL)-6 and IL-8 as a result of adrenocorticotropic hormone (ACTH) stimulation, implicating that IBS is characterized by an overactivation of the HPA axis and increased proinflammatory cytokines.[62]

The disruption of the GBA has been shown to be associated to the pathogenesis of several disorders, ranging from IBS, liver disease, and chronic abdominal pain syndromes to psychiatric disorders such as depression and anxiety as well as neurodegenerative disorders such as Parkinson's and Alzheimer's.[31,48,63,64]

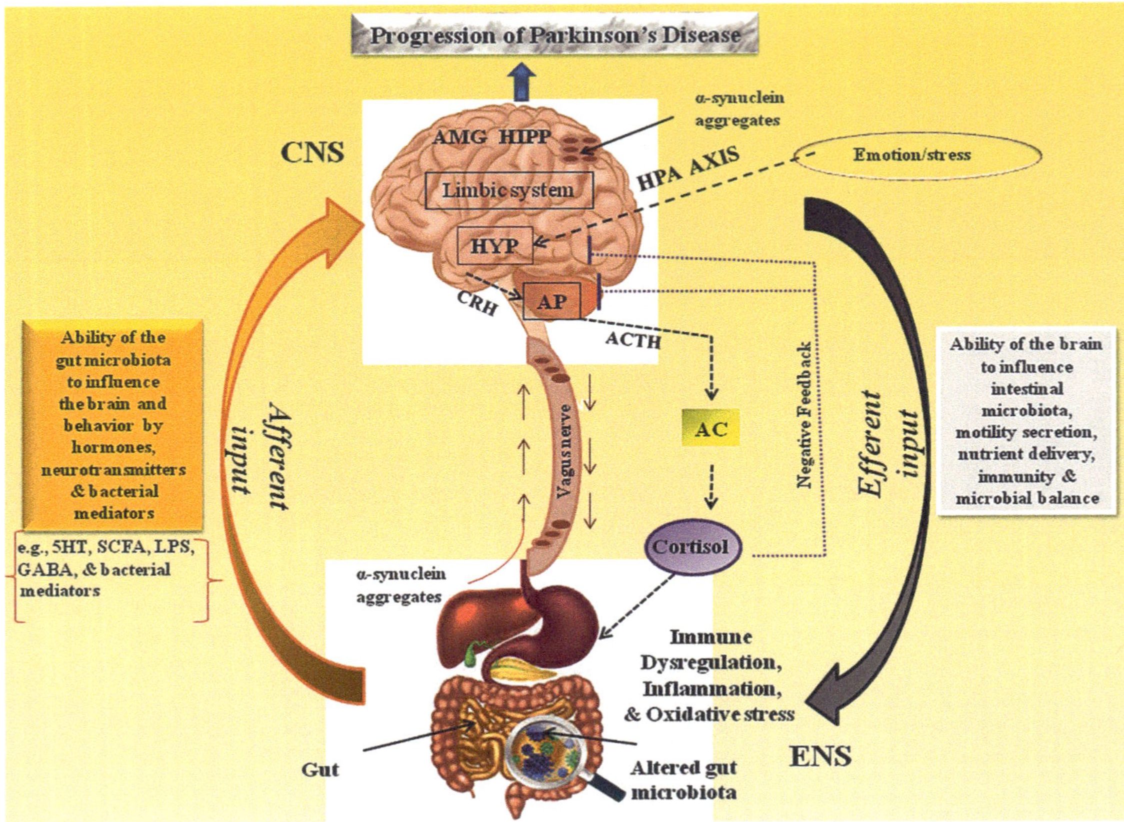

Progression of Parkinson's Disease
CNS
α-synuclein aggregates
AMG HIPP
Limbic system
HPA AXIS
Emotion/stress
HYP
CRH
AP
ACTH
Ability of the gut microbiota to influence the brain and behavior by hormones, neurotransmitters & bacterial mediators
e.g., 5HT, SCFA, LPS, GABA, & bacterial mediators
Afferent input
Vagus nerve
AC
Negative Feedback
Efferent input
Ability of the brain to influence intestinal microbiota, motility secretion, nutrient delivery, immunity & microbial balance
Cortisol
α-synuclein aggregates
Immune Dysregulation, Inflammation, & Oxidative stress
Gut
Altered gut microbiota
ENS

Fig. 4, Cont'd

4 Synuclein propagation from gut to brain

Another hypothesis for the pathogenesis of Parkinson's disease is an abnormal accumulation of α-synuclein.[65] Briefly, α-synuclein is abundantly expressed in the CNS and involved in the regulation of the presynaptic release of the neurotransmitter dopamine. Moreover, α-synuclein is also expressed within the ENS and can be detected in intestinal submucosal neuronal structures of neurologically healthy individuals with increasing age,[66,67] implicating that it may not be a pathological correlate. The pattern of α-synuclein staining in Parkinson's disease was distinct from healthy subjects and those with inflammatory bowel disease. Furthermore, the absence of this pattern in subjects with inflammatory bowel disease suggests it is not a sequence of inflammation or oxidative stress. Parkinson's disease patients are found to be α-synuclein positive, with higher prevalence and grade than age-matched controls; however, Alzheimer's disease patients were not α-synuclein positive,[68] implicating that absence of increased α-Syn expression in Alzheimer's disease points to differences in pathogenesis. Moreover, either Parkinson's disease develops selectively in the enterically α-synuclein-positive population subset or Parkinson's disease induces α-Syn expression. Through interactions with environmental factors and other proteins and

---

Fig. 4 The disruption of gut-brain axis may result in the progression of Parkinson's disease. Afferent input includes hormonal, chemo, and mechanosensory pathways by which microbiota influences neurotransmitter release affecting stress, behavior and mood. Efferent input initiates from brain, and influences intestinal microbiota and modulates motility secretion, nutrient delivery, immunity, and microbial balance. HPA axis is the central stress response system that is a complex set of direct influences and feedback interactions among three components: the HYP, the pituitary gland (a pea-shaped structure located below the thalamus), and the adrenal gland (small, conical organs on top of the kidneys). The HPA axis is involved in the neurobiology of mood disorders and functional illnesses, including neurological and gastrointestinal disorders. Alpha synuclein deposition starts in the ENS, finally propagating to the CNS. Gut microbiota imbalance can break intestinal epithelial barrier, which significantly influences the gut-brain crosstalk, leaving a marked impact on digestive processes, immune responses, emotional status, perception and cognitive functions, contributing to digestive disorders (IBS, IBD) and neurological disorders such as depression, anxiety and even PD. Corticotrophin releasing hormone (CRH), Anterior pituitary (AP), Adrenocorticotropic hormone (ACTH), Adrenal cortex (AC), Hypothalamus (HYP), Hypothalamic pituitary adrenal axis (HPA), Enteric nervous system (ENS), Central nervous system (CNS), 5-Hydroxytryptamine (5-HT), Short chain Fatty acids (SCFAs), Lipopolysaccharide (LPS), γ-Amino butyric acid (GABA), Inflammatory bowel disease (IBD), Irritable bowel syndrome (IBS), Parkinson's disease (PD).

small molecules, α-synuclein follows a β-sheet structure formation and loses its physiologic membrane-binding capacity, leading to the aggregation of misfolded proteins forming so-called Lewy neurites and Lewy bodies in dopaminergic neurons of substantia nigra and noradrenergic neurons of the locus coeruleus.[69,70] Aggregated α-synuclein impairs mitochondrial function by inhibiting mitochondrial complex I activity,[71] causing mitochondrial dysfunction and ultimately leading to oxidative stress in dopaminrgic neurons. Both *α*-synuclein aggregation and mitochondrial dysfunction have been proposed to be important for the loss of dopaminergic neurons in substantia nigra in Parkinson's disease.[72–74]

Thus the spread of α-synuclein from the ENS to the CNS by trans-synaptic cell-to-cell transmission in both sympathetic and parasympathetic nervous systems[75] is the foundation for the "prion concept" in Parkinson's disease pathophysiology.[16,76] In addition to α-syn aggregation, inflammation also plays a key role in the progression of Parkinson's disease. Chemical or viral exposure may give rise to immune activation in the gastrointestinal tract or olfactory system, triggering misfolding, aggregation, and subsequent propagation of α-syn. The dysregulated action of glial cells and astrocytes triggered by the neurotoxic α-syn can induce inflammation in the brain, but neuroinflammation can prompt the prion-like pathology of α-syn by increasing its release, uptake, or both. However, this hypothesis has not yet been fully confirmed.[77]

5 Targeting of GBA in Parkinson's disease by probiotics, prebiotics, and/or symbiotics

There is no cure for Parkinson's disease. Available palliative treatments focus only on restoring dopamine deficits and controlling symptoms. Since there is evidence for a role of the intestinal microbiome in Parkinson's disease, a dietary approach may be helpful in preventing or modifying progression of this disease. Dietary membrane precursors and cofactors can increase neuronal membrane formation and function and reduce pro-inflammatory response. Several studies have suggested that dietary interventions (e.g., probiotics, prebiotics, and/or symbiotics) may influence the GBA by changing gut microbiota composition and therefore affect neuronal functioning in both the ENS and CNS, which may alter the pathogenesis of Parkinson's disease.[78–82]

Probiotics are specific microorganisms that when administered in adequate amounts can exert a health benefit on the host by restoring microbiota and maintaining immune homeostasis. The most common probiotic bacteria currently used are representatives of *Lactobacilli*, *Enterococci*, *Bifidobacteria*, yeasts, and mixtures of different beneficial bacteria. The beneficial effects of probiotics, including enhancement of intestinal epithelial integrity, protection from barrier disruption, stimulation of mucosal immune system, and suppression of pathogenic bacterial growth, as well as modulation of brain function, improve anxiety and depression.[83] One study reported that Parkinson's disease patients suffering from chronic constipation receiving fermented milk containing *Lactobacillus casei* Shirota for 5 weeks experienced improved stool consistency and reduced bloating and abdominal pain.[84] Improving gastrointestinal function by supplementing with probiotics may lead to better functionality and protection of the intestine as well as improve levodopa absorption and reduce behavioral and cognitive deficits (such as anxiety, depression, and memory problems),[85,86] which are common symptoms in Parkinson's disease patients. At present the studies regarding the use of probiotics for the treatment of Parkinson's disease are very limited, but in the long run probiotics may turn out to be a powerful tool for altering Parkinson's disease-associated microbiota composition and improving gut leakiness, homeostasis, and associated neuroinflammation in the ENS and CNS.

Prebiotics are nondigestible oligosaccharides that are known to beneficially affect the host by selectively stimulating the growth and/or activity of a limited number of bacteria in the gut.[87] Prebiotic fibers are shown to have beneficial effects on immune function, bowel motility, and constipation, which may be very relevant for inflammation and gastrointestinal-related symptoms in Parkinson's disease patients. Two well-known, nondigestible carbohydrates are the galacto-oligosaccharides (GOS), based on lactose, and fructo-oligosaccharides (FOS), synthesized from fructose. These have been shown to increase the levels of brain-derived neurotrophic factor (BDNF) in the dentate gyrus of the hippocampus in rats.[88] As BDNF signaling is critical for neuronal protection, survival, and plasticity, GOS and FOS supplementation may provide brain neuroprotection in Parkinson's disease.[89] However, despite the evidence supporting the use of prebiotics for gastrointestinal dysfunction, immune function, and neuroprotection, their use has never been investigated in Parkinson's disease patients. Future clinical research is required for precise understanding of intracellular and extracellular actions of BDNF in response to prebiotics in Parkinson's disease patients.

Gibson and Roberfroid also speculated the concept of additional benefits of synbiotics.[87] By combining the rationale of pro- and prebiotics, the concept of synbiotics is proposed to characterize some colonic foods with interesting nutritional properties that make these compounds candidates for classification as health-enhancing functional food ingredients. The term symbiotic is used for those products in which a prebiotic component selectively favors a probiotic microorganism.[90] The probiotic strains used in symbiotic formulations include *Lactobacilli*, *Bifidobacteria* spp., *Saccharomyces boulardii*, and *Bacillus coagulans* etc., while the major prebiotics used include oligosaccharides like FOS, GOS, and xyloseoligosaccharide (XOS), inulin, and prebiotics from natural sources like chicory and yacon roots. The health benefits of symbiotic consumption include elevated levels of *Lactobacilli* and *Bifidobacteria* and balanced gut microbiota, improved liver function in cirrhotic patients, improved immunomodulating ability, and prevention of bacterial translocation and reduced incidences of nosocomial infections in surgical patients.[91,92]

Chronic constipation is a common symptom that influences the quality of life in Parkinson's disease patients. To evaluate whether synbiotic intake may efficiently restore the balance of gut microbiota and improve gastrointestinal functions, a randomized, placebo-controlled trial was performed with dietary supplementation with the synbiotic (BIFICOPEC) containing 0.63 g of bifid triple viable capsules (BIFICO) and 8 g of soluble dietary fiber.[93] Results showed that 12 weeks of supplementation with probiotics and soluble dietary fiber increased bowel movements, improved Patient Assessment of Constipation-Symptoms (PAC-SYM) and Gastrointestinal Quality of Life Index (GIQLI) scores, and relieved constipation-related symptoms in patients with slow transit constipation.[93] Moreover, another group independently reported that the consumption of fermented milk (containing probiotics and prebiotics) led to an increase in the number of bowel movements and relieved constipation.[94] Thus symbiotics may be another potential therapeutic approach to prevent and/or treat Parkinson's disease by regulating intestinal microbiota composition and exerting beneficial effects on immune function, dysbiosis, and bowel function.

6 Conclusion

Increased evidence indicates that many factors, such as unbalanced nutrition, antibiotic use, age, and infection, result in increased intestinal permeability, causing alterations in microbial metabolites, triggering α-synuclein accumulation in the intestinal mucosa cells, leading to intestinal inflammation producing pro-inflammatory cytokines. Pathologic α-synuclein can be spread from intestine to the brain via the vagal nerve and pro-inflammatory cytokines and immune cells transmitting to the brain through the humoral system. The diversity and amount of microorganisms in the gut may affect both the ENS and CNS. Therefore, a better understanding of the GBA may aid in preventing the pathogenesis of disease. Since amount, type, and balance of dietary macronutrients carbohydrates, proteins, and fats may have a great protective impact on Parkinson's disease, nutrition-based approaches may provide opportunities to complement traditional Parkinson's disease therapies. At present, the studies conducted on the GBA in Parkinson's disease patients are inadequate, therefore randomized controlled trials are needed to evaluate nutritional or probiotic, prebiotic, and/or synbiotic supplementation to treat gut microbiota dysbiosis.

References

1. de Lau LM, Breteler MM. Epidemiology of Parkinson's disease. *Lancet Neurol.* 2006;5(6):525–535.
2. Hirtz D, Thurman DJ, Gwinn-Hardy K, Mohamed M, Chaudhuri AR, Zalutsky R. How common are the "common" neurologic disorders? *Neurology.* 2007;68:326–337.
3. Moustafa AA, Chakravarthy S, Phillips JR, et al. Motor symptoms in Parkinson's disease: a unified framework. *Neurosci Biobehav Rev.* 2016;68:727–740.
4. Pfeiffer RF. Non-motor symptoms in Parkinson's disease. *Parkinsonism Relat Disord.* 2016;22:S119–S122.
5. Poewe W. Non-motor symptoms in Parkinson's disease. *Eur J Neurol.* 2008;15(Suppl. 1):14–20.
6. Martinez-Martin P, Rodriguez-Blazquez C, Kurtis MM, Chaudhuri KR, NMSS Validation Group. The impact of non-motor symptoms on health-related quality of life of patients with Parkinson's disease. *Mov Disord.* 2011;26(3):399–406.
7. Aarsland D, Kramberger MG. Neuropsychiatric symptoms in Parkinson's disease. *J Parkinsons Dis.* 2015;5(3):659–667.
8. Yang D, Zhao D, Shah SZA, et al. The role of the gut Microbiota in the pathogenesis of Parkinson's disease. *Front Neurol.* 2019;10:1155.
9. Dias V, Junn E, Mouradian MM. The role of oxidative stress in Parkinson's disease. *J Parkinsons Dis.* 2013;3(4):461–491.
10. Fortin DL, Nemani VM, Voglmaier SM, Anthony MD, Ryan TA, Edwards RH. Neural activity controls the synaptic accumulation of α-synuclein. *J Neurosci.* 2005;25:10913–10921.
11. Emanuele M, Chieregatt E. Mechanisms of alpha-synuclein action on neurotransmission: cell-autonomous and non-cell autonomous role. *Biomolecules.* 2015;5(2):865–892.
12. Atik A, Stewart T, Zhang J. Alpha-synuclein as a biomarker for Parkinson's disease. *Brain Pathol.* 2016;26(3):410–418.
13. Oueslati A, Fournier M, Lashuel HA. Role of post-translational modifications in modulating the structure, function and toxicity of alpha-synuclein: implications for Parkinson's disease pathogenesis and therapies. *Prog Brain Res.* 2010;183:115–145.
14. Smith WW, Margolis RL, Li X, et al. Alpha-synuclein phosphorylation enhances eosinophilic cytoplasmic inclusion formation in SH-SY5Y cells. *J Neurosci.* 2005;25(23):5544–5552.

15. Datta SK, Verma S, Jain V, et al. Parkinson's disease: the emerging role of gut dysbiosis, antibiotics, probiotics, and fecal microbiota transplantation. *J Neurogastroenterol Motil.* 2019;25(3):363–376.
16. Braak H, de Vos RA, Bohl J, Del Tredici K. Gastric alpha-synuclein immunoreactive inclusions in Meissner's and Auerbach's plexuses in cases staged for Parkinson's disease-related brain pathology. *Neurosci Lett.* 2006;396(1):67–72.
17. Brundin P, Melki R. Prying into the prion hypothesis for Parkinson's disease. *J Neurosci.* 2017;37(41):9808–9818.
18. Braak H, Del Tredici K, Rub U, de Vos RA, Jansen Steur EN, Braak E. Staging of brain pathology related to sporadic Parkinson's disease. *Neurobiol Aging.* 2003;24(2):197–211.
19. Goetze O, Wieczorek J, Mueller T, Przuntek H, Schmidt WE, Woitalla D. Impaired gastric emptying of a solid test meal in patients with Parkinson's disease using 13C-sodium octanoate breath test. *Neurosci Lett.* 2005;375(3):170–173.
20. Noyce AJ, Bestwick JP, Silveira-Moriyama L, et al. Meta-analysis of early nonmotor features and risk factors for Parkinson disease. *Ann Neurol.* 2012;72(6):893–901.
21. Hawkes CH, Del Tredici K, Braak H. A timeline for Parkinson's disease. *Parkinsonism Relat Disord.* 2010;16(2):79–84.
22. Katzenschlager R, Lees AJ. Olfaction and Parkinson's syndromes: its role in differential diagnosis. *Curr Opin Neurol.* 2004;17(4):417–423.
23. Hirayama M, Hakusui S, Koike Y, et al. A scintigraphical qualitative analysis of peripheral vascular sympathetic function with meta-[123I] iodobenzylguanidine in neurological patients with autonomic failure. *J Auton Nerv Syst.* 1995;53(2–3):230–234.
24. Abbott RD, Petrovitch H, White LR, et al. Frequency of bowel movements and the future risk of Parkinson's disease. *Neurology.* 2001;57(3):456–462.
25. Lin C-H, Lin J-W, Liu Y-C, Chang C-H. Risk of Parkinson's disease following severe constipation: a nationwide population-based cohort study. *Parkinsonism Relat Disord.* 2014;20(12):1371–1375.
26. Liddle RA. Parkinson's disease from the gut. *Brain Res.* 2018;1693(Pt B):201–206.
27. Hooper LV, Littman DR, Macpherson AJ. Interactions between the microbiota and the immune system. *Science.* 2012;336:1268–1273.
28. Sonnenburg JL, Bäckhed F. Diet-microbiota interactions as moderators of human metabolism. *Nature.* 2016;535:56–64.
29. Mayer EA, Knight R, Mazmanian SK, Cryan JF, Tillisch K. Gut microbes and the brain: paradigm shift in neuroscience. *J Neurosci.* 2014;34:15490–15496.
30. Carabotti M, Sciroccoa A, Masellib MA, Severia C. The gut-brain axis: interactions between enteric microbiota, central and enteric nervous systems. *Ann Gastroenterol.* 2015;28:203–209.
31. Jiang C, Li G, Huang P, Liu Z, Zhao B. The gut microbiota and Alzheimer's disease. *J Alzheimers Dis.* 2017;58:1–15.
32. Kowalski K, Mulak A. Brain-gut-microbiota axis in Alzheimer's disease. *J Neurogastroenterol Motil.* 2019;25(1):48–60.
33. Berer K, Mues M, Koutrolos M, et al. Commensal microbiota and myelin autoantigen cooperate to trigger autoimmune demyelination. *Nature.* 2011;479:538–541.
34. Lee YK, Menezes JS, Umesaki Y, Mazmanian SK. Proinflammatory T-cell responses to gut microbiota promote experimental autoimmune encephalomyelitis. *Proc Natl Acad Sci U S A.* 2011;108(Suppl. 1):4615–4622.
35. Xu R, Wu B, Liang J, et al. Altered gut microbiota and mucosal immunity in patients with schizophrenia. *Brain Behav Immun.* 2020;85:120–127.
36. Severance EG, Yolken RH, Eaton WW. Autoimmune diseases, gastrointestinal disorders and the microbiome in schizophrenia: more than a gut feeling. *Schizophr Res.* 2016;176(1):23–35.

37. Strati F, Cavalieri D, Albanese D, et al. New evidences on the altered gut microbiota in autism spectrum disorders. *Microbiome*. 2017;5:24.
38. Fattorusso A, Genova LD, Dell'Isola GV, Mencaroni E, Esposito S. Autism spectrum disorders and the gut microbiota. *Nutrients*. 2019;11:521.
39. Zheng P, Zeng B, Zhou C, et al. Gut microbiome remodeling induces depressive-like behaviors through a pathway mediated by the host's metabolism. *Mol Psychiatry*. 2016;21:786–796.
40. Cheung SG, Goldenthal AR, Uhlemann A-C, Mann JJ, Miller JM, Sublette E. Systematic review of gut microbiota and major depression. *Front Psych*. 2019;10:34.
41. Jiang H, Ling Z, Zhang Y, et al. Altered fecal microbiota composition in patients with major depressive disorder. *Brain Behav Immun*. 2015;48:186–194.
42. Unger MM, Spiegel J, Dillmann KU, et al. Short chain fatty acids and gut microbiota differ between patients with Parkinson's disease and age-matched controls. *Parkinsonism Relat Disord*. 2016;32:66–72.
43. Bedarf JR, Hildebrand F, Coelho LP, et al. Functional implications of microbial and viral gut metagenome changes in early stage L-DOPA-naive Parkinson's disease patients. *Genome Med*. 2017;9(1):39.
44. Hopfner F, Kunstner A, Muller SH, et al. Gut microbiota in Parkinson disease in a northern German cohort. *Brain Res*. 2017;1667:41–45.
45. Keshavarzian A, Green SJ, Engen PA, et al. Colonic bacterial composition in Parkinson's disease. *Mov Disord*. 2015;30(10):1351–1360.
46. Hill-Burns EM, Debelius JW, Morton JT, et al. Parkinson's disease and Parkinson's disease medications have distinct signatures of the gut microbiome. *Mov Disord*. 2017;32(5):739–749.
47. Scheperjans F, Aho V, Pereira PA, et al. Gut microbiota are related to Parkinson's disease and clinical phenotype. *Mov Disord*. 2015;30(3):350–358.
48. Sampson TR, Debelius JW, Thron T, et al. Gut microbiota regulate motor deficits and neuroinflammation in a model of Parkinson's disease. *Cell*. 2016;167:1469–1480.
49. Minato T, Maeda T, Fujisawa Y, et al. Progression of Parkinson's disease is associated with gut dysbiosis: two-year follow-up study. *PLoS One*. 2017;12:e0187307.
50. Pringsheim T, Jette N, Frolkis A, Steeves TD. The prevalence of Parkinson's disease: a systematic review and meta-analysis. *Mov Disord*. 2014;29(13):1583–1590.
51. Hasegawa S, Goto S, Tsuji H, et al. Intestinal dysbiosis and lowered serum lipopolysaccharide-binding protein in Parkinson's disease. *PLoS One*. 2015;10(11):e0142164.
52. Wallen ZD, Appah M, Dean MN, et al. Characterizing dysbiosis of gut microbiome in PD: evidence for overabundance of opportunistic pathogens. *NPJ Parkinsons Dis*. 2020;6:11.
53. Braak H, Rub U, Gai WP, Del Tredici K. Idiopathic Parkinson's disease: possible routes by which vulnerable neuronal types may be subject to neuroinvasion by an unknown pathogen. *J Neural Transm (Vienna)*. 2003;110:517–536.
54. Messaoudi M, Lalonde R, Violle N, Javelot H. Assessment of psychotropic-like properties of a probiotic formulation (*Lactobacillus helveticus* R0052 and *Bifidobacterium longum* R0175) in rats and human subjects. *Br J Nutr*. 2011;105:755–764.
55. Wüllner U, Schmitz-Hübsch T, Antony G, et al. Autonomic dysfunction in 3414 Parkinson's disease patients enrolled in the German network on Parkinson's disease (KNP e.V.): the effect of ageing. *Eur J Neurol*. 2007;14:1405–1408.
56. Fasano A, Bove F, Gabrielli M, et al. The role of small intestinal bacterial overgrowth in Parkinson's disease. *Mov Disord*. 2013;28:1241–1249.
57. Grenham S, Clarke G, Cryan JF, Dinan TG. Brain-gut-microbe communication in health and disease. *Front Physiol*. 2011;2:94.

58. Browning KN, Travagli RA. Central nervous system control of gastrointestinal motility and secretion and modulation of gastrointestinal functions. *Compr Physiol.* 2014;4:1339–1368.
59. Foster JA, Rinaman L, Cryan JF. Stress and the gut-brain axis: regulation by the microbiome. *Neurobiol Stress.* 2017;7:124–136.
60. Tsigos C, Chrousos GP. Hypothalamic-pituitary-adrenal axis, neuroendocrine factors and stress. *J Psychosom Res.* 2002;53:865–871.
61. Zheng G, Wu S-P, Hu Y, Smith DE, Wiley JW, Hong S. Corticosterone mediates stress-related increased intestinal permeability in a region-specific manner. *Neurogastroenterol Motil.* 2013;25:e127–e139.
62. Dinan TG, Quigley EMM, Ahmed SMM, et al. Hypothalamic-pituitary-gut axis dysregulation in irritable bowel syndrome: plasma cytokines as a potential biomarker? *Gastroenterology.* 2006;130:304–311.
63. Houser MC, Tansey MG. The gut-brain axis: is intestinal inflammation a silent driver of Parkinson's disease pathogenesis? *NPJ Parkinsons Dis.* 2017;3, 3.
64. Ambrosini YK, Borcherding D, Kanthasamy A, et al. The gut-brain axis in neurodegenerative diseases and relevance of the canine model: a review. *Front Aging Neurosci.* 2019;11:130.
65. Wong YC, Krainc D. α-Synuclein toxicity in neurodegeneration: mechanism and therapeutic strategies. *Nat Med.* 2017;23(2):1–13.
66. Böttner M, Zorenkov D, Hellwig I, et al. Expression pattern and localization of α-synuclein in the human enteric nervous system. *Neurobiol Dis.* 2012;48:474–480.
67. Shannon KM, Keshavarzian A, Mutlu E, et al. α-Synuclein in colonic submucosa in early untreated Parkinson's disease. *Mov Disord.* 2012;27:709–715.
68. Gold A, Turkalp ZT, Munoz DG. Enteric α-synuclein expression is increased in Parkinson's disease but not Alzheimer's disease. *Mov Disord.* 2013;28:237–240.
69. Hasegawa M, Fujiwara H, Nonaka T, et al. Phosphorylated α-synuclein is ubiquitinated in α-synucleinopathy lesions. *J Biol Chem.* 2002;277:49071–49076.
70. Breydo L, Wu JW, Uversky VN. α-Synuclein misfolding and Parkinson's disease. *Biochim Biophys Acta.* 2012;822:261–285.
71. Reeve AK, Ludtmann MHR, Angelova PR, et al. Aggregated α-synuclein and complex I deficiency: exploration of their relationship in differentiated neurons. *Cell Death Dis.* 2015;6(7):e1820.
72. Fitzgerald E, Murphy S, Martinson HA. Alpha-synuclein pathology and the role of the microbiota in Parkinson's disease. *Front Neurosci.* 2019;13:369.
73. Jenner P. Oxidative stress in Parkinson's disease. *Ann Neurol.* 2003;53:S26–S38.
74. Prigione A, Isaias IU, Galbussera A, et al. Increased oxidative stress in lymphocytes from untreated Parkinson's disease patients. *Parkinsonism Relat Disord.* 2009;15:327–328.
75. Danzer KM, Kranich LR, Ruf WP, et al. Exosomal cell-to-cell transmission of α synuclein oligomers. *Mol Neurodegener.* 2012;7:42.
76. Brundin P, Ma J, Kordower JH. How strong is the evidence that Parkinson's disease is a prion disorder? *Curr Opin Neurol.* 2016;29:459–466.
77. Ma J, Gao J, Wang J, Xie A. Prion-like mechanisms in Parkinson's disease. *Front Neurosci.* 2019;13:552.
78. Clemente JC, Ursell LK, Parfrey LW, Knight R. The impact of the gut microbiota on human health: an integrative view. *Cell.* 2012;148:1258–1270.
79. Cryan JF, Dinan TG. Mind-altering microorganisms: the impact of the gut microbiota on brain and behaviour. *Nat Rev Neurosci.* 2012;13:701–712.
80. Cryan JF, Dinan TG. More than a gut feeling: the microbiota regulates neurodevelopment and behavior. *Neuropsychopharmacology.* 2015;40:241–242.
81. Maslowski KM, Mackay CR. Diet, gut microbiota and immune responses. *Nat Immunol.* 2011;12:5–9.

82. Perez-Pardoa P, Kliesta T, Dodiyab HB, et al. The gut-brain axis in Parkinson's disease: possibilities for food-based therapies. *Eur J Pharmacol.* 2017;817:86–95.
83. Varankovich NV, Nickerson MT, Korber DR. Probiotic-based strategies for therapeutic and prophylactic use against multiple gastrointestinal diseases. *Front Microbiol.* 2015;6:685.
84. Cassani E, Privitera G, Pezzoli G, et al. Use of probiotics for the treatment of constipation in Parkinson's disease patients. *Minerva Gastroenterol Dietol.* 2011;57:117–121.
85. Liang S, Wang T, Hu X, et al. Administration of *Lactobacillus helveticus* NS8 improves behavioral, cognitive, and biochemical aberrations caused by chronic restraint stress. *Neuroscience.* 2015;310:561–577.
86. Maes M, Kubera M, Leunis J-C. The gut-brain barrier in major depression: intestinal mucosal dysfunction with an increased translocation of LPS from gram negative enterobacteria (leaky gut) plays a role in the inflammatory pathophysiology of depression. *Neuro Endocrinol Lett.* 2008;29:117–124.
87. Gibson GR, Roberfroid MB. Dietary modulation of the human colonic microbiota: introducing the concept of prebiotics. *J Nutr.* 1995;125:1401–1412.
88. Savignac HM, Corona G, Mills H, et al. Prebiotic feeding elevates central brain derived neurotrophic factor, N-methyl-D-aspartate receptor subunits and D-serine. *Neurochem Int.* 2013;63:756–764.
89. Numakawa T, Suzuki S, Kumamaru E, Adachi N, Richards M, Kunugi H. BDNF function and intracellular signaling in neurons. *Histol Histopathol.* 2010;25:237–258.
90. De Vrese M, Schrezenmeir J. Probiotics, prebiotics, and synbiotics. *Adv Biochem Eng Biotechnol.* 2008;111:1–66.
91. Pandey KR, Naik SR, Vakil BV. Probiotics, prebiotics and synbiotics—a review. *J Food Sci Technol.* 2015;52(12):7577–7587.
92. Zhang MM, Cheng JQ, Lu YR, Yi ZH, Yang P, Wu XT. Use of pre-, pro-and synbiotics in patients with acute pancreatitis: a meta-analysis. *World J Gastroenterol.* 2010;16(31):3970.
93. Ding C, Ge X, Zhang X, et al. Efficacy of synbiotics in patients with slow transit constipation: a prospective randomized trial. *Nutrients.* 2016;8:605.
94. Barichella M, Pacchetti C, Bolliri C, et al. Probiotics and prebiotic fiber for constipation associated with Parkinson disease: an RCT. *Neurology.* 2016;87:1274–1280.

CHAPTER 8

The contribution of microbiota, cerebral blood flow, and sleep deprivation in the pathogenesis of Alzheimer's disease

Akhlaq A. Farooqui
Department of Molecular and Cellular Biochemistry, The Ohio State University, Columbus, OH, United States

1 Introduction

Alzheimer's disease (AD) is a progressive neurodegenerative disease characterized by the accumulation of extracellular β-amyloid (Aβ) plaques (senile plaques) and intracellular neurofibrillary tangles (NFTs), loss of memory, dramatic changes in character and behavior, and inability to carry out normal daily activities in the latter stages of the disease.[1–5] To this end, there have been many reports indicating the presence of Aβ plaques in brains from aged individuals without cognitive dysfunction and memory loss,[6,7] suggesting that the association of Aβ with AD is collateral rather than causal. In addition, anti-Aβ therapies (solanezumab and gantenerumab) have failed to preserve or rescue cognitive function, indicating that Aβ may not be universally neurotoxic.[8,9] In addition, attempts to treat AD using BACE1 inhibitors have worsened cognitive function. However, BACE1 inhibitor treatment results in Aβ reduction as judged by positron emission tomography (PET) studies, and significantly improves the cerebrospinal fluid (CSF) Aβ42/40 ratio and total tau pTau-181. This infers that reducing Aβ levels in the brain has a positive effect on downstream pathological pathways.[10] Finally, neither Aβ plaques nor phosphotau NFTs are specific for AD. Other neuropathological conditions such as Parkinson's disease,[11] which is characterized by monoaminergic dysfunction, Lewy body pathology, and cerebrovascular disease also show deposition of Aβ in the brain, suggesting that Aβ alone is insufficient to cause neuronal loss and cognitive symptoms observed in AD.[12] These observations and emerging data on the involvement of cerebral blood flow (CBF),[13] mitochondrial dysfunction and synaptic network dysfunction, insulin resistance, gut microbiome, and neuroinflammation,[14]

Gut Microbiota in Neurologic and Visceral Diseases
https://doi.org/10.1016/B978-0-12-821039-0.00001-0

genetic, epigenetic, and environmental factors, are driving the interest of investigators in new directions on the pathogenesis of AD.[14–16] These processes are closely linked not only with consumption of diet, vascular risk factors, sleep, and regulation of CBF by arterioles and the capillary neurovascular unit, but also with genetic, epigenetic, and environmental factors.[14,17–24] Collective evidence suggests that other mechanisms directly or indirectly related to Aβ may contribute to the pathogenesis of AD.

2 Factors affecting human gut microbiota

The mammalian gut microbiota is a dynamic and complex entity that consists mainly of bacteria, both symbiotic and potentially pathogenic species (~ 1000 species and 7000 strains of bacteria; 10^8 or 10^{11} in the small and large intestine, respectively).[25] More than 90% of gut microbiota are symbiotic bacteria of which a major portion consists of anaerobes able to modulate metabolic processes of the host.[26] The most common bacteria are Firmicutes (51%) and Bacteroidetes (48%).[27] Firmicutes include both Gram-positive and Gram-negative species such as those belonging to the genera *Lactobacillus* (Gram-positive), *Eubacterium* (Gram-positive), and *Clostridium* (Gram-positive). Bacteroidetes comprise Gram-negative bacteria of the genera *Bacteroides* and *Prevotella*. The remaining 1% of bacteria belong to other divisions such as Proteobacteria (Gram-negative, genus *Escherichia* in particular), Actinobacteria (Gram-positive, genus *Bifidobacterium* in particular), *Fusobacteria* (Gram-negative), Spirochaetes (Gram-negative), Verrucomicrobia (Gram-negative), and Lentisphaerae (Gram-negative).[28] The genes of bacteria in the gut are 150 times larger than those in the human genome, and the mass of bacteria in the body can reach 1.5 kg, or ~ 2% of the weight of an average 75-kg person. Both animal and human studies have demonstrated that diet can influence the composition and function of the gut microbiome. Thus, gut microbiota population is affected by environment, diet, medications (especially antibiotics), age, mode of delivery at birth, method of infant feeding, geographic factors, surgical interventions, and host genetics, particularly genes related to the immune system and metabolism (Fig. 1). Alterations in the relative abundance of each microbiota species in the gut microbial community not only influences immune performance and physiological homeostasis, but also healthy aging.[29,30] Gut microbiota composition is modulated by several factors including aging, genetics, lifestyle (low physical activity and lack of sleep), diet, and antibiotic therapy.[31,32] Among these factors,

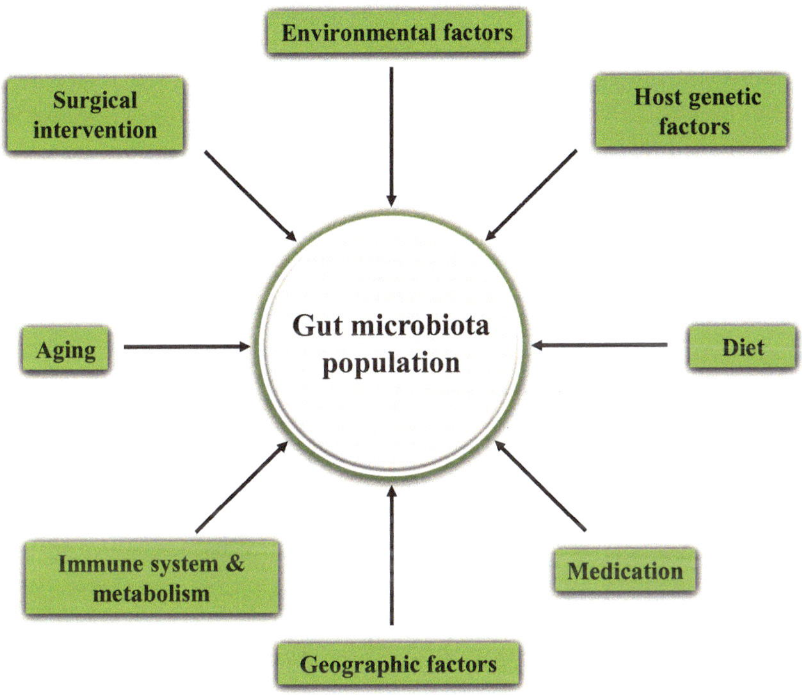

Fig. 1 Factors affecting gut microbiota population.

long-term consumption of a Western diet (high-fat diet), limited physical activity, and lack of sleep are known to alter the composition of the gut microbiome. These processes decrease the number of beneficial bacteria, such as *Lactobacillus* and *Bifidobacteria* and increase the number of "unhealthy" proinflammatory bacteria, such as *Propionibacteria*, *Fusobacteria*, *Shigella*, and *Clostridia*. There is a bidirectional communication occurs between the gut and brain through autonomic, neuroendocrine, enteric, and immune system pathways.[33] Gut microbiota communicate with the brain by producing a number of neuroactive and immunocompetent substances (hormones, neurotransmitters, neuropeptides, and short-chain fatty acids (SCFAs)). These substances are transported from gut to the brain through lymphatic subsystem, which allows for direct or indirect transport.[34,35] Gut microbiota also modulate afferent sensory nerves, for example, through the inhibition of calcium-dependent potassium channels,[36] and regulate the mucosal immune function.[37] Furthermore, microbiota influence the specific phenotypes and physiological functions of T and B cells in the gut mucosa layer.[38] These cells play a pivotal role in immune homeostasis by defending against foreign antigens and by maintaining the integrity of the gut mucosal barrier. Collective evidence suggests that microbiota-gut-brain axis is a complex multidirectional cross-talk system between the gut microbiota, the enteric nervous system (ENS), and the brain. Consumption of unhealthy diet increases the number of proinflammatory gut microbiota and promotes the secretion of neurotoxic substances such as bacterial amyloid and lipopolysaccharides (LPS) leading to dysbiosis, a state in which the homeostasis of the gut microbiome is disrupted, leading to impairment in barrier functions of the intestinal wall and the blood-brain barrier (BBB) (Fig. 2).

3 Effect of microbiota on immune system

Gut microbiota are known to modulate immune cells in the gut and neurons in the central nervous system (CNS) via microbial products[39–41] including LPS,[42] metabolites, and SCFAs,[43,44] all of which can contribute to the pathogenesis of chronic visceral and neurological disorders.[14] LPS produces low-grade inflammation, which is mediated by the induction of inflammatory cytokines by immune cells. SCFAs are end products of the microbial fermentation of macronutrients. They not only modulate immune cell function but also reduce proinflammatory cytokine and chemokine production by monocytes and glial cells (Fig. 3).[14,45] They can also promote regulatory T-cell generation, thereby suppressing the function of inflammatory T cells.[46] Thus, these immune regulatory SCFAs act not only on immune cells systemically but also influence the metabolism of intestinal tissue cells locally. Collective evidence suggests that diet-mediated changes play a predominant role in modulating the composition of gut microbiota. Since elderly people experience impaired barrier functions of the intestinal wall and the BBB, bacterial amyloid and LPS are able to penetrate from the intestinal cavity into the bloodstream, and further into the brain tissue, and have a direct and/or systemic negative effect on the structure and functions of the brain and promote the development of neuroinflammation and neurodegeneration.[47,48]

4 Contribution of microbiota in the pathogenesis of Alzheimer's disease

Once established, the composition of the gut microbiota remains stable throughout adult life. However, there are some differences between the gut microbiota of elderly and young adults.[49] Elderly subjects have predominant

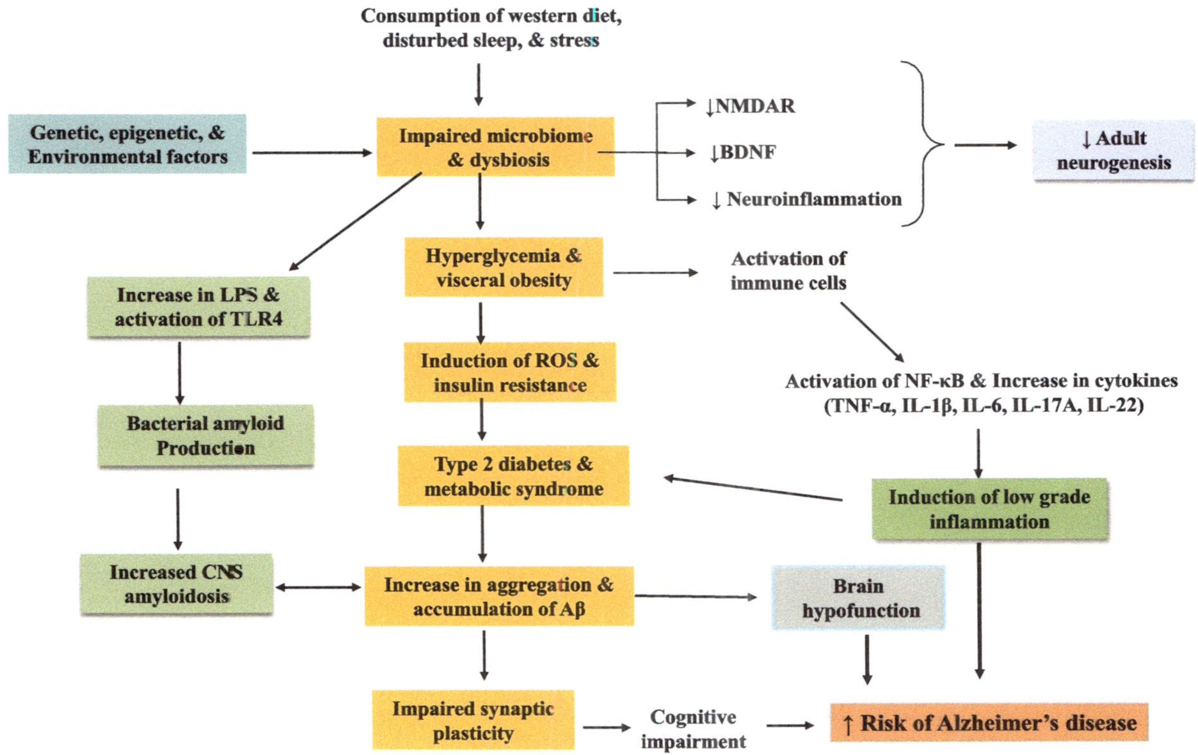

Fig. 2 Involvement of gut microbiome in the pathogenesis of Alzheimer's disease.

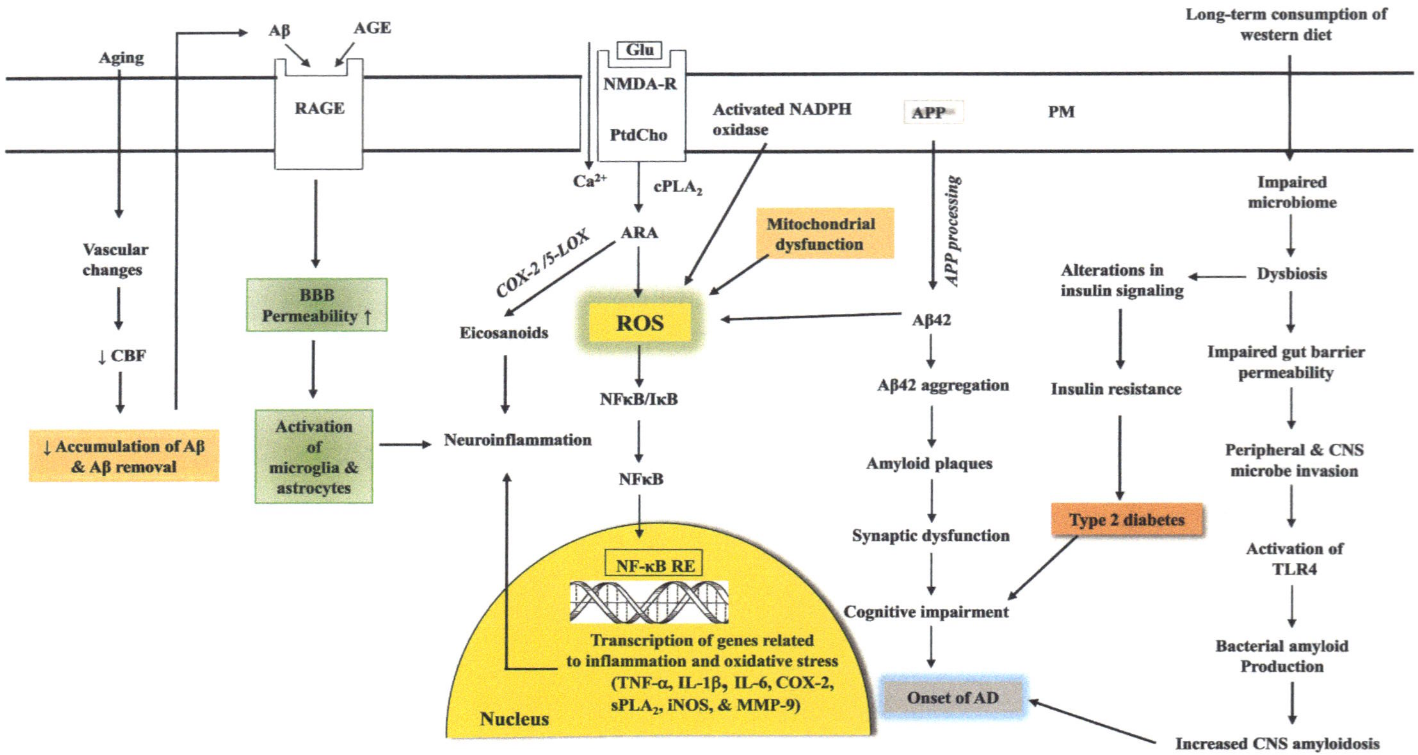

Fig. 3 Hypothetical diagram showing neurochemical processes contributing to the pathogenesis of Alzheimer's disease. Amyloid precursor protein (APP); beta-amyloid (Aβ); Alzheimer's disease (AD); neurofibrillary tangles (NFTs); Aβ-derived diffusible ligands (ADDLs); advanced glycation end products (AGE); receptor for advanced glycation end products (RAGE); jun amino-terminal kinases (JNK); extracellular-signal-regulated kinase (ERK); tumor necrosis factor-α (TNF-α); interleukin-1β (IL-1β); and blood-brain barrier (BBB).

population of the *Bacteroides* and *Clostridium* genera, whereas the guts of young adults predominantly contain Firmicutes.[50] Gut microbiota modulates the expression of many genes in the human intestinal tract,[51] including genes involved in immunity, nutrient absorption, energy metabolism, and intestinal barrier function. Recent studies have hypothesized that AD may be associated with a dysbiosis of microbes in the gut.[52,53] This hypothesis is linked to the fact that the intestinal flora is able to influence the activity of the brain and cause its dysfunctions.[54,55] The imbalance between the microbiota and host organism leads to immune imbalance. Dysbiosis also contributes to the dormant blood microbiome (atopobiosis), a process that directly promotes systemic inflammation via the amyloidogenic formation and shedding of potent inflammagens such as LPS. Thus, alterations in the gut microbiota may also be associated with elevation in the permeability of the gut barrier and immune activation leading to systemic inflammation. This process may in turn impair the BBB and promote neuroinflammation and oxidative stress. Triggering receptors expressed on myeloid cells (TREMs) are key players in dysbiosis, leaky gut, and inflammation. As previously stated, during gut dysbiosis, the intestines release several compounds that activate TREMs on macrophages, leading to an exuberant proinflammatory reaction up to damage in the gut barrier.[56] In turn, TREM-positive activated macrophages along with inflammatory mediators may reach the brain through the blood, glymphatic system, circumventricular organs, or the vagus nerve via the microbiota-gut-brain axis. This leads to a systemic inflammatory response, which, in turn, impairs the BBB while promoting further TREM-dependent neuroinflammation and, ultimately, neural injury.[56] Abnormalities in the composition of the microbiota, and translocation of bacterial antigens into the systemic circulation and the brain, have also become areas of intense research in the pathogenesis of AD.[57,58] Based on these studies it is suggested that gut microbiota can secrete large amounts of amyloids and LPS, which may contribute to the modulation of signaling pathways and the production of proinflammatory cytokines associated with the pathogenesis of AD[59] (Figs. 2 and 3).

Aβ also has antimicrobial properties and participates in the innate immune responses. However, in the dysregulated state, Aβ production may induce harmful effects. Importantly, bacterial amyloids through molecular mimicry may elicit cross-seeding of misfolding and induce microglial priming.[60] The Aβ seeding and propagation may occur at different levels of the microbiota-gut-brain axis. The potential mechanisms of amyloid spreading include neuron-to-neuron or distal neuron spreading, direct BBB crossing, and activation of other cells such as astrocytes, fibroblasts, microglia, and immune system cells. It is suggested that gut microbiota-host interactions and gut dysbiosis play an important role in neurodegeneration. Based on studies, it is suggested that induction of gut-mediated inflammatory responses and poor diet in the elderly (seniors) may contribute to chronic neurodegeneration in the pathogenesis of AD.[52,60]

5 Association of cerebral blood flow in Alzheimer's disease

Healthy cardiovascular and cerebrovascular systems are required for brain development, function, and survival. To meet metabolic demands efficiently, the blood flow to the brain needs to be approximately 750 mL per minute or about 50 mL/100 mL of brain tissue per minute.[61] Global cerebral blood flow (CBF) declines with age,[62–64] which may be due to regional atrophy (although reduced CBF has been shown independent of regional atrophy),[62] lower cerebral metabolism, increased cerebrovascular resistance, or vascular dysfunction.[64] Older age is accompanied by less blood flow to the brain probably due to the onset of atherosclerosis (Fig. 4).[65] Prolonged reduction in blood flow due to aging and atherosclerosis not only results in hypertension-mediated damage in occipito-temporal, prefrontal, and medial temporal lobe regions, but also impacts blood vessel function.[66] It is hypothesized that prolonged reduction in blood flow may be one of the very early events in the pathogenesis of AD.[67] This hypothesis is based on the idea that: (1) greater levels of CBF may prevent or slow the accumulation of neuropathology, in particular amyloid-β[68] and (2) greater levels of CBF may allow for greater cognitive reserve for a given level of neuropathological burden.[69,70] CBF has been reported to be decreased in patients with diabetes and AD (30%–40%) compared to age-matched control subjects[71] causing cerebrovascular hypoperfusion and hypometabolism.[72,73] Brain hypometabolism may be caused by defect in glucose transport, BBB, decrease in glycolysis, and/or mitochondrial dysfunction. The decrease in CBF due to atherosclerosis may negatively affect the synthesis of proteins required for learning and memory and eventually lead to neuritic injury and neuronal death. Prolonged hypertension in old age may result in cognitive dysfunction[74] and dementia.[75] This makes hypertension a major risk factor for vascular cognitive impairment in neurological disorders. Hypertension contributes to both the development and progression of cerebrovascular disease.[76] There is growing evidence that hypertension is the most powerful modifiable risk factor for cerebral vessel dysfunction and it is closely

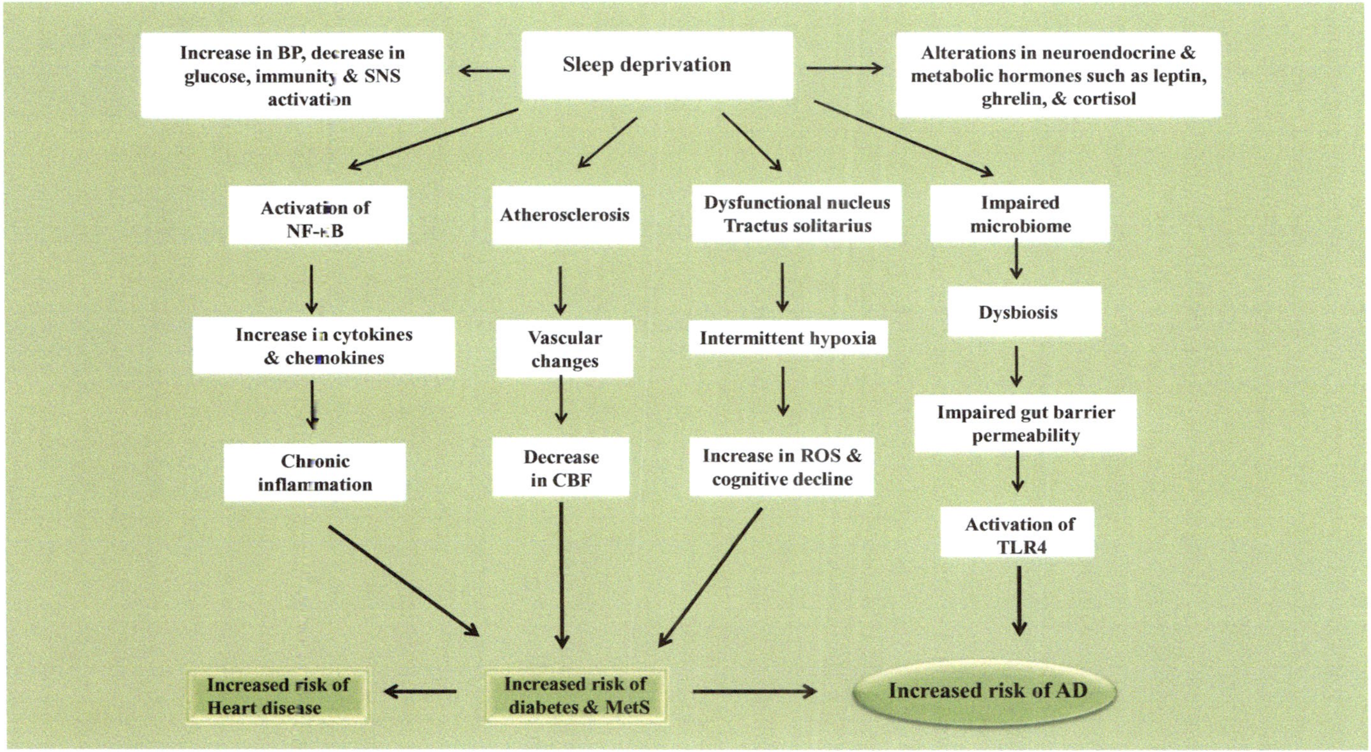

Fig. 4 Effect of microbiota, cerebral blood flow, and sleep deprivation on the pathogenesis of Alzheimer's disease. Nuclear factor-κB (NF-κB); sympathetic nervous system (SNS); reactive oxygen species (ROS); cerebral blood flow (CBF); metabolic syndrome (MetS); and Toll-like receptor 4 (TLR4).

associated with cognitive decline.[77] The relationship among aging, hypertension, and cognitive function is complex and not completely understood. Nevertheless, changes in CBF and blood pressure are considered markers of cerebrovascular health.[78] In seniors, consumption of a Western diet containing high fat and salt not only increases oxidative stress, elevates circulating catecholamines,[79] and decreases nitric oxide (NO) bioavailability, but also enhances renin-angiotensin-aldosterone system (RAAS) levels and levels of the potent vasoconstrictor angiotensin II (Ang II) in a dose-dependent manner.[79,80] The brain RAAS contains several functional components to produce the active ligands Ang II, Ang III, Ang IV, Ang 1–7, and Ang 3–7. These ligands interact with several receptor proteins including AT_1, AT_2, AT_4, and Mas, which are distributed within the central and peripheral nervous systems[81] and modulate blood pressure. Among the aforementioned ligands, Ang II is the best candidate for the mechanistic link between hypertension and AD.[82] It is well known that vasoconstrictor and pro-oxidant effects of Ang II contribute to pathogenesis of essential hypertension.[83,84] Furthermore, increased circulating levels of Ang II accelerate development of AD pathology by promoting β-secretase activity.[85,86] Existing literature supports the concept that Ang II increases superoxide production by activation of angiotensin II type 1 receptor (AT_1R) in cerebral microvascular endothelium thereby causing endothelial dysfunction.[87]

6 Contribution of sleep deprivation in the pathogenesis of Alzheimer's disease

Sleep is a complex biobehavioral process that is an essential part of human life. Sleep is required for optimal health and performance. Obstructive sleep apnea is a common sleep disorder characterized by recurrent partial or complete upper airway obstruction during sleep, leading to either stoppage of breathing (apnea) or very shallow breathing (hypopnea). Obstructive sleep apnea not only leads to intermittent hypoxia due to cyclical oxygen desaturation but also to frequent arousals and sleep fragmentation causing fatigue, bad mood, and depression.[88,89] Due to the presence of polyunsaturated fatty acids and low activities of antioxidant enzymes, the human brain is extremely sensitive to hypoxia, ischemia, and glucose depletion. Impaired delivery of oxygen in obstructive sleep apnea alters neuronal homeostasis, induces pathology, and triggers neuronal degeneration/death leading to neurocognitive impairment, with negative influence on vigilance, attention, executive functioning, and memory.[90] Sleep apnea-mediated pathological changes are known to occur in loss of gray matter in the frontal, parietal, temporal, and occipital cortices, the thalamus, hippocampus, and key brainstem nuclei including the nucleus tractus solitarius. Human and animal model studies propose that intermittent hypoxic insult-mediated changes in sleep apnea result in increased platelet aggregation, metabolic dysregulation, oxidative stress, neuroinflammation, atherosclerosis, endothelial dysfunction, and hypertension.[91] Sleep apnea also stimulates the activation and translocation of NF-κB in specific brain regions involved in sleep regulation,[92] which promotes an increase in expression of inflammatory cytokines such as IL-2, IL-4, IL-5, IL-6, IL-8, IFN-γ, and TNF-α.[93] Other neuro-physiological changes in sleep apnea include increased blood pressure,[91] decreased leptin levels (Fig. 4), and increased levels of the appetite-stimulating hormone ghrelin (28%).[94,95] These changes may predispose animals to increased cardiovascular disease.[96] The severity of sleep apnea also correlates with the degree of insulin resistance,[97] a condition that promotes the pathogenesis of metabolic syndrome (MetS), a major risk factor for stroke, AD, and depression.[98] Collectively these studies indicate that AD is a multifactorial disease mediated by neuroinflammation and neurotoxic protein accumulation[99] that can be associated in part with sleep deprivation, disruption of the glymphatic system, BBB dysfunction,[100–102] oxidative stress, and microglia dysfunction.

7 Conclusion

Due to the failure of anti-Aβ therapies in AD, investigators attempt to explain the pathogenesis of AD on the basis of long-term consumption of a Western diet, alterations in gut microbiome, vascular risk factors, sleep deprivation, and decrease in CBF. These factors usually coexist in MetS, a condition characterized by brain hypofunction, insulin resistance, decreased CBF, hypertension, dyslipidemia, and vascular syndrome.[103] Hypoperfusion of the brain in old age may promote the deposition of amyloid plaques by damaging the BBB and upregulating amyloid precursor proteins. Because neurodegeneration in AD starts in midlife, it is important to identify early risk factors for the disease. Identification of other risk factors may not only shed some light on the pathophysiology of AD but also provide new potential avenues for prevention and treatment. Long-term presence of these factors in the body may result in induction of severe oxidative stress and induction of chronic neuroinflammation. These processes lead to neuroglial network dysfunction, neuroglial energy crisis, and cognitive impairment along with onset of AD.

References

1. Braak H, Del Tredici K. Alzheimer's pathogenesis: is there neuron-to-neuron propagation? *Acta Neuropathol.* 2011;121:589–595.
2. Braak H, Thal DR, Ghebremedhin E, Del Tredici K. Stages of the pathologic process in Alzheimer disease: age categories from 1 to 100 years. *J Neuropathol Exp Neurol.* 2011;70:960–969.
3. Farooqui AA. *Neurochemical Aspects of Alzheimer Disease: Risk Factors, Pathogenesis, Biomarkers, and Treatment Strategies.* Academic Press; 2017.
4. Hardy J. The amyloid hypothesis for Alzheimer's disease: a critical reappraisal. *J Neurochem.* 2009;110:1129–1134.
5. Jahn H. Memory loss in Alzheimer's disease. *Dialogues Clin Neurosci.* 2013;15:445–454.
6. Dickson DW, Crystal HA, Mattiace LA, et al. Identification of normal and pathological aging in prospectively studied nondemented elderly humans. *Neurobiol Aging.* 1992;13:179–189.
7. Tomlinson BE, Blessed G, Roth M. Observations on the brains of demented old people. *J Neurol Sci.* 1970;11:205–242.
8. Robakis NK. Mechanisms of AD neurodegeneration may be independent of Abeta and its derivatives. *Neurobiol Aging.* 2011;32:372–379.
9. Tayeb HO, Murray ED, Price BH, Tarazi FI. Bapineuzumab and solanezumab for Alzheimer's disease: is the 'amyloid cascade hypothesis' still alive? *Expert Opin Biol Ther.* 2013;13:1075–1084.
10. Lewczuk P, Riederer P, O'Bryant SE, et al. Cerebrospinal fluid $A\beta_{42/40}$ corresponds better than $A\beta_{42}$ to amyloid PET in Alzheimer's disease. *World J Biol Psychiatry.* 2018;19:244–328.
11. Petrou M, Dwamena BA, Foerster BR, et al. Amyloid deposition in Parkinson's disease and cognitive impairment: a systematic review. *Mov Disord.* 2015;30:928–935.
12. Coria F, Castaño EM, Frangione B. Brain amyloid in normal aging and cerebral amyloid angiopathy is antigenically related to Alzheimer's disease beta-protein. *Am J Pathol.* 1987;129:422–428.
13. Bangen KJ, Nation DA, Clark LR, Harmell AL, Wierenga CE, et al. Interactive effects of vascular risk burden and advanced age on cerebral blood flow. *Front Aging Neurosci.* 2014;6:159.
14. Farooqui AA. *Insulin Resistance as a Risk Factor for Visceral and Neurological Disorders.* Academic Press; 2020.
15. Wijesinghe P, Shankar SK, Yasha TC, et al. Vascular contributions in Alzheimer's disease-related neuropathological changes: first autopsy evidence from a South Asian aging population. *J Alzheimers Dis.* 2016;54:1607–1618.
16. Zlokovic BV. Neurovascular pathways to neurodegeneration in Alzheimer's disease and other disorders. *Nat Rev Neurosci.* 2011;12:723–738.
17. Askarova S, Umbayev B, Masoud AR, et al. The links between the gut microbiome, aging, modern lifestyle and Alzheimer's disease. *Front Cell Infect Microbiol.* 2020;10:104.
18. Girouard H, Iadecola C. Neurovascular coupling in the normal brain and in hypertension, stroke, and Alzheimer disease. *J Appl Physiol.* 2006;100:328–335.
19. Gupta A, Iadecola C. Impaired Abeta clearance: a potential link between atherosclerosis and Alzheimer's disease. *Front Aging Neurosci.* 2015;7:115.
20. Iadecola C. The pathobiology of vascular dementia. *Neuron.* 2013;80:844–866.
21. Qiu C. Preventing Alzheimer's disease by targeting vascular risk factors: hope and gap. *J Alzheimers Dis.* 2012;32:721–731.
22. Rius-Perez S, Tormos AM, Perez S, Talens-Visconti R. Vascular pathology: cause or effect in Alzheimer disease? *Neurologia.* 2018;33:112–120.
23. Stampfer MJ. Cardiovascular disease and Alzheimer's disease: common links. *J Intern Med.* 2006;260:211–223.

24. Yarchoan M, Xie SX, Kling MA, et al. Cerebrovascular atherosclerosis correlates with Alzheimer pathology in neurodegenerative dementias. *Brain.* 2012;135:3749–3756.
25. Sender R, Fuchs S, Milo R. Revised estimates for the number of human and bacteria cells in the body. *PLoS Biol.* 2016;14:e1002533.
26. Kang JX. Gut microbiota and personalized nutrition. *J Nutrigenet Nutrigenomics.* 2013;6:I–II.
27. The Human Microbiome Project Consortium, Huttenhower C, Gevers D, et al. Structure, function and diversity of the healthy human microbiome. *Nature.* 2012;486:207.
28. Westfall S, Lomis N, Kahouli I, Yuan Dia S, Singh S, Prakash S. Microbiome, probiotics and neurodegenerative diseases: deciphering the gut brain axis. *Cell Mol Life Sci.* 2017;74:3769–3787.
29. Candela M, Biagi E, Brigidi P, O'Toole PW, De Vos WM. Maintenance of a healthy trajectory of the intestinal microbiome during aging: a dietary approach. *Mech Ageing Dev.* 2014;136:70–75.
30. Ottman N, Smidt H, de Vos WM, Belzer C. The function of our microbiota: who is out there and what do they do? *Front Cell Infect Microbiol.* 2012;2:104.
31. Chwalba A, Otto-Buczkowska E. Participation of the microbiome in the pathogenesis of diabetes mellitus. *Clin Diabetol.* 2017;6:178–181.
32. Miele L, Giorgio V, Alberelli MA, De Candia E, Gasbarrini A, Grieco A. Impact of gut microbiota on obesity, diabetes, and cardiovascular disease risk. *Curr Cardiol Rep.* 2015;17:120.
33. Zhao L, Xiong Q, Stary CM, et al. Bidirectional gut-brain-microbiota axis as a potential link between inflammatory bowel disease and ischemic stroke. *J Neuroinflammation.* 2018;15:339.
34. Winek K, Dirnagl U, Meisel A. The gut microbiome as therapeutic target in central nervous system diseases: implications for stroke. *Neurotherapeutics.* 2016;13:762–774.
35. Winek K, Meisel A, Dirnagl U. Gut microbiota impact on stroke outcome: fad or fact? *J Cereb Blood Flow Metab.* 2016;36:891–898.
36. Kunze WA, Mao YK, Wang B, et al. *Lactobacillus reuteri* enhances excitability of colonic AH neurons by inhibiting calcium-dependent potassium channel opening. *J Cell Mol Med.* 2009;13:2261–2270.
37. McDermott AJ, Huffnagle GB. The microbiome and regulation of mucosal immunity. *Immunology.* 2014;142:24–31.
38. Honda K, Littman DR. The microbiota in adaptive immune homeostasis and disease. *Nature.* 2016;535:75–84.
39. Burcelin R. Gut microbiota and immune crosstalk in metabolic disease. *Mol Metab.* 2016;5:771–781.
40. Pomié C, Blasco-Baque V, Klopp P, et al. Triggering the adaptive immune system with commensal gut bacteria protects against insulin resistance and dysglycemia. *Mol Metab.* 2016;5:392–403.
41. Saita D, Ferrarese R, Foglieni C, et al. Adaptive immunity against gut microbiota enhances apoE-mediated immune regulation and reduces atherosclerosis and western-diet-related inflammation. *Sci Rep.* 2016;6:29353.
42. Cani PD, Osto M, Geurts L, Everard A. Involvement of gut microbiota in the development of low-grade inflammation and type 2 diabetes associated with obesity. *Gut Microbes.* 2012;3:279–288.
43. Alvarez-Curto E, Milligan G. Metabolism meets immunity: the role of free fatty acid receptors in the immune system. *Biochem Pharmacol.* 2016;114:3–13.
44. Meijer K, de Vos P, Priebe MG. Butyrate and other short-chain fatty acids as modulators of immunity: what relevance for health? *Curr Opin Clin Nutr Metab Care.* 2010;13:715–721.
45. Usami M, Kishimoto K, Ohata A, et al. Butyrate and trichostatin A attenuate nuclear factor kappaB activation and tumor necrosis factor alpha secretion and increase

prostaglandin E2 secretion in human peripheral blood mononuclear cells. *Nutr Res.* 2008;28:321–328.
46. Arpaia N, Campbell C, Fan X, et al. Metabolites produced by commensal bacteria promote peripheral regulatory T-cell generation. *Nature.* 2013;504:451–455.
47. Cândido FG, Valente XF, Grze´skowiak LM, Moreira AP, Rocha DM, Alfenas R. Impact of dietary fat on gut microbiota and low-grade systemic inflammation: mechanisms and clinical implications on obesity. *Int J Food Sci Nutr.* 2017;69:125–143.
48. Harris K, Kassis A, Major G, Chou CJ. Is the gut microbiota a new factor contributing to obesity and its metabolic disorders? *J Obes.* 2012;2012, 879151.
49. Clemente JC, Ursell LK, Parfrey LW, Knight R. The impact of the gut microbiota on human health: an integrative view. *Cell.* 2012;148:1258–1270.
50. Jones ML, Ganopolsky JG, Martoni CJ, Labbé A, Prakash S. Emerging science of the human microbiome. *Gut Microbes.* 2014;5:446–457.
51. Dahiya DK, Renuka Puniya M, Shandilya UK, et al. Gut microbiota modulation and its relationship with obesity using prebiotic fibers and probiotics: a review. *Front Microbiol.* 2017;8:563.
52. Jiang C, Li G, Huang P, Liu Z, Zhao B. The gut microbiota and Alzheimer's disease. *J Alzheimers Dis.* 2017;58:1–15.
53. Wang Y, Kasper LH. The role of microbiome in central nervous system disorders. *Brain Behav Immun.* 2014;38:1–12.
54. Gareau MG. *Advances in Experimental Medicine and Biology, Microbiota-Gut-Brain Axis and Cognitive Function.* New York, NY: Springer New York; 2014:357–371.
55. Jagust W. Imaging the evolution and pathophysiology of Alzheimer disease. *Nat Rev Neurosci.* 2018;19:687–700.
56. Natale G, Biagioni F, Busceti CL, Gambardella S, Limanaqi F, Fornai F. TREM receptors connecting bowel inflammation to neurodegenerative disorders. *Cell.* 2019;8. pii: E1124.
57. Pretorius L, Kell DB, Pretorius E. Iron dysregulation and dormant microbes as causative agents for impaired blood rheology and pathological clotting in Alzheimer's type dementia. *Front Neurosci.* 2018;12:851.
58. Vogt NM, Kerby RL, Dill-McFarland KA, et al. Gut microbiome alterations in Alzheimer's disease. *Sci Rep.* 2017;7:13537.
59. Sperling RA, Aisen PS, Beckett LA, et al. Toward defining the preclinical stages of Alzheimer's disease: recommendations from the National Institute on Aging-Alzheimer's Association workgroups on diagnostic guidelines for Alzheimer's disease. *Alzheimers Dement.* 2011;7:280–292.
60. Kowalski K, Mulak A. Brain-gut-microbiota axis in Alzheimer's disease. *J Neurogastroenterol Motil.* 2019;25:48–60.
61. Ito H, Kanno I, Fukuda H. Human cerebral circulation: positron emission tomography studies. *Ann Nucl Med.* 2005;19:65–74.
62. Chen JJ, Rosas HD, Salat DH. Age-associated reductions in cerebral blood flow are independent from regional atrophy. *Neuroimage.* 2011;55:468–478.
63. Liu W, Lou X, Ma L. Use of 3D pseudo-continuous arterial spin labeling to characterize sex and age differences in cerebral blood flow. *Neuroradiology.* 2016;58:943–948.
64. Tarunmi T, Ayaz Khan M, Liu J, et al. Cerebral hemodynamics in normal aging: central artery stiffness, wave reflection, and pressure pulsatility. *J Cereb Blood Flow Metab.* 2014;34:971–978.
65. Dennis NA, Cabeza R. Neuroimaging of healthy cognitive aging. In: Craik FIM, Salthouse TA, eds. *The Handbook of Aging and Cognition.* 3rd ed. New York: Psychology Press; 2008:1–54.
66. Beason-Held LL, Moghekar A, Zonderman AB, Kraut MA, Resnick SM. Longitudinal changes in cerebral blood flow in the older hypertensive brain. *Stroke.* 2007;38:1766–1773.

67. Iadecola C. Neurovascular regulation in the normal brain and in Alzheimer's disease. *Nat Rev Neurosci.* 2004;5:347–360.
68. Mattsson N, Tosun D, Insel PS, et al. Association of brain amyloid-beta with cerebral perfusion and structure in Alzheimer's disease and mild cognitive impairment. *Brain.* 2014;137:1550–1561.
69. Davenport MH, Hogan DB, Eskes GA, Longman RS, Poulin MJ. Cerebrovascular reserve: the link between fitness and cognitive function? *Exerc Sport Sci Rev.* 2012;40:153–158.
70. Stern Y. Cognitive reserve in ageing and Alzheimer's disease. *Lancet Neurol.* 2012;11:1006–1012.
71. Erol A. An integrated and unifying hypothesis for the metabolic basis of sporadic Alzheimer's disease. *J Alzheimers Dis.* 2008;13:241–253.
72. Ngwa JS, Fungwe TV, Ntekim O, Allard JS, Johnson SM. Alzheimer's disease neuroimaging initiative associations of pulse and blood pressure with hippocampal volume by APOE and cognitive phenotype: the Alzheimer's disease neuroimaging initiative (ADNI). *Dement Geriatr Cogn Disord.* 2018;45:66–78.
73. Roher AE, Debbins JP, Malek-Ahmadi M, et al. Cerebral blood flow in Alzheimer's disease. *Vasc Health Risk Manag.* 2012;8:599–611.
74. Giordano N, Tikhonoff V, Palatini P, Bascelli A, Boschetti G, et al. Cognitive functions and cognitive reserve in relation to blood pressure components in a population-based cohort aged 53 to 94 years. *Int J Hypertens.* 2012;2012, 274851.
75. Goldstein FC, Levey AI, Steenland NK. High blood pressure and cognitive decline in mild cognitive impairment. *J Am Geriatr Soc.* 2013;61:67–73.
76. MacMahon S, Peto R, Cutler J, et al. Blood pressure, stroke, and coronary heart disease. Part 1, prolonged differences in blood pressure: prospective observational studies corrected for the regression dilution bias. *Lancet.* 1990;335:765–774.
77. Nelson PT, Alafuzoff I, Bigio EH, Bouras C, Braak H, et al. Correlation of Alzheimer disease neuropathologic changes with cognitive status: a review of the literature. *J Neuropathol Exp Neurol.* 2012;71:362–381.
78. Barodka VM, Joshi BL, Berkowitz DE, Hogue CW, Nyhan D. Review article: implications of vascular aging. *Anesth Analg.* 2011;112:1048–1060.
79. Tran LT, MacLeod KM, McNeill JH. Chronic etanercept treatment prevents the development of hypertension in fructose-fed rats. *Mol Cell Biochem.* 2009;330:219–228.
80. Wright JW, Kawas LH, Harding JW. A role for the brain RAS in Alzheimer's and Parkinson's diseases. *Front Endocrinol.* 2013;4:158.
81. Wright JW, Harding JW. The brain RAS and Alzheimer's disease. *Exp Neurol.* 2010;223:326–333.
82. Kehoe PG, Miners S, Love S. Angiotensins in Alzheimer's disease—friend or foe? *Trends Neurosci.* 2009;32:619–628.
83. Coffman TM. Under pressure: the search for the essential mechanisms of hypertension. *Nat Med.* 2011;17:1402–1409.
84. Reudelhuber TL. Where hypertension happens. *J Clin Invest.* 2013;123:1934–1936.
85. Cifuentes D, Poittevin M, Dere E, Broquères-You D, Bonnin P, et al. Hypertension accelerates the progression of Alzheimer-like pathology in a mouse model of the disease. *Hypertension.* 2015;65:218–224.
86. Faraco G, Park L, Zhou P, Luo W, Paul SM, et al. Hypertension enhances Aβ-induced neurovascular dysfunction, promotes β-secretase activity, and leads to amyloidogenic processing of APP. *J Cereb Blood Flow Metab.* 2016;36:241–252.
87. Xiao X, Zhang C, Ma X, Miao H, Wang J, et al. Angiotensin-(1-7) counteracts angiotensin II-induced dysfunction in cerebral endothelial cells via modulating Nox2/ROS and PI3K/NO pathways. *Exp Cell Res.* 2015;336:58–65.
88. Ancoli-Israel S. Sleep apnea in older adults- is it real and should age be the determining factor in the treatment decision matrix? *Sleep Med Rev.* 2007;11:83–85.

89. Levy P, Bonsignore MR, Eckel J. Sleep, sleep-disordered breathing and metabolic consequences. *Eur Respir J.* 2009;34:243–260.
90. Beebe DW, Groesz L, Wells C, Nichols A, McGee K. The neuropsychological effects of obstructive sleep apnea: a meta-analysis of norm-referenced and case-controlled data. *Sleep.* 2003;26:298–307.
91. Daulatzai MA. Pathogenesis of cognitive dysfunction in patients with obstructive sleep apnea: a hypothesis with emphasis on the nucleus tractus solitaries. *Sleep Disord.* 2012;2012, 251096.
92. Ramesh V, Thatte HS, McCarley RW, Basheer R. Adenosine and sleep deprivation promote NF-kappaB nuclear translocation in cholinergic basal forebrain. *J Neurochem.* 2007;100:1351–1363.
93. Irwin MR, Wang M, Ribeiro D, et al. Sleep loss activates cellular inflammatory signaling. *Biol Psychiatry.* 2008;64:538–540.
94. Morselli L, Leproult R, Balbo M, Spiegel K. Role of sleep duration in the regulation of glucose metabolism and appetite. *Best Pract Res Clin Endocrinol Metab.* 2010;24:687–702.
95. Spiegel K, Tasali E, Penev P, Van Cauter E. Brief communication: sleep curtailment in healthy young men is associated with decreased leptin levels, elevated ghrelin levels, and increased hunger and appetite. *Ann Intern Med.* 2004;141:846–850.
96. Caro JF, Sinha MK, Kolaczynski JW, Zhang PL, Considine RV. Leptin: the tale of an obesity gene. *Diabetes.* 1996;45:1455–1462.
97. Wallace AM, McMahon AD, Packard CJ, et al. Plasma leptin and the risk of cardiovascular disease in the West of Scotland Coronary Prevention Study (WOSCOPS). *Circulation.* 2001;104:3052–3056.
98. Farooqui AA. *Metabolic syndrome: An Important Risk Factor for Stroke, Alzheimer, and Depression.* New York: Springer; 2013.
99. Boland B, Yu WH, Corti O, et al. Promoting the clearance of neurotoxic proteins in neurodegenerative disorders of ageing. *Nat Rev Drug Discov.* 2018;17:660–688.
100. Cordone S, Annarumma L, Rossini PM, De Gennaro L. Sleep and β-amyloid deposition in Alzheimer disease: insights on mechanisms and possible innovative treatments. *Front Pharmacol.* 2019;10:695.
101. Jessen NA, Munk AS, Lundgaard I, Nedergaard M. The glymphatic system: a beginner's guide. *Neurochem Res.* 2015;40:2583–2599.
102. Sweeney MD, Zlokovic BV. A lymphatic waste-disposal system implicated in Alzheimer's disease. *Nature.* 2018;560:172–174.
103. Clemente-Postigo M, Queipo-Ortuno MI, Murri M, et al. Endotoxin increase after fat overload is related to postprandial hypertriglyceridemia in morbidly obese patients. *J Lipid Res.* 2012;53:973–978.

CHAPTER 9

Implications of microbiota in the pathogenesis of diabetes mellitus and cardiovascular disease

Daniela Maria Tanase[a], Alina Mihaela Leustean[b], and Evelina Maria Gosav[a]
[a]Department of Internal Medicine, "Grigore T. Popa" University of Medicine and Pharmacy, Internal Medicine Clinic, "St. Spiridon" County Clinical Emergency Hospital, Iasi, Romania
[b]Department of Gastroenterology, "Grigore T. Popa" University of Medicine and Pharmacy, Institute of Gastroenterology and Hepatology "St. Spiridon" County Clinical Emergency Hospital, Iasi, Romania

1 Introduction

The increased incidence and prevalence of metabolic diseases is a global concern. Alteration of the gut microbiota has been linked with metabolic disorders such as obesity, insulin resistance, diabetes, cardiovascular diseases, and inflammatory bowel disease.[1–7] Under different external or internal factors like dietary habits, medication, age, individual genotype, physical activity, infections, antibiotic history, and even changes in the circadian rhythm, the gut microbiota pattern can alter and lead to different metabolic disorders.[8, 9] Over the years, researchers have studied the pivotal role of gut microbiota in cardiovascular disease and diabetes and provided new insights regarding pathophysiological, biochemical, and immune system mechanisms with hope of discovering new predictive and treatment tools for better outcomes.

2 Microbiota and diabetes

The gut microbiota is a complex microbial organ that starts developing in utero or during birth, when early exposure to external environment defines its future composition. In adult subjects the gut microbiome weighs approximatively 100–200 grams, and it is inhabited by trillions of microorganisms such as viruses, bacteria, fungi, archaea, and protozoans.[10] Each person stores more than 150 bacterial species with the vast majority residing in the colon. In addition, almost 90% of the bacterial species identified in the human gut belong to the phyla Bacteroidetes spp. (Gram-negative, e.g. *Bacteroides*,

Gut Microbiota in Neurologic and Visceral Diseases
https://doi.org/10.1016/B978-0-12-821039-0.00005-8

Prevotella) and Firmicutes spp. (Gram-positive, e.g., *Clostridium*, *Lactobacillus*, and *Ruminiococcus*), while the rest belong to Tenericutes, Proteobacteria, Cyanobacteria, and Actinobacteria spp. Among the habitual species, the gastrointestinal tract also includes anaerobes (e.g., *Escherichia coli*), and aerobic bacteria (e.g., *Pseudomonas* spp.).[11, 12] The enterotype hypothesis suggests that gut microbiota diversify continuously over time, is linked to dietary patterns, and is composed of three main enterotypes: *Bacteroides*, *Prevotella*, and *Rumincoccus*.[13, 14]

The gut microbiota has crucial roles in the host. It protects against pathogens, regulates gastrointestinal hormone secretion and gut epithelial sheathing, metabolizes drugs, and mediates immune system functions.[15] In addition, some microorganism can produce short-chain fatty acids (SCFAs) by fermentation, and can synthesize various vitamins (e.g., vitamin B12, vitamin K) and micronutrients.[16] Imbalance in the gut composition, bacterial variety deficiency, and pathobacteria multiplication lead to gut dysbiosis. Dysbiosis of the gut microbiome map may increase risk of opportunistic bacteria and insulin resistance (IR) with metabolic alteration, which in turn increases the risk of metabolic disease such as diabetes.[17] It is still debatable whether microbiome imbalance causes or is a consequence of disease.

2.1 Type 2 diabetes

Type 2 diabetes mellitus (T2D) is a metabolic disease defined by altered blood glucose level (hyperglycemia) and lipid metabolism, and is linked to obesity and high-risk hypertension. According to the World Health Organization (WHO) and International Diabetes Federation (IDF), almost 400 million people worldwide suffer from metabolic diseases such as diabetes. Because of external environmental factors, TD2 may soon be in the top 10 causes of death worldwide.[18–20] T2D frequently appears in the fourth and fifth decades of life and is associated with metabolic syndromes like atherosclerotic disease, hypertension, low high-density lipoprotein cholesterol (HDLc), high circulating level of low-density lipoprotein cholesterol (LDLc), and overweight/obesity status.[21] Among the risk factors for T2D, the most notable are fetal programming, genetic predisposition, and obesity. In T2D the development mechanism is represented by insulin insufficiency and non-responsiveness due to failure of cells to sense normal insulin levels. Other mechanisms include incretin deficiency, high glucagon plasma levels, and elevated lipid catabolism along with other metabolic irregularities.[22] The first evidence that linked gut microbiota alteration and glucose metabolism was published by Backhed et al. in 2004.[23] Since then, researchers have continued to investigate the two-way street that connects intestinal dysbiosis and diabetes.

Interestingly, in patients with T2D the gut microbiome is marked by increased ratio of Bacteroides: Firmicutes and reduction of Clostridia, whereas obese people without T2D exhibit a decreased ratio of Bacteroides:Firmicutes. These results raise the question of whether these two entities have the same or different risk factors. *Lactobacillus* was positively correlated to glucose and HBA1c levels and *Clostridium leptum* and *C. coccoides* (main butyrate-producing bacteria) were correlated to a decrease in T2D.[24,25] These Gram-negative bacteria could generate a pro-inflammatory state through lipopolysaccharides (LPS) alteration, energy homoeostasis, and glucose deregulation, leading to IR and T2D. Therefore, three main mechanisms that relate the interchange between dysbiosis and glucose metabolism are the elevation of LPS in the systemic circulation, the production of SCFAs, and metabolization of bile acids.

Studies showed that LPS shed from the outer membranes of Gram-negative bacteria can promote endotoxemia. High-fat diet elevates LPS circulating levels via elevated intestinal permeability with increased adiposity and subsequently inflammation.[26] High circulating LPS activate the immune system through toll-like receptor 2 (TLR-2) and toll-like receptor 4 (TLR-4) signaling pathways (especially the TLR4/CD14). TLR-4 together with other complex systems mediates IR and LPS translocation into intestinal vessels. Subsequently, the nitric oxide-induced synthase (iNOS) generated reacts with cysteine and forms nitrosothiol. Then, nitrosothiol provokes phosphorylation of the substrate of insulin receptor 1 (IRS-1) in serine and induces hormone resistance. It appears that the TLR2 and TLR5 system plays a key role in identifying microbial cells and in activating certain pro-inflammatory pathways. Evidence show that inflammation, immune system, and metabolism have a synergic and dependent function in metabolism disease development.[27,28]

SCFAs like butyrate, acetate, and propionate contribute to intestinal hormone production, gluconeogenesis, lipogenesis, and hepatic metabolism. SCFAs such as *Faecalibacterium prausnitzi*, a butyrate-producer, and *Bifidobacterium* along with hydrogen are implicated in colonic fermentation and energy output. Butyrate is considered an essential energy substrate for colonic mucosa cells as it enhances glucagon-like peptide-2 (GLP-2) release, promotes gut integrity, and prevents translocation and LPS-related endotoxemia.[29,30] In addition, the signals between SCFAs, which are linked to G protein-coupled receptors (GPR41 and GPR43) and specific immune cells, enhance GLP-1 and YY peptide (PYY) levels with subsequently raised insulin secretion. As mentioned, only few G-protein-coupled receptors inhabit immune cells and activate certain pro-inflammatory cytokines (e.g. IL-1, IL-6, and TNF alpha) and chemokines modulating the systemic immune response.[31, 32]

Bile acids have emulsification properties and help with digestion and absorption of certain vitamins and lipids. Evidence shows that bile acids are also strong signaling molecules and have a codependent relationship with the intestinal microbiota family.[33] In hepatocytes the cholesterol is transformed into primary bile acids, which are then conjugated to glycine and taurine (acid, 3-ketodeoxycholic acid, deoxycholic acid) and stored in the gallbladder and transported to enterohepatic circulation by re-absorption in the distal ileum.[34] Primary bile acids enhance the nuclear farnesoid X receptor (FXR) in ileal enterocytes, which activates the fibroblast growth factor expression that further activates hepatic fibroblast growth factor receptor 4 (FGFR4) with secondary bile acid inhibiton. Secondary bile acids (e.g., lithocholic and deoxycholic acid) modulated by Firmiticus bacteria, connect with the TGR5 (G protein-coupled receptor signaling pathway known also as GPBAR-1 or GPCR19). Research showed that FXR and THR5 are implicated in energy metabolism and glucose homeostasis. They were described to have protective roles against diet-induced obesity by activating GLP1 secretion from L cells.[35, 36]

Even if the precise process that explains how the gut bacteria family can modulate gut permeability is still unclear, the common ingredient for most of these signaling pathways is represented by impairment of the gut epithelial barrier with increased cellular permeability, dysbiosis, macromolecule intake, immune responses, activation of signaling pathways that interfere with glucose and lipid metabolism, and low-degree inflammation, which finally instigates IR, atherosclerosis, and eventually T2D or cardiovascular disease development.[37]

2.2 Type 1 diabetes

Statistics show that incidence and prevalence of type 1 diabetes (T1D) is increasing independent of genetic causes, which means that environmental factors may have a decisive role in its appearance. T1D represents around 10% of diabetes and has been linked to gut microbiota disorder.[38, 39] T1D is characterized by complete insulin insufficiency resulting from the destruction of pancreatic β cells in the islets of Langerhans mediated by T cells.[40] A high-gluten diet may be one factor implicated in the development of T1D and may induce microbiota alterations. Regarding microbiota changes in T1D, studies demonstrated that large amounts of *Bacteroides* are linked to T1D onset. In addition, they found that low gut *Bacteroides* spp. levels are associated with obese individuals, and that *Lactobacillus* spp. levels are dependent on fasting glucose and HbA1c.[26,41] Also, T1D microbiota harbor lower levels of butyrate-producing microorganisms such as *Faecalibacterium* and *Roseburia*, and mucin degraders like *Akkermansia muciniphila* and *Prevotella*[42] Interestingly, authors investigated infants who were genetically predisposed to develop TID and found after tests that their gut microbiota diversity reduced by a quartile compared to normal subjects.[43] Even if the conducted studies so far have several differences like ethnicity, intake of medication, and diet, they all show the fundamental connection between gut dysbiosis, glucose metabolism, and obesity or TID/T2D development (Table 1).[44–58]

3 Microbiota and cardiovascular disease

Changing the composition of the intestinal microbiota and the activity of the multiple metabolism products play an important role in the appearance of cardiovascular diseases (CVDs) such as hypertension, heart failure, and atrial fibrillation, all of which are associated with high morbidity and mortality.[59] Environmental and internal factors like smoking, hypercholesterolemia, obesity, and diabetes, along with dietary patterns, participate in CVD appearance and increased incidence.[60] Each individual has a unique gut microbial spectrum composed of abundant species and subspecies of bacterial communities. Dysbiosis resulting from changes in the gut microbiota diversity can lead to metabolic systemic disturbance.[61–64] The intestinal microbiome is a key factor at the intersection of diet and cardiovascular disease through its role in metabolizing dietary components and thus synthesis and release of important metabolites such as SCFAs and trimethylamine-N-oxide (TMAO). It is known that the intestinal microbiota is involved in the synthesis of proatherogenic metabolites, namely TMAO. The breakdown of foodstuffs rich in phosphatidylcholine, choline, or carnitine (from red meat, fish, and eggs) forms trimethylamine (TMA), the TMAO precursor.[65] TMA is transported to the liver through the circulation of the portal vein, where it is oxidized by liver enzyme flavin monooxygenase (FMO), mainly FMO3 to TMAO, which then is excreted into the urine. There are five members of the FMO family (FMO1-5); FMO1, FMO2 and FMO3 are relatively the most active. In particular, FMO3 exhibits activity approximately 10-fold greater than FMO1.[66] TMA can also be converted to dimethylamine

Table 1 Summary of different studies that link altered microbiota and diabetes.

Disease	Human/ animal study	Method	Salient results	Ref
Type 2 diabetes	Human	Shotgun sequencing metagenomic	Less butyrate-producing bacteria (*Roseburia* and *Faecalibacterium prausnitzii*), and increased *Escherichia coli*	44
	Human	Shotgun sequencing metagenomic	Less butyrate-producing bacteria (*Roseburia* and *F. prausnitzii*)	45
	Human	(qPCR) and 6S rRNA gene	Less phylum Firmicutes and class Clostridia increased *Betaproteobacteria* and *Proteobacteria*	46
	Human	16S rDNA-based high-throughput sequencing	Less *F. praunitzii* and *R. intestinalis* and increased *Lactobacillus*	47
	Human	qPCR	Less *C. coccoides* and *C. leptum* and increased *Lactobacillus*	25
	Human	Sequencing of 16S rRNA Gene Amplicons	Abundance of Firmicutes and *Lactobacillus*, decreased butyrate-producing bacterial families like Ruminococcaceae and Lachnospiraceae	21
	Human	16 rDNA qRT-PCR	Increased Gram-negative (*C. leptum*, *Butyricicoccus* spp.)	48
	Human	16S rRNA sequencing and imputed metagenomic	Less *Bacteroides*, *Prevotella*, *Roseburia*, *Lachnospira*, *Faecalibacterium* and increased *Enterobacteriaceae*, *Collinsella*, *Streptococcus*, *Lactobacillus*, *Lachnospiraceae:Ruminococcus*	49
	Human	RT-qPCR method	Less *Prevotella* spp. and more *Lactobacillus*	50
	Human	qPCR, pyrosequencing of the V4 region of the 16S rRNA gen	Less Firmicutes, Clostridia and *Lactobacillus*, increased *Betaproteobacteria*	51
	Mice	16S ribosomal RNA	Altered ratio Firmicutes to Bacteroidetes	52
	Mice	PCR amplification and DNA polymerase	Increased *Lactobacillus* spp.	53

Type 1 diabetes	Rats	FISH and fluorescent microscopy targeted against the 16S rRNA	Less *Lactobacillus* and *Bifidobacterium* and increased *Bacteroides*	54
	Human	DNA extraction	Increased *Christensenella* and *Bifidobacteria*	55
	Human	High-throughput 16S rRNA gene sequencin	Increased *B. dorei* and *B.vulgatu*	56
	Human	DNA was extracted, the V4 region of the 16S rRNA gene amplified and 454-pyrosequenced	Increased *Bacteroides*	57
	Human	DNA extraction, 16S rRNA amplification and pyrosequencing	Increased ratio of Bacteroidetes to Firmicutes	42
	Human	Radiobinding assays	Abundance of butyrate producers, such as *Faecalibacterium*, and mucin degraders, such as *Prevotella*, *Akkermansia Bifidobacterium adolescentis* and *B. pseudocatenulatum*	58

Real-time quantitative PCR (qPCR), reverse transcription-quantitative PCR (RT-qPCR), fluorescent in situ hybridization (FISH), ribosomal ribonucleic acid (rRNA), ribosomal deoxyribonucleic acid (rDNA).

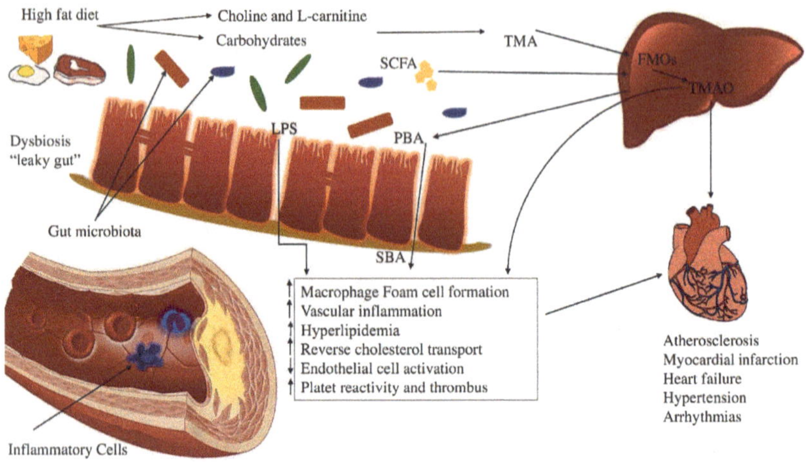

Fig. 1 The interplay between gut microbiota and CVD development. TMA trimethylamine; FMAO liver enzyme flavin monooxygenase; TMAO trimethylamine -N-oxide; SCFA short-chain fatty acid; LPS lipopolysaccharides; PBA primary bile acid; SBA secondary bile acid.

(DMA) and formaldehyde by direct N-demethylation, which may be influenced by FMO3 genetic variation. TMAO is a biologically active molecule and has been shown to be a predictor of acute coronary syndrome and stroke mostly through proatherogenic pathways.[67] There is a mechanical and functional link between the gut microbiota and the physiological processes in the body that influence the risk of CVD. TMAO is also considered to have predictive roles for cardiovascular risk that correlates with atherosclerosis development and other metabolic diseases such as diabetes.[68–71] So far, the intricate mechanism that links gut microbiota dysbiosis and CVD development includes bile acid, LPS (SCFAs), TMA, TMAO metabolism, and immune system (toll-like receptors 2, 4, 9, NF-κB), all of which are influenced by exogenous and endogenous triggers. (Fig. 1).[72] Further studies are needed in order to elucidate an exact cause and effect between gut microbiota alteration and human cardiovascular diseases, how they can be modulated, and whether there will be new therapeutic approaches discovered in the near future.

3.1 Atherosclerosis

Atherosclerosis is a chronic inflammatory disorder that includes risk factors such as hypercholesterolemia, high-fat diet, hypertension, diabetes, obesity, and sedentary lifestyle.[73] It is characterized by lipid migration to the inner layer of the arterial wall (endothelium), adhesion of monocytes and circulating lymphocytes to endothelial cells, migration of monocytes into the sub-endothelial space, and acquisition of macrophage characteristics and transformation into foam cells. Smooth vascular muscle cells also migrate to the endothelium and proliferate, with the additional accumulation of modified LDL macrophages and inflammatory cells, together forming atherosclerotic plaque.[74] The mechanism behind the microbiota–atherosclerosis axis is associated with the ability of the healthy gut microbiota to maintain intestinal permeability by strengthening the tight junctions of the intestinal epithelium. Dysbiosis disrupts these tight junctions, increasing intestinal permeability and the uptake of endotoxin, especially LPS. These endotoxins can then enter the systemic circulation, causing chronic systemic inflammation, which can be modulated by the gut microbiota ecosystem.[75–77]

Recently, several studies have reported mechanical links between TMAO and the pathogenesis of atherosclerosis.[78, 79] TMAO, mainly produced by FMO3 containing flavin from the microbial precursor TMA, has been designated as a key unit in bridge dysbiosis and atherosclerosis.[80] TMAO seems to be capable of activating the multiprotein inflammasome complex in vascular cells that promote reactive oxygen species production (ROS) and lipid deposition. The influence of the host microbial state on atherosclerotic progression is also evident from the fact that the infiltration of immune cells into the vascular atherosclerotic plaques depends on the composition of the intestinal microbiota.[63, 79] Dietary patterns influence TMAO plasma variability the most.[66] Studies showed that a long-term, high-fat diet, especially one with exaggerated consumption of animal products, leads to high levels of TMAO with atherosclerosis stimulation and further raises CVD risks.[79–82]

As shown, many studies demonstrated that gut dysbiosis may be responsible for atherosclerosis development,[83, 84] and the most frequent findings regarding gut microbiome alteration reveal an increase in Firmicutes-to-Bacteroidetes ratio[84] and enhanced levels of *Clostridia*, *Enterobacteriaceae*, and *Streptococcus* spp.[85] Patients with atherosclerotic stroke or transient ischemic attack have developed dysbiosis, with an increase in opportunistic bacteria such as *Enterobacter*, *Megasphaera*, *Oscillibacter*, and *Desulfovibrio*, and a decrease in commensal bacteria such as *Bacteroides*, *Prevotella*, and *Fecalibacterium*.[64–67] In addition, researchers found that *A. muciniphila* has beneficial effects on the gut wall functions and can provide protective effects against atherosclerosis.[86] Further human studies are needed to elucidate the mechanism behind host–gut microbiota and the atherosclerotic–CVD interplay.

3.2 Acute coronary syndrome

Acute coronary syndrome (ACS) is an important cause of death worldwide[87] and is accompanied by clinical symptoms, ECG changes, and raised cardiac biomarkers.[88] Circulating TMAO was associated with the presence of a vulnerable coronary plaque, plaque rupture, and long-term risks of cardiovascular events occurring in patients with ACS. Significant progress in the microbiome study highlights the proatherogenic effect of TMAO. TMAO has been shown to induce macrophage foam cell formation and plaque formation in the aorta and coronary arteries. High levels of TMAO in patients are positively correlated with increased mortality and morbidity risks for myocardial infarction.[89–91] Currently, several mechanisms have been proposed for the pro-atherosclerotic role of TMAO in ACS. For example, TMAO can inhibit the reverse transport of cholesterol by lowering cholesterol 7a-hydroxylase (Cyp7a1), the enzyme that limits bile acid synthesis rate. TMAO can also regulate macrophage shedding receptors, SRA (scavenger receptor), and CD36 expression, which results in the deposition of cholesterol into macrophages and the generation of foam cells.[67] It can induce endothelial dysfunction by activating nuclear factor-κB and protein kinase C by activating monocyte adhesion and promoting the expression of vascular endothelial inflammatory factors including vascular cell adhesion molecule.[92, 93] Researchers suggest that TMAO could be a marker for coronary plaque vulnerability and progression and a direct participant at increased risk of myocardial infarction.[89, 94] Although these observations point out that plasma levels of TMAO correlate with CVD risk, there are still some questions regarding the causal effects of TMAO and the basic mechanical link that explains how TMAO could directly or indirectly promote CVD.

3.3 Atrial fibrillation

Atrial fibrillation (AF) is a frequent abnormal heart rhythm characterized by rapid and irregular beating of the atria with globally increased mortality and morbidity.[95] Among the risk factors for AF, hypertension, diabetes, and obesity has been correlated with dietary patterns that can lead to gut microbiome impairment.[96] Adrenic and palmitic acids belonging to the gut microbiome-derived free fatty acids can provide arrhythmogenic effects contributing to AF development.[97] Alternatively, TMAO can alter heart function by aggravating myocardial interstitial fibrosis and perivascular fibrosis. TMAO has been demonstrated to trigger vascular inflammation through PYD domains containing protein 3 (NLRP3) inflammasome, activation of caspase 1, and the production and secretion of powerful pro-inflammatory cytokines such as IL-1β and IL-18. Inflammasome activation has been linked to adverse cardiac remodeling and cytokine-mediated systolic dysfunction, which might also translate into an arrhythmogenic substrate in the atrium.[98–101] Yu et al.[97] demonstrated in an experimental study that TMAO could be responsible of increased fluctuation of atrial electrophysiology with subsequent atrial fibrillation development, possibly through activation of the p65 NF-κB signaling pathway. In addition, other studies showed that TMAO predicts the risk of atrial fibrillation and is positively correlated with the occurrence of AF.[99, 100] In patients with diagnosed AF, researchers found specific alteration of gut microbiota and metabolites that include abundance of *Enterococcus*, *Ruminococcus*, and *Streptococcus* and decreased levels of *Faecalibacterium*, *Oscillibacter*, *Alistipes*, and *Bilophila*.[101] The microbial enterotype examined in AF patients with persistent atrial fibrillation for more than 12 months revealed similar and dynamic changes in gut microbiota dysbiosis. The two subgroups had reassembled increased levels of *B. vulgatus*, *P. copri*, and *F. prausnitzii*, and distinctive genera changes such as decreased *Butyricicoccus* (SCFA-producing bacteria) and abundance of *Coprococcus* (high risk of artery coronary heart disease in subjects with chronic schizophrenia).[102, 103] The relationship between the intestinal microbiome dysbiosis and atrial fibrillation needs further clinical and practical studies for identification of new therapeutic strategies that could delay AF progression snd to determine if gut microbiota represent a prognostic biomarker for AF.

3.4 Heart failure

Over the years, substantial proof of the key role of gut microbiota in the pathogenesis of heart failure has led to the "gut hypothesis of heart failure." In patients with heart failure, decreased cardiac output and redistribution of blood causes a decrease in intestinal perfusion and disturbance of the intestinal barrier. Disturbance of intestinal barrier function could increase intestinal permeability bacterial translocation and growth of circulating endotoxins, which can contribute to the specific inflammation.[4, 104] Activation of TLR4 induces the release of cytokines, such as TNF-α, IL-1, and IL-6. These mediators of inflammation are associated with cardiac apoptosis, hypertrophy, and fibrosis. The immune activation mechanism marked by elevated inflammatory cytokines plays a crucial role in the development of chronic heart failure. The composition of the intestinal microbiota may change rapidly during ischemia and intestinal reperfusion or as a result of increased portal vein pressure, due to the activation of bacterial virulence in the microbiota by intestinal hypoxia, hypercapnia, changes in local pH, redox status, and norepinephrine. Hypoperfusion and congestion in the intestine can reduce cardiac output and further disrupt the barrier function of the intestine and promote systemic inflammation through bacterial translocation, which can lead to further exacerbations of heart failure.[105, 106]

TMAO levels are raised in chronic heart failure patients, but the mechanism of their increase is multifactorial. TMAO levels are likewise primarily affected by TMAO substrate intakes, such as choline and betaine.[107] Moreover, chronic heart failure patients have impaired intestinal mucosal barrier and increased permeability, which makes it easier for TMAO to enter the bloodstream through the intestinal mucosal barrier, and can also up-regulate vascular cell adhesion molecule-1 expression, promote monocyte adherence, and activate protein kinase C and NF-kB. It can also encourage the development of chronic heart failure by accelerating endothelial dysfunction, including decreasing endothelial self-repair and activating the inflammatory response.[108] Other metabolites implicated in the heart failure–gut microbiota mechanism are represented by secondary bile acid metabolites,[109] indoxy,[110] p-cresil sulfate (PCS),[111] and phenylacetylglutamine (PAG).[112] Pasini et al.[4] showed that fungi (*Candida* spp.) and bacteria (*Campylobacter* and *Shigella* spp.) are positively correlated with the severity of heart failure and dysfunction of intestinal barrier function. Patients with long-term heart failure seem to be more susceptible to *C. difficile* infections.[113] Additional cohort studies revealed that heart failure subjects have variable decrease of microbial pattern and depletion of *F. prasusnitzii*, Lachnospiraceae family, and *Eubacterium hallii*, which are bacterial butyrate producers.[114–116]

3.5 Arterial hypertension

The mechanisms of hypertension are complex and multifactorial. The gut microbiota is considered a risk factor for certain diseases, including atherosclerosis, obesity, metabolic syndrome, and diabetes, all of which have been shown to be linked to hypertension.[116] Researchers found associations between gut microbiota and high blood pressure with their main founding being substantial decrease in microbial richness and diversity and increased Firmicutes-to-Bacteroidetes ratio.[117, 118] Compared to other metabolites of intestinal microflora, several studies have shown the physiological functions of SCFAs in regulating blood pressure. The most common SCFAs are acetate, propionate, and butyrate, which are known to mediate multiple biological effects on the endocrine system, nervous system, CVD, inflammation, and intestinal homeostasis by binding to receptors such as G41 protein-coupled receptor (GPR41), G43 protein-coupled receptor (GPR43), the receptor 109A coupled to G protein (GPR 109A), and vascular olfactory receptor 78 (Olfr78).[119] Olfr78 is expressed in olfactory neurons, renal arterioles, and vascular smooth muscle cells where they play a role in regulating blood pressure. Olfr78 has been found to increase renin level, leading to increased blood pressure, while GPR41 had an antagonistic effect.[120, 121] Recent data suggest that hypertensive subjects have more opportunistic pathogenic species (e.g., *Streptococcus* spp. and *Klebsiella* spp.) and that gut levels correlate with the severity of the disease.[122] Also, overgrowth of bacteria like *Prevotella* and *Klebsiella* was linked to microbial dysfunction in both pre-hypertensive and hypertensive populations.[123]

Communication between the intestinal enteric nervous system and the central nervous system appeared similarly as a potential link to blood pressure. The microbial products have been implicated in sympathetic activation and maintenance of lymphocyte flow to intestinal tissue. Thus, the gut microbiota is potentially functionally interwoven to control blood pressure, and their dysfunctions may be associated with hypertension. In fact, a beneficial role for *Lactobacillus* probiotics in regulating blood pressure has been reported.[124] The interaction of the microbiota with the host's immune system is well established, and disruption of the homeostasis of the immune system can influence CVD. The gut microbiota–hypertension axis requires more attention and further extensive investigation.[125]

4 Investigative tools

Over the last years, new DNA-based techniques for commensal gut microorganism detection have been researched. Among these techniques, we mention the 16S rRNA gene sequencing and the whole-genome shotgun, which helps gather information about bacteria presence, number and phylogenetic connections. Metagenomic sequencing gives an exhaustive characterization of the gut microbiota genome and its functional properties. Research suggests that in the near future techniques like this need to be more specific for diagnosis and should include age as well as ethnical and geographical data. The most used tools for microbiota detection include fluorescence in situ hybridization, denaturing gradient gel electrophoresis and temperature gradient gel electrophoresis, gnotobiosis, and polymerase chain reactions (PCRs).[126–129] Chromatography–tandem mass spectrometry

(LC/MS/MS) is a new validated method for the quantitative analysis of bile acids in both mice and humans. This method could serve as a vigorous investigative instrument for research of bile acid remodulation in the gut microbiota of diabetic patients.[130] Advanced techniques, such as shotgun metagenomic sequencing, metabolomics, and transcriptomes analysis are future approaches that need extensive investigation to clear up the molecular interplay in the gut microbiome in individuals at risk for or with already developed diabetes-related disorders.

5 Treatment and future perspective

5.1 Diabetes

Depending on the mother's diet during intrauterine life, to childhood and adulthood, the gut microbiota is continuously changing. Our personal dietary lifestyle is one of the key modulators of the intestinal microbiota which shapes future metabolic disease development.[131, 132] Studies showed that individuals who consume diets rich in meat-based foods have gut microbiota composed mostly of *Bacteroides*, whereas individuals who consume vegetable-based diets have greater prevalence of *Pevotella*. Dietary pattern plays a crucial role in gut microbial composition and diversity. High-protein diets exert diminished levels of *E. rectale* and *Roseburia* and increased levels of insulin-like growth factor 1 (IGF-1), which is linked to a high risk of diabetes and mortality. Data shows that consuming meat leads to raised catabolic processes (deterioration of amino acids and glycans) and that gut microbiota can recover after short-term dietary restriction.[133–136] Conversely, plant-based diets seem to have protective effects in inflammatory/metabolic disease such as diabetes. A robust systematic review and methodology revealed that consumption of botanical fermented foods (BFFs) such as miso, nattō, fermented olives, seaweed, and red yeast rice have beneficial effects in T2D subjects.[137] The use of artificial sweeteners both in rodents and in humans led to glucose intolerance with increased abundance of *Bacteroides* and decreased *Lactobacillus* spp., which were previously associated with type 2 diabetes. These results may imply that the consumption of artificial sweeteners could coincide with the dramatic increase in obesity and diabetes epidemics.[138]

Nutritional approaches that include prebiotics, synbiotics, arabinoxylans, and arabinoxylans continue to be implemented in current studies. Probiotics that contain *A. muciniphil* help repair the mucosal wall, while *Christensenella minut*, *P. copri*, and *B. fragilis* have anti-inflammatory effects. Gut modulation using probiotics improves lipid and glucose levels, and could represent an attractive therapeutic approach to prevent and treat metabolic disorders.[139,140] Most recently, transplantation of *F. prausnitzii* (butyrate-producing bacterium) has been used as a therapeutic approach in diabetes dysbiosis with encouraging results.[141]

Antidiabetic drugs have an anti-inflammatory effect and interact with the gut microbiota patterns. For example, patients treated with metformin showed an abundance of SCFA-producing bacteria (*Bifidobacterium*, *Prevotella*), *Lactobacillus*, *Akkermansia* spp., and *Clostridium* spp.[142,143] In addition, animals treated with the SGLT2 inhibitor dapagliflozin had increased gut levels of *A. muciniphila* and a decreased Firmicutes-to-Bacteroidetes ratio.[144] Administration of GLP-1 agonists like liraglutide decreased gut abundance of weight-promoting bacteria such as *Ruminiclostridium* and *Romboutsia* and increased weight-loss-related bacteria levels of *Coprococcus* and *Blautia*.[145] Additionally, *Akkermansia* seems to increase considerably after liraglutide treatment compared to metformin treatment.[146] Sitagliptin, vildagliptin, and saxagliptin are DPP-4 inhibitors that are included in the therapeutic panel for diabetes.[147] Increase of *Bifidobacterium* spp. and overall improvement of the intestinal microbiota was observed in pregnant females after sitagliptin use. Meanwhile, saxagliptin has been shown to enrich Lactobacillaceae and Erysipelotrichaceae gut levels, and repress the *Bacteroides* and *Prevotella*.[148] Vildagliptin is another DPP inhibitor that influences the gut microbiota architecture and has anti-inflammatory effects. Vildagliptin therapy decreased *Ruminococcaceae* levels and enhanced phylotypes of Bacteroidaceae and Streptococcaceae.[149] Even if peroxisome proliferator-activated receptor gamma (PPARγ) has many important adverse effects, it continues to be used in clinical practice. Studies show that certain bacteria such as *Atopobium parvulum* and *P. copri*, which are linked to other inflammatory diseases, enhance PPARγ phosphorylation.[150] These discrepancies found in the gut microbiota after treatment with anti-diabetic molecules thus proving how these therapies have common but distinct effects.

5.2 Cardiovascular disease

The numerous links between the modified gut microbial community, metabolites, CVD sensitivity, and metabolic diseases have shone a light on the gut microbiome as a potential new target for therapeutics. Currently, diet modulation is the main therapeutic tool used in clinical practice to affect chronic metabolic diseases and although this lifestyle interaction may have a clear impact on the structure and function of the gut microbial community, there are few studies exploring the impact of dietary interventions on the gut microbiome in humans.[151, 152] Long-term consumption of diets rich in red meat or high L-carnitine supplementation leads to enhanced TMA/TMAO generation.[153] In such cases, statin or aspirin treatment seems to have a beneficial effect, as they can partially attenuate TMAO levels.[154, 155]

The specific diet and intake of prebiotics and probiotics has beneficial effects in optimizing the structure of the microbiome. Further approaches for promising gut microbiome manipulation, although not yet developed for metabolic disorders, include prebiotics, probiotics, and small molecule pathway inhibitors of defined microbial enzymes. Probiotics and prebiotics also play an important role in the treatment of cardiovascular disease. Usual probiotics include *Lactobacillus*, *Bifidobacterium*, *Lactococcus*, *Streptococcus*, and *Enterococcus*. Prebiotics can stimulate the activity of one or more beneficial bacteria selectively and have beneficial effects on the host. Most prebiotics are carbohydrates that are present in natural products such as fruits, vegetables, and cereals.[156] The utility of probiotics and prebiotics in changing the gut microbiome ecosystem has become a new target for CVD prevention and treatment. A number of studies have shown that both probiotics and prebiotics have beneficial effects on lipid modulation and therefore indirect benefits on CVD.[157, 158] Thus, CVD patients can supplement their diet with probiotics and prebiotics depending on their own condition to regulate the gut microbiota and to improve bowel dysbiosis. Probiotics have been shown to reduce the size of myocardial infarction, the area of the atherosclerotic plaque, and the incidence rates of post-infarct myocardial hypertrophy and heart failure. Probiotics can alleviate hypertrophy and cardiac remodeling, delay heart failure, and increase ejection fraction by reducing leptin and increasing tissue taurine after coronary syndrome.[159, 160] *Lactobacillus* probiotic has been shown to be useful in controlling blood pressure by reducing vascular inflammation and protecting endothelial function; it is also believed to reduce the myocardial infarction area.[160,161] Prebiotics are effective in controlling both blood glucose and plasma lipid profiles. A high-fiber diet significantly reduces blood pressure in patients with hypertension, improves cardiac function, and alleviates cardiac fibrosis.[162]

Fecal microbiota transplant (FMT) is effective in treating irritable bowel syndrome, recurrent *C. difficile* infection, and inflammatory bowel disease.[163] Currently, FMT remains a promising therapy for CVD, although more studies are needed.[164] The FMT, which transfers the intestinal flora from a donor to the intestinal tract of a recipient, can increase the diversity of the intestinal microbiota and change its composition. The use of FMT has been proposed as one of the ways to overcome resistance to intestinal microbiota at colonization, even under dysbiosis conditions. This intervention has already been tested in humans with severe obesity, indicating an increased insulin sensitivity to recipients of a transplanted microbiota from weak donors.[165]

As a new strategy against TMAO precursor formations and adverse cardiovascular outcomes, scientists have developed selectively targeting microbial synthetic enzymes such as DMB (3,3-dimethyl,1-butanol). DMB can alleviate enhanced macrophage foam cell formation in choline diet and atherosclerosis and suppress TMA production via inhibition of several carnitine derivative nutrients.[166] Thus, small molecule antimicrobial enzyme therapeutics could be innovative tools for treatment of cardiometabolic phenotypes in individuals. Interestingly, studies linked use of antibiotics in experimental animal studies and showed that they can lower TMAO levels and oxidative stress and can reverse endothelial dysfunction, and even reduce infarct size.[159, 167]

6 Conclusion

There is substantial published literature that affiliates intestinal flora dysbiosis and metabolic diseases such as diabetes and cardiovascular diseases. Nonetheless, the complete mechanism of this intricate connection continues to raise questions, as it is still not fully understood. Dietary interventions, prebiotics and probiotics, fecal microbial transplant, and bariatric surgery can change the gut microbial composition. However, the question still remains whether the gut microbiota is the generator and/or a repercussion of T1D/T2D or cardiovascular disease. The answer will direct future identification of the human microbiome atlas and function, paving the way for potential new biomarkers or instruments for prevention and treatment of metabolic disease.

References

1. Sharma S, Tripathi P. Gut microbiome and type 2 diabetes: where we are and where to go? *J Nutr Biochem*. 2019;63:101–108. https://doi.org/10.1016/j.jnutbio.2018.10.003.
2. Ramos-Romero S, Hereu M, Atienza L, et al. Mechanistically different effects of fat and sugar on insulin resistance, hypertension and gut microbiota in rats. *Am J Physiol Endocrinol Metab*. 2018;314:E552–E563. https://doi.org/10.1152/ajpendo.00323.20.
3. Dasinger JH, Fehrenbach DJ, Abais-Battad JM. Dietary Protein: Mechanisms Influencing Hypertension and Renal Disease. *Curr Hypertens Rep*. 2020;22(2). https://doi.org/10.1007/s11906-020-1018-8.

4. Pasini E, Aquilani R, Testa C, Baiardi P, Angioletti S, et al. Pathogenic gut flora in patients with chronic heart failure. *JACC Heart Fail.* 2016;4:220–227.
5. Moreno-Navarrete JM, Serino M, Blasco-Baque V, et al. Gut microbiota interacts with markers of adipose tissue browning, insulin action and plasma acetate in morbid obesity. *Mol Nutr Food Res.* 2018;62. https://doi.org/10.1002/mnfr.201700721, 1700721.
6. Liu A, Lv H, Wang H, Yang H, Li Y, Qian J. Aging Increases the severity of colitis and the related changes to the gut barrier and gut microbiota in humans and mice. *J Gerontol: Ser A.* 2020. https://doi.org/10.1093/gerona/glz263.
7. Guo XY, Liu XJ, Hao JY. Gut microbiota in ulcerative colitis: insights on pathogenesis and treatment. *J Dig Dis.* 2020 Feb 10. https://doi.org/10.1111/1751-2980.12849.
8. Voigt RM, Forsyth CB, Green SJ, Engen PA, Keshavarzian A. Circadian rhythm and the gut microbiome. *Int Rev Neurobiol.* 2016;131:193–205.
9. Luo D, Chen K, Li J, Fang Z, Pang H, Yin Ye. Gut microbiota combined with metabolomics reveals the metabolic profile of the normal aging process and the anti-aging effect of FuFang Zhenshu TiaoZhi (FTZ) in mice. *Biomed Pharmacother.* 2020;121. https://doi.org/10.1016/j.biopha.2019.109550, 109550.
10. Duvallet C, Gibbons SM, Gurry T, Irizarry RA, Alm EJ. Meta-analysis of gut microbiome studies identifies disease-specific and shared responses. *Nat Commun.* 2017;8(1):1784.
11. Mejía-León ME, de la Barca AM. Diet, microbiota and immune system in type 1 diabetes development and evolution. *Nutrients.* 2015;7(11):9171–9184.
12. Karen Beer PA-C, MSPAS, RD, LD, CDE. The gut microbiome in Type 2 diabetes. *Clin Rev.* 2018;28(2):13–14. 16.
13. Knights D, Ward TL, McKinlay CE, et al. Rethinking 'enterotypes'. *Cell Host Microbe.* 2014;16:433–437.
14. Zhao L, Zhang F, Ding X, et al. Gut bacteria selectively promoted by dietary fibers alleviate type 2 diabetes. *Science.* 2018;359:1151–1156. https://doi.org/10.1126/science.aao5774.
15. Roglic G, Varghese C, Thamarangsi T. Diabetes in South-East Asia: burden, gaps, challenges and ways forward. WHO South East Asia. *J Public Health.* 2016;5:1–4.
16. Grarup N, Sandholt CH, Hansen T, Pedersen O. Genetic susceptibility to type 2 diabetes and obesity: from genome-wide association studies to rare variants and beyond. *Diabetologia.* 2014;57:1528–1541. https://doi.org/10.1007/s00125-014-3270-4.
17. Lazar V, Ditu L-M, Pircalabioru GG, et al. Gut Microbiota, Host Organism, and Diet Trialogue in Diabetes and Obesity. *Front Nutr.* 2019;6:21. https://doi.org/10.3389/fnut.2019.00021.
18. World Health Organization. *Global report on diabetes: executive summary*; 2016. http://www.who.int/diabetes/global-report.
19. Kelly T, Yang W, Chen CS, Reynolds K, He J. Global burden of obesity in 2005 and projections to 2030. *Int J Obes (Lond).* 2008;32:1431–1437.
20. International Diabetes Federation, ed. *IDF Diabetes Atlas.* 8th ed. International Diabetes Federation; 2017.
21. Bhute SS, Suryavanshi MV, Joshi SM, Yajnik CS, Shouche YS, Ghaskadbi SS. Gut microbial diversity assessment of indian Type-2-diabetics reveals alterations in eubacteria, archaea, and eukaryotes. *Front Microbiol.* 2017;8:214.
22. Nauck MA, Meier JJ. Incretin hormones: their role in health and disease. *Diabetes Obes Metab.* 2018;20(Suppl. 1):5–21. https://doi.org/10.1111/dom.13129T2D.
23. Backhed F, Ding H, Wang T, et al. The gut microbiota as an environmental factor that regulates fat storage. *PNAS.* 2004;101:15718–15723. https://doi.org/10.1073/pnas.0407076101.
24. Hartsra AV, Bouter KEC, Backhed F, Nieuwdorp M. Insights into the role of the microbiome in obesity and type 2 diabetes. *Diabetes Care.* 2015;38(1):159–165.

25. Pei-Chi C, Yi-Wen C, Suh-Ching Y. The alteration of gut microbiota in newly-diagnosed type 2 diabetic patients. *Nutrition*. 2018. https://doi.org/10.1016/j.nut.2018.11.019.
26. Hersoug LG, Møller P, Loft S. Gutmicrobiota-derived lipopolysaccharide uptake and trafficking to adipose tissue: implications for inflammation and obesity. *Obes Rev.* 2016;17(4):297–312.
27. Zhang X, Zhao Y, Xu J, et al. Modulation of gut microbiota by berberine and metformin during the treatment of high-fat diet-induced obesity in rats. *Sci Rep.* 2015;5:14405.
28. Lissé AD, Valmore BP, Waldo D, et al. The Microbiome and the Epigenetics of Diabetes Mellitus. In: *Diabetes Food Plan*. Viduranga Waisundara: IntechOpen; 2018.
29. Xu YH, Gao CL, Guo HL, et al. Sodium butyrate supplementation ameliorates diabetic inflammation in db/db mice. *J Endocrinol.* 2018;238:231–244. https://doi.org/10.1530/JOE-18-0137.
30. Sanna S, van Zuydam NR, Mahajan A, et al. Causal relationships among the gut microbiome, short-chain fatty acids and metabolic diseases. *Nat Genet*. 2019;51(4):600–605.
31. Drucker A. DJ. Mechanisms of action and therapeutic application of glucagonlike peptide-1. *Cell Metab.* 2018;27:740–756. https://doi.org/10.1016/j.cmet.2018.03.001.
32. Larsen MP, Torekov SS. Glucagon-like peptide 1: a predictor of type 2 diabetes? *J Diabetes Res.* 2017;2017. https://doi.org/10.1155/2017/7583506, 7583506.
33. Di Ciaula A, Garruti G, Lunardi Baccetto R, et al. Bile acid physiology. *Ann Hepatol.* 2017;16:s4–14. https://doi.org/10.5604/01.3001.0010.5493. Suppl. 1:s3–105.
34. Chiang JYL, Ferrell JM. Bile acid metabolism in liver pathobiology. *Gene Expr.* 2018;18:71–87. https://doi.org/10.3727/105221618X15156018385515.
35. Thomas C, Gioiello A, Noriega L, et al. TGR5- mediated bile acid sensing controls glucose homeostasis. *Cell Metab.* 2009;10:167–177. https://doi.org/10.1016/j.cmet.2009.08.001.
36. König J, Wells J, Cani PD, et al. Human intestinal barrier function in health and disease. *Clin Transl Gastroenterol.* 2016;7:e196. https://doi.org/10.1038/ctg.2016.54.
37. Ding R, Goh W-R, Wu R, et al. Revisit gut microbiota and its impact on human health and disease. *J Food Drug Anal.* 2019. https://doi.org/10.1016/j.jfda.2018.12.012.
38. Adeloye D, et al. Global and regional estimates of the morbidity due to type I diabetes among children aged 0–4 years: a systematic review and analysis. *J Glob Health*. 2018;8, 021101.
39. Thomas RM, Jobin C. Microbiota in pancreatic health and disease: the next frontier in microbiome research. *Nat Rev Gastroenterol Hepatol.* 2020;17:53–64. https://doi.org/10.1038/s41575-019-0242-7.
40. Marino E, Richards JL, McLeod KH, et al. Gut microbial metabolites limit the frequency of autoimmune T cells and protect against type 1 diabetes. *Nat Immunol.* 2017. https://doi.org/10.1038/ni.3713. Nature Publishing Group, a division of Macmillan Publishers Limited. All Rights Reserved, Available:.
41. Vaarala O. Gut microbiota and type 1 diabetes. *Rev Diabet Stud.* 2012;9:251–259. https://doi.org/10.1900/RDS.2012.9.251.
42. Giongo A, Gano KA, Crabb DB, et al. Toward defining the autoimmune microbiome for type 1 diabetes. *ISME J.* 2011;5(1):82–91.
43. Kostic AD, Gevers D, Siljander H, et al. The dynamics of the human infant gut microbiome in development and in progression toward type 1 diabetes. *Cell Host Microbe.* 2015;17:260–273. https://doi.org/10.1016/j.chom.2015.01.001.
44. Qin J, Li Y, Cai Z, et al. A metagenome-wide association study of gut microbiota in type 2 diabetes. *Nature.* 2012;490:55–60. https://doi.org/10.1038/nature11450.
45. Karlsson FH, Tremaroli V, Nookaew I, et al. Gut metagenome in European women with normal, impaired and diabetic glucose control. *Nature.* 2013;498:99–103. https://doi.org/10.1038/nature12198.

46. Larsen N, Vogensen FK, van den Berg FW, et al. Gut microbiota in human adults with type 2 diabetes differs from non-diabetic adults. *PLoS One.* 2010;5:e9085. https://doi.org/10.1371/journal. pone.0009085.
47. Zhang X, Shen D, Fang Z, Jie Z, Qiu X, et al. Human gut microbiota changes reveal the progression of glucose intolerance. *PLoS One.* 2013;8:e71108.
48. Suceveanu AI, Pantea Stoian A, Parepa RI, et al. Gut microbiota patterns in obese and type 2 diabetes. (T2D) patients from romanian black sea coast region. *Rev Chim.* 2018;69:2260–2267.
49. Candela M, et al. Modulation of gut microbiota dysbiosis in type 2 diabetic patients by macrobiotic Ma-Pi2diet. *Br J Nutr.* 2016;116(1):80–93.
50. Sato J, Kanazawa A, Ikeda F, Yoshihara T, Goto H, et al. Gut dysbiosis and detection of "live gut bacteria" in blood of Japanese patients with type 2 diabetes. *Diabetes Care.* 2014;37:2343–2350.
51. Sedighi M, Razavi S, Navab-Moghadam F, Khamseh ME, Alaei-Shahmiri F. Comparison of gut microbiota in adult patients with type 2 diabetes and healthy individuals. *Microb Pathog.* 2017;111:362–369.
52. Ley RE, Turnbaugh PJ, Klein S, Gordon JI. Microbial ecology: human gut microbes associated with obesity. *Nature.* 2006;444:1022–1023. https://doi.org/10.1038/4441022a.
53. Horie M, et al. Comparative analysis of the intestinal flora in type 2 diabetes and non-diabetic mice. *Exp Anim.* 2017;66(4):405–416.
54. Brugman S, Klatter FA, Visser JT, et al. Antibiotic treatment partially protects against type 1 diabetes in the bio-breeding diabetes-prone rat. Is the gut flora involved in the development of type 1 diabetes? *Diabetologia.* 2006;49(9):2105–2108.
55. De Groot PF, Belzer C, Aydin Ö, et al. Distinct fecal and oral microbiota composition in human type 1 diabetes, an observational study. *PLoS One.* 2017;12. https://doi.org/10.1371/journal.pone.0188475, e0188475.
56. Davis- Richardson AG, et al. Bacteroides dorei dominates gut microbiome prior to autoimmunity in Finnish children at high risk for type 1 diabetes. *Front Microbiol.* 2014;5:678.
57. Mejía- León ME, Petrosino JF, Ajami NJ, Domínguez- Bello MG, de la Barca AMC. Fecal microbiota imbalance in Mexican children with type 1 diabetes. *Sci Rep.* 2014;4:3814.
58. de Goffau MC, Luopajarvi K, Knip M, et al. Fecal microbiota composition differs between children with beta-cell autoimmunity and those without. *Diabetes.* 2013;62:1238–1244. https://doi.org/10.2337/db12-05.
59. Lynch SV, Pedersen O. The human intestinal microbiome in health and disease. *N Engl J Med.* 2016;375:2369–2379.
60. Rothschild D, Weissbrod O, Barkan E, et al. Environment dominates over host genetics in shaping human gut microbiota. *Nature.* 2018;555(7695):210–215.
61. Gabriele GS, Anna S, Giovanni E. Diagnostics and therapeutic implications of gut microbiota alterations in cardiometabolic diseases. *Trends Cardiovasc Med.* April 2019;29(3):141–147.
62. Jieting P, Xun X. MinHu Interaction between gut microbiome and cardiovascular disease. *Life Sci.* 2018;214(1):153–157.
63. Blacher E, Levy M, Tatirovsky E, Elinav E, et al. Microbiome-modulated metabolites at the interface of host immunity. *J Immunol.* 2017;198(2):572–580.
64. Lau K, Srivatsav V, Rizwan A, et al. Bridging the gap between gut microbial dysbiosis and cardiovascular diseases. *Nutrients.* 2017;9:859–874.
65. Rath S, Heidrich B, Pieper DH, Vital M. Uncovering the trimethylamine-producing bacteria of the human gut microbiota. *Microbiome.* 2017;5(1):54.
66. Chen K, Zheng X, Feng MC, Li DL, Zhang HQ. Gut microbiota-dependent metabolite trimethylamine N-oxide contributes to cardiac dysfunction in western diet induced obese mice. *Front Physiol.* 2017;8:139.

67. Schiattarella GG, Sannino A, Toscano E, et al. Gut microbe-generated metabolite trimethylamine-N-oxide as cardiovascular risk biomarker: a systematic review and dose-response meta-analysis. *Eur Heart J.* 2017;38(39):2948–2956.
68. Zhu W, Gregory JC, Org E, et al. Gut microbial metabolite tmao enhances platelet hyperreactivity and thrombosis risk. *Cell.* 2016;165:111–124.
69. Shafi T, Powe NR, Meyer TW, et al. Trimethylamine N-oxide and cardiovascular events in hemodialysis patients. *J Am Soc Nephrol.* 2017;28:321–331.
70. Stubbs JR, House JA, Ocque AJ, et al. Serum trimethylamine-Noxide is elevated in CKD and correlates with coronary atherosclerosis burden. *J Am Soc Nephrol.* 2016;27:305–313.
71. Tang WH, Wang Z, Li XS, et al. Increased trimethylamine N-oxide portends high mortality risk independent of glycemic control in patients with type 2 diabetes mellitus. *Clin Chem.* 2017;63:297–306.
72. Canyelles M, Tondo M, Cedó L, Farràs M, Escolà-Gil J, Blanco-Vaca F. Trimethylamine N-oxide: a link among diet, gut microbiota, gene regulation of liver and intestine cholesterol homeostasis and HDL function. *Int J Mol Sci.* 2019;19(10):3228. https://doi.org/10.3390/ijms19103228.
73. Camaré C, Pucelle M, Nègre-Salvayre A, Salvayre R. Angiogenesis in the atherosclerotic plaque. *Redox Biol.* 2017;12:18–34. https://doi.org/10.1016/j.redox.2017.01.007.
74. Mach F, Baigent C, Catapano AL, et al. 2019 ESC/EAS Guidelines for the management of dyslipidaemias: lipid modification to reduce cardiovascular risk. *Eur Heart J.* 2019;00:178. https://doi.org/10.1093/eurheartj/ehz455.
75. Singh V, Yeoh BS, Vijay-Kumar M. Gut microbiome as a novel cardiovascular therapeutic target. *Curr Opin Pharmacol.* 2016;27:8–12.
76. Yamashita T. Intestinal immunity and gut microbiota in atherogenesis. *J Atheroscler Thromb.* 2017;24(2):110–119.
77. Ma J, Li H. The role of gut microbiota in atherosclerosis and hypertension. *Front Pharmacol.* 2018;9. https://doi.org/10.3389/fphar.2018.01082.
78. Hein MT, Frederick CL. Role of gut microbiota in cardiovascular disease that links to host genotype and diet the gut microbiome. *Implicat Human Dis.* 2016;1(1):1–15.
79. Senthong V, Li XS, Hudec T, et al. Plasma trimethylamine N-Oxide, a gut microbe–generated phosphatidylcholine metabolite, is associated with atherosclerotic burden. *J Am Coll Cardiol.* 2016;67:2620. https://doi.org/10.1016/j.jacc.2016.03.546.
80. Skagen K, et al. The carnitine-butyrobetaine-trimethylamine-oxide pathway and its association with cardiovascular mortality in patients with carotid atherosclerosis. *Atherosclerosis.* 2016;247:64–69.
81. Koeth R, Wang Z, Levison B, et al. Intestinal microbiota metabolism of L-carnitine, a nutrient in red meat, promotes atherosclerosis. *Nat Med.* 2013;19:576–585. https://doi.org/10.1038/nm.3145.
82. Rune I, Rolin B, Lykkesfeldt J, et al. Long-term Western diet fed apolipoprotein E-deficient rats exhibit only modest early atherosclerotic characteristics. *Sci Rep.* 2018;8(1). https://doi.org/10.1038/s41598-018-23835-z.
83. Gregory JC, Buffa JA, Org E, et al. Transmission of atherosclerosis susceptibility with gut microbial transplantation. *J Biol Chem.* 2015;290:5647–5660. https://doi.org/10.1074/jbc.M114.618249.
84. Brown JM, Hazen SL. Microbial modulation of cardiovascular disease. *Nat Rev Microbiol.* 2018;16(3):171–181. https://doi.org/10.1038/nrmicro.2017.149.
85. Jie Z, Xia H, Zhong SL, et al. The gut microbiome in atherosclerotic cardiovascular disease. *Nat Commun.* 2017;845. https://doi.org/10.1038/s41467-017-00900-14.
86. Li J, Lin S, Vanhoutte PM, Woo CW, Xu A. Akkermansia muciniphila protects against atherosclerosis by preventing metabolic endotoxemia-induced inflammation in apoe-/- mice. *Circulation.* 2016;133:2434–2446. https://doi.org/10.1161/circulationaha.115.019645.

87. Benjamin EJ, et al. Heart Disease and Stroke Statistics-2018 Update: A Report From the American Heart Association. *Circulation*. 2018;137(12):e67–e492.
88. Thygesen K, et al. Fourth universal definition of myocardial infarction (2018). *Eur Heart J.* 2019;40(3):237–269.
89. McMillan A, Hazen SL. Gut microbiota involvement in ventricular remodeling post-myocardial infarction. *Circulation*. 2019;139(5):660–662.
90. Emoto T, Yamashita T, Sasaki N, et al. Analysis of gut microbiota in coronary artery disease patients: A possible link between gut microbiota and coronary artery disease. *J Atheroscler Thromb.* 2016;23(8):908–921.
91. Senthong V, Wang Z, Li XS, et al. Intestinal microbiota-generated metabolite trimethylamine-N-oxide and 5-year mortality risk in stable coronary artery disease: the contributory role of intestinal microbiota in a COURAGE-like patient cohort. *J Am Heart Assoc.* 2016;5. https://doi.org/10.1161/JAHA.115.002816, e002816.
92. Haghikia A, Li XS, Liman TG, et al. Gut Microbiota-dependent trimethylamine N-oxide predicts risk of cardiovascular events in patients with stroke and is related to proinflammatory monocytes. *Arterioscler Thromb Vasc Biol.* 2018;38:2225–2235.
93. Liu H, Chen X, Hu X, et al. Alterations in the gut microbiome and metabolism with coronary artery disease severity. *Microbiome.* 2019;7:68. https://doi.org/10.1186/s40168-019-0683-9.
94. Li XS, Obeid S, Klingenberg R, et al. Gut microbiota dependent trimethylamine N-oxide in acute coronary syndromes: a prognostic marker for incident cardiovascular events beyond traditional risk factors. *Eur Heart J.* 2017;38(11):814–824.
95. Brugada J, Katritsis DG, Arbelo E, et al. 2019 ESC Guidelines for the management of patients with supraventricular tachycardia The Task Force for the management of patients with supraventricular tachycardia of the European Society of Cardiology (ESC). *Eur Heart J.* 2019. https://doi.org/10.1093/eurheartj/ehz467.
96. Du X, Dong J, Ma C. Is atrial fibrillation a preventable disease? *J Am Coll Cardiol.* 2017;69(15):1968–1982.
97. Kirchhof P, Benussi S, Kotecha D, et al. ESC Guidelines for the management of atrial fibrillation developed in collaboration with EACTS. *Eur Heart J.* 2016;37(38):2893–2962.
98. Lin T, Li K, He W, Chen L, Wang T, Wang N. Trimethylamine N-oxide: A new therapeutic target for atrial fibrillation? *Int J Cardiol.* 2019;274:194. https://doi.org/10.1016/j.ijcard.2018.07.147.
99. Yu L, Meng G, Huang B, et al. A potential relationship between gut microbes and atrial fibrillation: Trimethylamine N-oxide, a gut microbe-derived metabolite, facilitates the progression of atrial fibrillation. *Int J Cardiol.* 2018;255:92–98.
100. Mishima RS, Elliott AD, Sanders P, Linz D. Microbiome and atrial fibrillation. *Int J Cardiol.* 2018;255:103–104. https://doi.org/10.1016/j.ijcard.2017.12.091.
101. Svingen GFT, Zuo H, Ueland PM, et al. Increased plasma trimethylamine-N-oxide is associated with incident atrial fibrillation. *Int J Cardiol.* 2018;267:100–106.
102. Zuo K, Li J, Li K, et al. Disordered gut microbiota and alterations in metabolic patterns are associated with atrial fibrillation. *GigaScience.* 2019;8(6). https://doi.org/10.1093/gigascience/giz058.
103. Nguyen TT, Kosciolek T, Maldonado Y, et al. Differences in gut microbiome composition between persons with chronic schizophrenia and healthy comparison subjects. *Schizophr Res.* 2019;4:23–29. https://doi.org/10.1016/j.schres.2018.09.014.
104. Sandek A, Swidsinski A, Schroedl W, Watson A, Valentova M, Herrmann R, Scherbakov N, Cramer L, Rauchhaus M, Grosse-Herrenthey A, Krueger M, von Haehling S, Doehner W, Anker SD, Bauditz J. Intestinal blood flow in patients with chronic heart failure: a link with bacterial growth, gastrointestinal symptoms, and cachexia. J Am Coll Cardiol 2014; 64:1092–1102. PubMed: 25212642.

105. Ahmad AF, Dwivedi G, O'Gara F. The gut microbiome and cardiovascular disease: current knowledge and clinical potential. *Am J Physiol.* 2019;1:1–30.
106. Organ CL, Otsuka H, Bhushan S, Wang Z, Bradley J, Trivedi R, Polhemus DJ, Tang WH, Wu Y, Hazen SL, Lefer DJ. Choline diet and its gut microbe-derived metabolite, trimethylamine n-oxide, exacerbate pressure overload-induced heart failure. Circ Heart Fail 2016; 9:e002314. PubMed: 26699388.
107. Suzuki T, Heaney LM, Bhandari SS, Jones DJ, Ng LL. Trimethylamine N-oxide and prognosis in acute heart failure. *Heart.* 2016;102:841–848.
108. Dantzer R, Cohen S, Russo SJ, Dinan TG. Resilience and immunity. *Brain Behav Immun.* 2018;74:28–42.
109. Wu CC, Hsieh MY, Hung SC, et al. Serum indoxyl sulfate associates with postangioplasty thrombosis of dialysis grafts. *J Am Soc Nephrol.* 2016;27:1254–1264.
110. Gryp T, Vanholder R, Vaneechoutte M, Glorieux G. p-Cresyl sulfate. *Toxins (Basel).* 2017;9(2):E52.
111. Poesen R, Claes K, Evenepoel P, et al. Microbiota-derived phenylacetylglutamine associates with overall mortality and cardiovascular disease in patients with CKD. *J Am Soc Nephrol.* 2016;27:3479-3487.
112. Mamic P, Heidenreich PA, Hedlin H, Tennakoon L, Staudenmayer KL. Hospitalized patients with heart failure and common bacterial infections: a nationwide analysis of concomitant clostridium difficile infection rates and in-hospital mortality. *J Card Fail.* 2016;22:891–900.
113. Kamo T, Akazawa H, Suda W, et al. Dysbiosis and compositional alterations with aging in the gut microbiota of patients with heart failure. *PLoS One.* 2017;12, e0174099.
114. Cui X, Ye L, Li J, et al. Metagenomic and metabolomic analyses unveil dysbiosis of gut microbiota in chronic heart failure patients. *Sci Rep.* 2018;8:635.
115. Kummen M, Mayerhofer CCK, Vestad B, et al. Gut microbiota signature in heart failure defined from profiling of 2 independent cohorts. *J Am Coll Cardiol.* 2018;71:1184–1186.
116. Williams B, Mancia G, Spiering W, et al. ESC/ESH guidelines for the management of arterial hypertension. *Eur Heart J.* 2018. https://doi.org/10.1093/eurheartj/ehy339.
117. Yang T, Santisteban MM, Rodriguez V, et al. Gut dysbiosis is linked to hypertension. *Hypertension.* 2015;65:1331–1340.
118. Adnan S, Nelson JW, Ajami NJ, et al. Alterations in the gut microbiota can elicit hypertension in rats. *Physiol Genomics.* 2017;49:96–104.
119. Pluznick JL, Protzko RJ, Gevorgyan H, et al. Olfactory receptor responding to gut microbiota-derived signals plays a role in renin secretion and blood pressure regulation. *Proc Natl Acad Sci U S A.* 2013;110:4410–4415.
120. Pluznick JL. Renal and cardiovascular sensory receptors and blood pressure regulation. *Am J Physiol Renal Physiol.* 2013;305:F439–F444.
121. Karbach SH, Schonfelder T, Brandao I, et al. Gut microbiota promote angiotensin II-induced arterial hypertension and vascular dysfunction. *J Am Heart Assoc.* 2016;5, e003698.
122. Yan Q, Gu Y, Li X, Yang W, Jia L, Chen C, et al. Alterations of the gut microbiome in hypertension. Front Cell Infect Microbiol 2017 7:381.
123. Li, J., Zhao, F., Wang, Y., Chen, J., Tao, J., Tian, G., et al.. Gut microbiota dysbiosis contributes to the development of hypertension. Microbiome, 5(1). doi:https://doi.org/10.1186/s40168-016-0222-x.
124. Gomez-Guzman M, Toral M, Romero M, et al. Antihypertensive effects of probiotics lactobacillus strains in spontaneously hypertensive rats. *Mol Nutr Food Res.* 2015;59:2326–2336.
125. Santisteban MM, Qi Y, Zubcevic J, et al. Hypertension-linked pathophysiological alterations in the gut. *Circ Res.* 2017;120:312–323.

126. Kuczynski J, Lauber CL, Walters WA, et al. Experimental and analytical tools for studying the human microbiome. *Nat Rev Genet*. 2012;13:47–58. https://doi.org/10.1038/nrg3129.
127. Karlsson F, Tremaroli V, Nielsen J, Bäckhed F. Assessing the human gut microbiota in metabolic diseases. *Diabetes*. 2013;62:3341–3349. https://doi.org/10.2337/db13-0844.
128. Lloyd-Price J, Mahurkar A, Rahnavard G, et al. Strains, functions and dynamics in the expanded human microbiome project. *Nature*. 2017;550:61–66.
129. Forster SC, Kumar N, Anonye BO, et al. A human gut bacterial genome and culture collection for improved metagenomic analyses. *Nat Biotechnol*. 2019;37:186–192.
130. Choucair I, Nemet I, Li L, et al. Quantification of bile acids: a mass spectrometry platform for studying gut microbe connection to metabolic diseases. *J Lipid Res*. Feb 2020;61(2):159–177. https://doi.org/10.1194/jlr.RA119000311.
131. Gentile CL, Weir TL. The gut microbiota at the intersection of diet and human health. *Science*. 2018;362(6416):776–780. https://doi.org/10.1126/science.aau5812.
132. Hills RD, Pontefract BA, Mishcon HR, Black CA, Sutton SC, Theberge CR. Gut Microbiome: Profound Implications for Diet and Disease. *Nutrients*. 2019;11(7):1613. https://doi.org/10.3390/nu11071613.
133. Tosti V, Bertozzi B, Fontana L. Health benefits of the mediterranean diet: metabolic and molecular mechanisms. *J Gerontol: Ser A*. 2017;73(3):318–326. https://doi.org/10.1093/gerona/glx227.
134. DeFilippis F, Pellegrini N, Vannini L, et al. High-level adherence to a Mediterranean diet beneficially impacts the gut microbiota and associated metabolome. *Gut*. 2016;65:1812–1821.
135. Lin S, Wang Z, Lam K-L, Zeng S, Tan BK, Hu J. Role of intestinal microecology in the regulation of energy metabolism by dietary polyphenols and their metabolites. *Food Nutr Res*. 2019;63. https://doi.org/10.29219/fnr.v63.151.
136. Zeng T, Cui H, Tang D, et al. Short-term dietary restriction in old mice rejuvenates the aging-induced structural imbalance of gut microbiota. *Biogerontology*. 2019;20:837–848. https://doi.org/10.1007/s10522-019-09830-5.
137. Chan M, Baxter H, Larsen N, et al. Impact of botanical fermented foods on metabolic biomarkers and gut microbiota in adults with metabolic syndrome and type 2 diabetes: a systematic review protocol. *BMJ Open*. 2019;9. https://doi.org/10.1136/bmjopen-2019-029242, e029242.
138. Suez J, Korem T, Zeevi D, et al. Artificial sweeteners induce glucose intolerance by altering the gut microbiota. *Nature*. 2014;514(7521):181–186.
139. Lau E, Neves JS, Ferreira-Magalhães M, Carvalho D, Freitas P. Probiotic ingestion, obesity, and metabolic-related disorders: results from NHANES, 1999–2014. *Nutrients*. 2019;11(7):1482. https://doi.org/10.3390/nu11071482.
140. Chang C-J, Lin T-L, Tsai Y-L, et al. Next generation probiotics in disease amelioration. *J Food Drug Anal*. 2019. https://doi.org/10.1016/j.jfda.2018.12.011.
141. Ganesan K, Chung SK, Vanamala J, Xu B. Causal relationship between diet-induced gut microbiota changes and diabetes: a novel strategy to transplant Faecalibacterium prausnitzii in preventing diabetes. *Int J Mol Sci*. 2018;19(12):3720. https://doi.org/10.3390/ijms19123720.
142. Kyriachenko Y, Falalyeyeva T, Korotkyi O, Molochek N, Kobyliak N. Crosstalk between gut microbiota and antidiabetic drug action. *World J Diabetes*. 2019;10(3):154–168. https://doi.org/10.4239/wjd.v10.i3.154.
143. Lee H, Ko G. Effect of metformin on metabolic improvement and gut microbiota. *Appl Environ Microbiol*. 2014;80:5935–5943. https://doi.org/10.1128/AEM.01357-14. 25038099.
144. Kuo GH, Gaul MD, Liang Y, et al. Synthesis and biological evaluation of benzocyclobutane-C-glycosides as potent and orally active SGLT1/SGLT2 dual inhibitors. *Bioorg Med Chem Lett*. 2018;28:1182–1187. https://doi.org/10.1016/j.bmcl.2018.02.057. 29523385.

145. Tilg H, Moschen AR. Microbiota and diabetes: an evolving relationship. *Gut.* 2014;63:1513–1521. https://doi.org/10.1136/gutjnl-2014-306928. 24833634.
146. Wang Z, Saha S, Van Horn S. Gut microbiome differences between metformin and liraglutide treated T2DM subjects. *Endocrinol Diab Metab.* 2018;1. https://doi.org/10.1002/edm2.9. e9. PMID: 30815546.
147. Olivares M, Schüppel V, Hassan AM, et al. The potential role of the dipeptidyl peptidase-4-like activity from the gut microbiota on the host health. *Front Microbiol.* 2018;9:1900. https://doi.org/10.3389/fmicb.2018.01900. 30186247.
148. Wang L, Li P, Tang Z, Yan X, Feng B. Structural modulation of the gut microbiota and the relationship with body weight: compared evaluation of liraglutide and saxagliptin treatment. *Sci Rep.* 2016;6:33251. https://doi.org/10.1038/srep33251. 27633081.
149. Zhang Q, Xiao X, Li M, et al. Vildagliptin increases butyrateproducing bacteria in the gut of diabetic rats. *PLoS One.* 2017;12. https://doi.org/10.1371/journal.pone.0184735, e0184735. 29036231.
150. Nepelska M, de Wouters T, Jacouton E, et al. Commensal gut bacteria modulate phosphorylation-dependent PPARγ transcriptional activity in human intestinal epithelial cells. *Sci Rep.* 2017;7:43199. https://doi.org/10.1038/srep4319. 28266623.
151. Xu Z, Knight R. Dietary effects on human gut microbiome diversity. *Br J Nutr.* 2015;113(Suppl):S1–S5.
152. Voreades N, Kozil A, Weir TL. Diet and the development of the human intestinal microbiome. *Front Microbiol.* 2014;5:494.
153. Miller MJ, Bostwick BL, Kennedy AD, et al. Chronic oral L-carnitine supplementation drives marked plasma TMAO elevations in patients with organic acidemias despite dietary meat restrictions. *JIMD Rep.* 2016;30:39–44.
154. Li DY, Wang Z, Li XS, Hazen SL, Tang WH. Relationship between statin use and trimethylamine N-oxide in cardiovascular risk assessment (abstract). *J Am Coll Cardiol.* 2018;71:A115.
155. Vallance HD, Koochin A, Branov J, et al. Marked elevation in plasma trimethylamine-Noxide (TMAO) in patients with mitochondrial disorders treated with oral l-carnitine. *Mol Genet Metab Rep.* 2018;15:130–133.
156. Markowiak P, Śliżewska K. Effects of probiotics, prebiotics, and synbiotics on human health. *Nutrients.* 2017;9:E1021. https://doi.org/10.3390/nu9091021.
157. Lew LC, Choi SB, Khoo BY, et al. Lactobacillus plantarum DR7 reduces cholesterol via phosphorylation of AMPK that down-regulated the mRNA expression of HMG-CoA reductase. *Korean J Food Sci Anim Resour.* 2018;38:350–361.
158. Korcz E, Kerényi Z, Varga L. Dietary fibers, prebiotics, and exopolysaccharides produced by lactic acid bacteria: potential health benefits with special regard to cholesterol-lowering effects. *Food Funct.* 2018;9:3057–3068.
159. Lam V, Su J, Koprowski S, et al. Intestinal microbiota determine severity of myocardial infarction in rats. *FASEB J.* 2012;26:1727–1735.
160. Gan XT, Ettinger G, Huang CX, et al. Probiotic administration attenuates myocardial hypertrophy and heart failure after myocardial infarction in the rat. *Circ Heart Fail.* 2014;7:491–499.
161. David LA, Maurice CF, Carmody RN, et al. Diet rapidly and reproducibly alters the human gut microbiome. *Nature.* 2014;505:559–563.
162. Cani PD, Neyrinck AM, Fava F, et al. Selective increases of bifidobacteria in gut microflora improve high-fat-diet-induced diabetes in mice through a mechanism associated with endotoxaemia. *Diabetologia.* 2007;50:2374–2383.
163. Khan MY, Dirweesh A, Khurshid T, Siddiqui WJ. Comparing fecal microbiota transplantation to standard-of-care treatment for recurrent Clostridium difficile infection: a systematic review and meta-analysis. *Eur J Gastroenterol Hepatol.* 2018;30:1309–1317.

164. Smits LP, Kootte RS, Levin E, et al. Effect of vegan fecal microbiota transplantation on Carnitine- and Choline-derived trimethylamine-N-oxide production and vascular inflammation in patients with metabolic syndrome. *J Am Heart Assoc.* 2018;7. https://doi.org/10.1161/JAHA.117.008342, e008342.
165. D'Odorico I, Di Bella S, Monticelli J, et al. Role of fecal microbiota transplantation in inflammatory bowel disease. *J Dig Dis.* 2018;19:322–334.
166. Wang Z, Roberts AB, Buffa JA, et al. Non-lethal inhibition of gut microbial trimethylamine production for the treatment of atherosclerosis. *Cell.* 2015;163:1585–1595.
167. Brunt VE, Gioscia-Ryan RA, Richey JJ, et al. Suppression of the gut microbiome ameliorates age-related arterial dysfunction and oxidative stress in mice. *J Physiol.* 2019. [E-pub ahead of print.

CHAPTER 10

Link between gut microbiome and cardiometabolic diseases

Tahira Farooqui
Department of Entomology, The Ohio State University, Columbus, OH, United States

1 Introduction

The human body contains a diverse community of microbes dominated by bacteria but also including viruses and fungi. The largest and most complex microbial reservoir of these communities is located in the gastrointestinal (GI) system, and its associated genome is known as the gut microbiome, which is estimated to harbor approximately 50- to 100-fold more genes compared to the host. These extra genes in the gut microbiome result in encoding additional enzymes that were non-encoded by the host and play a critical role in facilitating host metabolism, thereby participating in the regulation of host physiology.[1] The diverse GI microbiota is predominantly composed of bacteria from three major phyla: Firmicutes, Bacteroidetes, and Actinobacteria.[2] The first two phyla, Bacteroidetes and Firmicutes, account for around 90% of overall gut microbiota.[3] The gut microbiota maintains several essential functions, such as colonic fermentation of dietary fibers, synthesis of certain vitamins and nutrients, prevention against colonization by pathogens, protection against epithelial injury, regulation of host metabolism and immune system, and development and regulation of the neuroendocrine hypothalamic-pituitary-adrenal (HPA) axis as well as the gut-brain axis (GBA), which is the biochemical signaling that takes place between the GI tract and the central nervous system to control host behavior (Fig. 1).[4–7] It is thus increasingly recognized that the configuration and activity of the gut microbiota, mainly influenced by host genotype and additional environmental factors, affect host biology in numerous ways. The microbiota field is still in the preliminary stages because the functional properties of the complex gut microbiome have not been fully understood. However, several promising findings exhibit a great potential toward revolutionizing disease etiology and medical treatments.[8]

Cardiovascular and metabolic diseases, together with some non-alcoholic liver diseases as a whole, are referred to as cardiometabolic diseases (CMDs),

Gut Microbiota in Neurologic and Visceral Diseases
https://doi.org/10.1016/B978-0-12-821039-0.00013-7

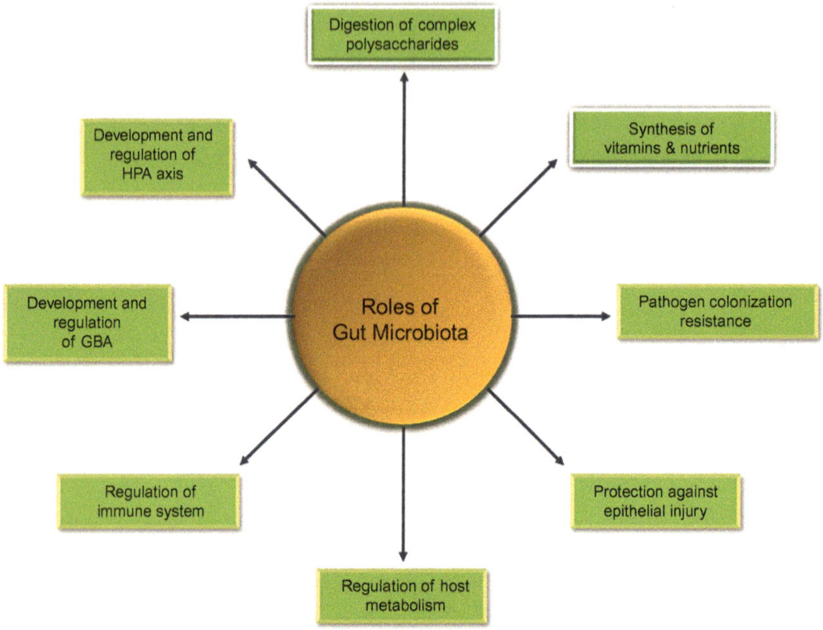

Fig. 1 Role of the gut microbiota in humans.

which are closely linked disorders that constitute an alarming global public health problem and are the leading causes of death worldwide. CMDs encompass various disorders like cardiovascular diseases (CVDs), which include conditions such as coronary artery disease (CAD), heart attack or myocardial infarction (MI), arrhythmia (malfunction of the heart's electrical system), heart failure (weakened pumping ability), and cardiomyopathy (abnormal enlargement of the heart); and metabolic diseases including type 2 diabetes (T2DM), obesity, chronic kidney disease (CKD), and non-alcoholic fatty liver disease (NAFLD) (Fig. 2).[9–12] Metabolic syndrome represents a group of cardiometabolic risk factors, including abdominal obesity combined with elevated blood pressure, fasting plasma glucose, and triglycerides, and reduced high-density lipoprotein cholesterol levels, implicating that the syndrome is associated with an increased risk of cardiovascular mortality.[13]

Over the last decades, the prevalence of many chronic diseases has increased significantly. It has been postulated that dysbiosis driven by environmental factors, such as consumption of an unhealthy diet (Western diet) or use of antibiotics that shifts the microbiome by increasing inflammation, results in the onset of chronic diseases. There have been advances in the characterization and analysis of the gut microbiota and associated bacterial

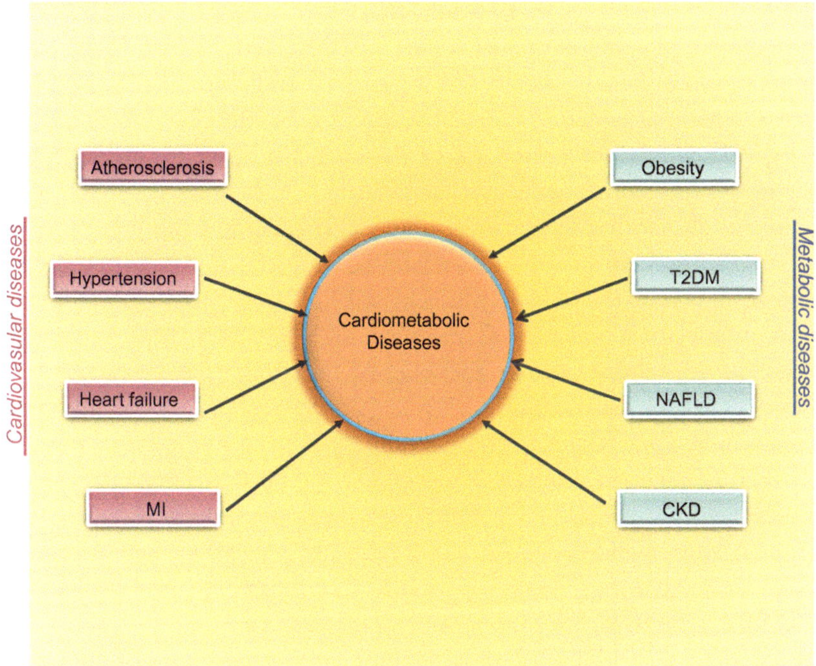

Fig. 2 Cardiometabolic diseases. Cardiovascular diseases (CVD) include atherosclerosis, hypertension, heart failure, and myocardial infarction (MI); and metabolic diseases include obesity, type 2 diabetes mellitus (T2DM), chronic kidney disease (CKD), cancer, and non-alcoholic fatty liver disease (NAFLD).

metabolites. However, the exact mechanisms through which they exert their actions in CMDs are not well understood. This chapter highlights the complex role of gut dysbiosis in the progression of CMDs.

2 CMDs

The common characteristics of CMDs include abdominal obesity (high body mass index and/or large waist circumference), insulin-resistant glucose metabolism (hyperinsulinemia, impaired fasting glucose, impaired glucose tolerance, T2DM); and dyslipidemia (high-serum triglyceride and low-serum high-density lipoprotein-cholesterol concentrations); and increased blood pressure (hypertension).

CMDs represent substantially larger economic and health burdens globally. According to the World Health Organization (WHO), 31% of all global deaths are related to CMDs.[14] Therefore, optimizing existing strategies and developing novel methods to treat and/or prevent CMDs is highly warranted.

CMDs are multifactorial diseases influenced by a wide spectrum of different factors, including environmental, socioeconomic, lifestyle (physical inactivity, smoking, and unhealthy dietary habits), genetic, and epigenetic factors.[15–23]

It is widely accepted that defective genes may contribute to the pathogenesis of many CMDs. Genetic variants are produced by altered exonic splicing leading to aberrant gene regulation.[18–20] Besides defective genes, epigenetic factors (such as DNA methylation, histone modification, and chromatin binding) also modulate gene expression in a heritable fashion without affecting the underlying genomic DNA sequence. In particular, DNA methylation and histone modifications have now been widely accepted to participate in diabetes, obesity, overweight, and cardiovascular disease.[23] Epigenetic factors have potential in accounting for additional phenotypic variation, along with providing insights into the causal relationship linking genetic variants to phenotypes.[21–23] Furthermore, increased inflammatory episodes also contribute to the pathophysiological driving force behind many cardiometabolic complications, including hyperlipidemia, dyslipidemia, obesity, T2DM, hypertension, nephropathy, and NAFLD.[24]

CMDs tend to have a number of risk factors in common. Poor diet and physical inactivity contribute to high cholesterol, hypertension, hyperglycemia, dyslipidemia, and unhealthy weight gain, which when combined result in developing cardiovascular and metabolic diseases. CMDs are estimated to cause more than 700,000 deaths per year in the United States, and approximately 50% of these deaths are directly related to diet.[25, 26]

3 Effects of diet on gut microbiota

The gut microbiota is fundamental to human health. Intestinal microbes metabolize dietary components, affecting human health by producing harmful or beneficial metabolites, which are involved in the incidence and progression of several intestinal-related and non-related diseases. Habitual diet (Western, vegetarian/vegan, Mediterranean, omnivore, and gluten-free) drives the composition of the gut microbiota and metabolome.[27]

Western diet refers to a high intake of animal protein, saturated fats, and sucrose, but low consumption of fiber. Western diet is also characterized by high intake of energy-dense processed and ultra-processed foods. These foods are industrial formulations made entirely from substances extracted from foods (e.g., oils, fats, sugar, starch, and proteins), derived from food constituents (e.g., hydrogenated fats and modified starch), or synthesized in laboratories from food substrates or other organic sources (e.g., flavor enhancers, colors, and several food additives used to make the product hyper-palatable).[28] Long-term consumption of a Western diet results in a marked decrease in numbers of total bacteria and beneficial *Bifdobacterium* and *Eubacterium* species.[29,30] Therefore, it is considered a risk factor for many CMDs, including diabetes, obesity, and cardiovascular diseases.[31] Western diet also has been shown to produce cancer-promoting nitrosamines.[32]

Vegetarian diets (plant-based) typically comprise plant foods such as grains, legumes, nuts, seeds, vegetables, and fruit, and exclude all kinds of animal food including meat, fish, and seafood. Vegetarian diets (vegan/vegetarian) offer significantly greater counts of certain Bacteroidetes-related operational taxonomic units compared to omnivore diets.[33] Both vegan and vegetarian diets are enriched in fermentable plant-based foods. Fibers (i.e., non-digestible carbohydrates found exclusively in plants) most consistently increase lactic acid bacteria, such as *Ruminococcus*, *Eubacterium rectale*, and *Roseburia*, and reduce *Clostridium* and *Enterococcus* species. Polyphenols, also abundant in plant foods, increase *Bifidobacterium* and *Lactobacillus*, which provide anti-pathogenic and anti-inflammatory effects and cardiovascular protection. High fiber intake encourages the growth of species that ferment fiber into metabolites as short-chain SCFAs, including acetate, propionate, and butyrate.[33] Thus the vegetarian diet (low-fat diet enriched with fruits and vegetables) appears to be beneficial for human health by promoting the development of more diverse and stable microbial systems due to the presence of fibers and polyphenols, and thereby can reduce risk factors for CMDs.[34–38]

A traditional Mediterranean diet including olive oil, assorted fruits, vegetables, cereals, legumes, and nuts; moderate consumption of fish, poultry, and red wine; and a lower intake of dairy products, red meat, processed meat and sweets, is considered a healthy, balanced diet.[39] It is distinguished by a beneficial fatty acid profile that is rich in both monounsaturated and polyunsaturated fatty acids, high levels of polyphenols, other antioxidants, high fiber, low glycemic carbohydrates, and both plant- and animal-based proteins.[39] DeFilippis et al. compared habitual omnivores, vegetarians, and vegans, and detected significant associations between degree of adherence to the Mediterranean diet and increased levels of fecal SCFAs, *Prevotella* bacteria, and other Firmicutes.[40] However, poor adherence to the Mediterranean diet has been shown to be associated with elevated urinary TMAO, implicating increased risk for cardiovascular risk.[40] Similarly, other studies have reported that foods comprising the typical Mediterranean diet improve obesity, lipid profile, and inflammation.[41–45] These changes may be mediated by diet-derived increase in *Lactobacillus*, *Bifidobacterium*, and *Prevotella*, and decrease in *Clostridium*.[45] In contrast, an omnivorous diet (intake of both plant and animal foods) fosters different microbiota when compared to vegetarian/vegan diets; however, there is only a marginal difference between vegans and vegetarians. Therefore, omnivorous diet consumption may result in greater chances of developing CMDs than consumption of vegan or vegetarian diets.[46,47]

A gluten-free diet (GFD) avoids wheat, rye, and barley, but includes many naturally gluten-free foods (such as fruits, vegetables, beans, seeds, legumes, and nuts in their natural unprocessed forms) as well as eggs, lean- and non-processed meats, fish, and poultry. GFD is mainly used for the treatment of celiac disease. In one study, Sanz took 10 healthy subjects who consumed GFD for 30 days and analyzed their fecal microbiota and dietary intake. Findings included a decrease in the numbers of healthy gut bacteria (*Bifidobacterium* and *Lactobacillus*), but an increase in numbers of unhealthy bacteria (*Escherichia coli* and total *Enterobacteriaceae*) parallel to reductions in the intake of polysaccharides after following a GFD.[48] Fecal samples of GFD-treated subjects exerted lower immune stimulatory effects on peripheral blood mononuclear cells than those of subjects on a regular diet.[48] Bonder et al. investigated the effect of a short-term GFD and noted reduction in *Ruminococcus bromii* and *Roseburia faecis* with an increase in *Victivallaceae* and *Clostridiaceae*, implicating that GFD changes the gut microbiome composition and alters the activity of microbial pathways.[49] Potter et al. provided evidence indicating that patients who consume a GFD are at risk of cardiovascular disease.[50] However, the GFD effect on health and gut microbiota cannot be extrapolated from one population to others, implicating that varieties of GFD do not necessarily lead to a healthier physiological state.[51]

Thus, habitual dietary contents and quantity have a major role in shaping the human gut microbiota composition and function, which may result either in positive or negative effects on human health.

4 Gut microbial dysbiosis

Trillions of microorganisms reside within the GI tract (including bacteria, archaea, fungi, and viruses). The collective genomes of whole microbial communities (microbiota) integrate the gut microbiome.[52] Gut microbiomes exert permanent pivotal functions by promoting food digestion, xenobiotic metabolism, and regulation of innate and adaptive immunological processes.

Proteins, peptides, and metabolites released locally and at distant sites trigger cell signaling and pathways, which allow for maintaining host-microbial homeostasis.[52] The microbiota functions in tandem with the host's defense and immune system to protect against pathogen colonization and invasion. It also performs an essential metabolic function, acting as a source of essential nutrients and vitamins and aiding in the extraction of energy and nutrients such as SCFAs and amino acids from diet. Ultimately, the host depends on its intestinal microbiota for a number of vital functions and thus the intestinal microbiota may contribute to health. Within the dietary components, polymers (mainly fibers, proteins, fat, and polyphenols) that are not hydrolyzed by human enzymes seem to be the main leads of the metabolic pathways of gut microbiota, which in turn directly influence the human metabolome.[53–55]

The microbiota have a commensal relation with the host and play a key role in human health and disease.[56] The homeostatic balance of the intestinal microbiome is considered extremely beneficial for the host.[57] In fact, diet-induced dysbiosis acts as a contributing factor in the development of GI diseases like inflammatory bowel disease, irritable bowel syndrome, and colorectal cancer as well as systemic diseases like obesity, diabetes, atherosclerosis, and NAFLD.[58]

Several factors, including diet, age, stress, lifestyle, and diseases, cause an increase or decrease in relative abundance and diversity of bacterial species of gut microbiota. A persistent imbalance of the gut's microbial community is known as dysbiosis, which is defined as a disturbance to microbiota homeostasis caused by disruption of the ecological equilibrium in the gut; changes in microbial gene richness, functional composition, and metabolic activities; or changes in local distribution. Dysbiosis has been related to three different phenomena that can occur at the same time: (1) loss of beneficial organisms, (2) excessive growth of potentially harmful bacteria, and (3) loss of overall microbial diversity.[57] Moreover, dysbiosis has been demonstrated to be associated with many diseases including inflammatory bowel, irritable bowel syndrome, and cardiovascular, metabolic, and brain diseases.

5 Gut microbial dysbiosis and CMDs

Human gastrointestinal tract harbors a complex and dynamic population of microorganisms called the gut microbiota. As mentioned earlier that human gut microbiota is mostly composed by two dominant bacterial phyla (Firmicutes and Bacteroidetes), representing more than 90% of the total community.[56] Gut microbiota composition remains relatively unaffected by acute perturbations, as its plasticity allows it to rapidly return to its initial composition.[59] However, in response to environmental changes and stress factors associated with modern lifestyles may chronically modify gut microbiota composition (dysbiosis), selecting more virulent microorganisms leading to deleterious effects on the host health.

Numerous research groups worldwide have shown that the gut microbiota perturbations and resulting dysbiosis are implicated in the progression and pathogenesis of CMDS. In the sections that follow, I briefly discuss and show schematically how external factors, such as Western diet, influence microbial dysbiosis resulting in CMDs (Fig. 2).

5.1 Microbial dysbiosis and obesity

Obesity is defined as the accumulation of excess body fat leading to a high body mass index (> 30) and is a growing epidemic in most countries around the world. It is known to be a major public health problem that affects more than 1.9 billion adults, suggesting that 39% of adults are considered obese and overweight. An imbalance in intestinal microbiota has been documented in obese individuals[60] and in metabolic syndrome-associated diseases.[3,61] A reduced Bacteroides-to-Firmicutes ratio, with greater relative abundance of Firmicutes, has been reported as a main feature of obesity-related intestinal dysbiosis,[60] as well as a factor that decreases total bacterial diversity and richness.[62,63] However, the idea that a greater Firmicutes-to-Bacteroidetes ratio is a hallmark of obesity is still controversial. Although most studies support this notion, others find no difference[64] or even support the opposite.[65] These controversies can be attributed to methodological differences.

Diets varying in caloric content (2400 kcal/day vs 3400 kcal/day) have shown altered nutrient load-induced rapid changes in the gut bacterial community.[4] Higher caloric intake associated with a 20% growth of Firmicutes and a 20% reduction in Bacteroidete, is directly related to body weight gain.[4]

Consumption of a Western diet is one of the major contributors to gut microbiome dysbiosis and obesity and derangement of intestinal tissues.[66] It promotes increased microbial dysbiosis-linked obesity by disrupting the intestinal barrier. Meanwhile, a reduced mucus layer thickness favors a microbiota encroachment, which increases the interaction of microbial-associated molecular patterns (MAMPs) with intestinal cells. Host cells recognize MAMPs through pattern recognition receptors (PRRs) such as Toll-like receptors (TLRs), nucleotide-binding oligomerization domain receptors

(NODs), and NOD-like receptors (NLRs). Disrupted intercellular junctions are associated with increased lipopolysaccharide (LPS) leakage called metabolic endotoxemia, which perturbs gut homeostasis, resulting in intestinal and systemic inflammation and impaired insulin sensitivity. Furthermore, addition of artificial sweeteners and emulsifiers in food products further worsens obesity-linked intestinal inflammation.[67, 68] Obesity-associated gut microbiota have an increased capacity to harvest energy from the diet, stimulate gene reprogramming in the colon, change polypeptide hormones and other bioactive molecules released by EC cells, decrease the intestinal barrier, and disturb immune homeostasis. SCFAs also participate in insulin-mediated fat accumulation in adipocytes via activation of their receptors GPR43 and GPR41, which inhibits lipolysis and encourages adipocyte differentiation.[69]

Bile acids (BAs) are produced in the liver from cholesterol, and are metabolized in the intestine by gut microbiota. These bioconversions modulate the signaling properties of BAs via the nuclear farnesoid X receptor (FXR) and G protein-coupled membrane receptor 5 (GPR5), which regulate numerous metabolic pathways in the host. Moreover, BAs can regulate gut microbial composition both directly and indirectly through the activation of innate immune response genes in the small intestine. Thus, host metabolism can be affected by both microbial modifications of BAs, which leads to altered signaling via bile acid receptors and by alterations in the composition of the microbiota.[70] Several studies in humans have demonstrated that the gut microbiota is altered in obesity and related metabolic diseases, which results in dysregulation of BAs signaling, leading to disruption of lipogenesis and gluconeogenesis (Fig. 3). However, the polyphenol-rich foods and extracts, due to their prebiotic effect, can reshape the gut microbiota, emphasizing the novel role of the mucin-degrading bacterium *Akkermansia muciniphila* in their metabolic benefits.[71]

5.2 Microbial dysbiosis and diabetes

T2DM is a highly prevalent metabolic disorder characterized by hyperglycemia in the context of insulin-resistance, accounting for about 90% of diabetes cases worldwide. Growing evidence has shown that gut microbial dysbiosis plays a critical role in the regulation of development of T2DM (Fig. 3). The microbiome of patients with T2DM is characterized by a decrease in Firmicutes and Clostridia as well as a non-significant increase in Bacteroidetes and Proteobacteria.[72] Among the commonly reported findings, the genera of *Bifidobacterium*, *Bacteroides*, *Faecalibacterium*, *Akkermansia*,

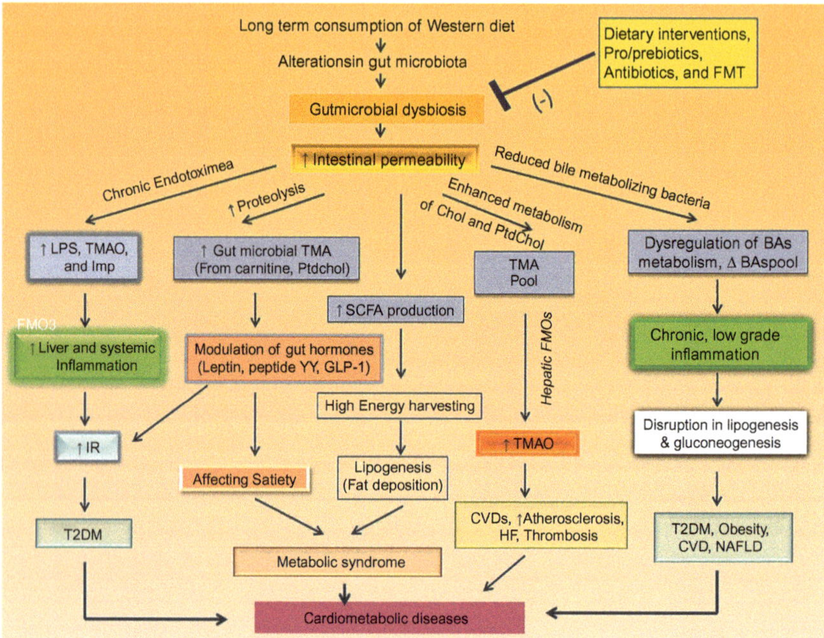

Fig. 3 Schematic representation of gut microbial dysbiosis-induced cardiometabolic diseases. Pancreatic peptide YY is a hormone secreted from endocrine cells called L cells in the small intestine (peptide YY), Glucagon-like peptide 1 (GLP-1), Type 2 diabetes mellitus (T2DM), short-chain fatty acids (SCFAs), trimethylamine (TMA), trimethylamine N-oxide (TMAO), hepatic flavin monooxygenase (FMO), cardiovascular diseases (CVDs), heart failure (HF), and imidazole propionate (ImP) from the breakdown of histidine.

and *Roseburia* were negatively associated with T2DM, while the genera of *Ruminococcus*, *Fusobacterium*, and *Blautia* were positively associated with T2DM.[73]

The endotoxins derived from the Gram-negative bacteria of the gut can be causes/mediators of low-grade systemic inflammation; their circulation, especially in T2DM patients, is influenced by alterations in diet.[74, 75] For example, consumption of a Western diet represents a growing health risk contributing to the increased occurrence of metabolic diseases, for example, diabetes. High plasmatic LPS is an endotoxin derived from the membrane of Gram-negative bacteria that binds to TLR4. Greater levels of LPS cause metabolic endotoxemia, triggering a low-grade inflammatory response, which is thought to underlie T2DM.[74,75] Changes in the intestinal homeostasis modulates BAs metabolism, decreases the abundance of some universal SCFA-producing bacteria, and metabolites that act synergistically on metabolic regulation systems contributing to insulin resistance. Butyrate promotes the development of the intestinal barrier. Patients with T2DM have decreased amounts of butyrate-producing bacteria in their microbiomes. However, feeding colonic cells with butyrate is associated with the maintenance of intestinal integrity and prevention of translocation, a process that moves Gram-negative intestinal bacteria across the lumen of the gut, causing endotoxemia. This process triggers a low-grade inflammatory response underlying T2DM.[76] Patients with T2DM and CKD have a greater proportion of TMA-producing microbiota and a positive correlation between TMAO and some biomarkers of inflammation and endothelial dysfunction.[77]

Moreover, alteration in intestinal permeability modulates BAs metabolism and increases various opportunistic pathogens and metabolites that act synergistically on metabolic regulation systems conferring sulphate reduction, oxidative stress resistance, and insulin resistance.[78] Dysregulation of bile acid metabolism contributes to dyslipidemia, hyperglycemia, and insulin resistance in T2DM (Fig. 3).

Therapeutic interventions such as dietary modifications, prebiotics, probiotics, antibiotics, and fecal transplantation can effectively alter the composition of gut bacteria.[79] As such, diet modification is a potential non-pharmacological prevention or treatment method for T2DM.

5.3 Microbial dysbiosis and NAFLD

NAFLD is a hepatic manifestation of metabolic syndrome that has emerged as one of the most common liver diseases worldwide. Its prevalence increases with increasing rates of obesity, insulin resistance, and diabetes mellitus.[80] NAFLD is characterized by simple hepatic steatosis and may progress to an inflammatory condition termed non-alcoholic steatohepatitis (NASH). NAFLD is less serious than NASH is, and usually does not progress to NASH or to liver cirrhosis. When NAFLD does progress to NASH, it may eventually lead to complications such as cirrhosis, liver cancer, liver failure, and cardiovascular disease.[81] It is well known that NAFLD patients usually die of extra-hepatic causes, frequently CVD, which highlights the importance of early diagnosis and prompt treatment of CVD risk factors. There is abundant evidence for a direct link between NAFLD and multiple CMDs including insulin resistance, hypertension, CKD, and cardiac arrhythmias.

Intestinal microbiota regulates metabolic functions and plays a role in the pathogenesis of NAFLD. An increased abundance of bacteria with pro-inflammatory features such as Proteobacteria or *E. coli* are dominantly present in these patients, whereas a reduction in the population of protective bacteria such as *Faecalibacterium prausnitzii* are decreased in NAFLD patients.[82, 83] The presence of a dysbiotic microbiome causes subsequent increases in gut permeability to bacteria-derived pathogens, including LPS and ethanol. In the liver, LPS causes inflammation by stimulating immune cells. Certain metabolites, such as BAs, SCFAs, and TMAO, also play a role in NAFLD pathophysiology, communication, and host metabolism. Bioactive molecules involved in this process include LPS, gut peptides, SCFAs, and lactate.[83] Moreover, individuals with a more aggressive NAFLD have a choline depletion linked to an increased choline metabolism.[84] Low choline may be important in NAFLD pathophysiology because it perturbs mitochondrial bioenergetics.[85, 86] Different approaches have been made to modulate intestinal microbiota, including pre/probiotics administration, fecal microbiota transportation, and lifestyle modification (Fig. 3). Intestinal microbiota have also been shown to be critically involved in cardiovascular health and diseases.[56, 87]

5.4 Microbial dysbiosis and atherosclerosis

Atherosclerosis is the major risk factor for CVDs, which are the leading causes of death in women among diverse populations worldwide. CVDs encompass a broad range of conditions and diseases, some of which are genetic, and many of which are the result of lifestyle choices. These conditions include: cardiomyopathy (an abnormally enlarged, thickened, and/or stiffened heart muscle that limits the heart's ability to pump blood effectively); congenital heart defects (problem with the heart structure that occurs while a fetus is developing that results in a leaky heart valve or malformations in the walls that separate the heart chambers); heart failure (weakened pumping resulting in inadequate supply of blood and oxygen to the body); arrhythmia (malfunction of the heart's electrical system resulting in irregular heartbeat); heart attack (the death of a segment of heart muscle caused by a loss of blood supply due to the hardening and narrowing of the coronary arteries from the build-up of fat, cholesterol, and other substances forming a plaque); and CAD (narrowing or blockage of coronary arteries caused by atherosclerosis).

Several studies have shown that gut microbial dysbiosis is critically involved in contributing to the development of atherosclerosis, including other CVDs (Fig. 3).[56, 88–92] The imbalances in host-microbial interactions impair homeostatic mechanisms that regulate health and can activate multiple pathways leading to CVDs risk factor progression. Dysbiosis has been implicated in the development of atherosclerosis through metabolism-independent and metabolite-dependent pathways. Metabolism-independent pathways deal with metabolic endotoxemia, leading to a pro-atherogenic inflammatory state, whereas metabolite-dependent pathways contribute to atherogenesis through dysbiosis-induced alterations in levels of TMAO, secondary BAs, and butyrate.[91] The enhanced metabolism of choline and phosphatidylcholine produces the pro-atherogenic compound TMAO.[92] There is a potential contribution of gut microbiota-derived production of TMAO from the metabolism of dietary choline and L-carnitine, which has been associated with an increased risk of major adverse events in humans. Several studies have also revealed a direct association of plasma concentrations of TMAO with dairy food consumption and low-grade inflammation due to expression of various pro-inflammatory cytokines such as TNF-α, IL-6, and C-reactive protein.[93] Since atherosclerosis is one of the major causes of CVD, it is plausible to assume that high plasma TMAO concentrations are related to the development of atherosclerosis. TMAO is therefore related to CVD as well (Fig. 3). Indeed, elevated plasma TMAO levels have been found in people at risk of CVD; these patients usually show greater concentrations of TMAO precursors, too.[94] TMAO concentrations can be influenced by several patient characteristics, including diet, renal function, liver function, excretory function of gut, general condition of the patient, as well as procedural factors, such as plasma collection time.

The secondary BAs are another group of gut microbiota-derived metabolites involved in CVDs and various metabolic diseases (Fig. 3). Furthermore, there is evidence that gut microbiota plays a causative role in atherosclerosis by modulating inflammation through production of microbial metabolites like SCFAs.[95] The abundance of dietary fiber consumption in diet (whole grain protein food, fruits, and vegetables) makes them attractive targets for disease prevention and reduction of risk of atherosclerosis and other CVDs.[96]

5.5 Microbial dysbiosis and hypertension

Apart from dyslipidemia, hypertension is another important risk factor for CVDs that is influenced by genetic as well as environmental factors. Gut dysbiosis has been linked with cerebrovascular events in addition to metabolic disorders.[97] Significant dysbiosis as a result of decreases in microbial richness, diversity, evenness, and increased Firmicutes-to-Bacteroidetes ratio has been reported in hypertensive animals.[98] Moreover, the gut microbiota has been shown to cause Ang-II-induced vascular dysfunction and hypertension.[99]

However, the causative relationship between gut dysbiosis and hypertension is complex and has not yet been fully evaluated.

It is well known that the groups of gut microbes that metabolize polysaccharides into different types of SCFAs are specific. *Streptococcus*, *Prevotella*, *Bifidobacterium*, *Clostridiums*, and *Akkermansi muciniphila* are the major acetate-producing bacteria, *Bacteroides*, *Salmonella*, *Dialister*, *Veillonella*, *Roseburia*, *Coprococcus*, and *Blautia* are propionate-producing bacteria. Lachnospiraceae, Ruminococcaceae, and Acidaminococcaceae are butyrate-producing bacteria.[97] The excess abundance of butyrate-producing bacteria has been reported to be associated with increased systolic and diastolic blood pressure in early pregnancy.[100] Greater dietary intake of fruit and vegetables has been shown to result in lower incidence of hypertension, implicating a modest beneficial effect in hypertension prevention.[101] Another group investigated the effect of a high-fiber diet and supplementation with SCFAs (acetate) on the gut microbiota and demonstrated its effect on the prevention of CVD.[102] High consumption of fiber modified the gut microbiota populations and increased the abundance of acetate-producing bacteria (*Bacteroides acidifaciens*), implicating improvement of gut dysbiosis and hypertension and heart failure in hypertensive mice.[102]

In addition to the altered regulation of various receptors through SCFAs, gut microbial dysbiosis may also contribute to hypertension through vasoconstriction mediated by oxidation of LDL.[103] Oxidized LDL (OX-LDL) is formed during oxidative stress when reactive oxygen species (ROS) interact with LDL. Greater levels of Ox-LDL decrease the production of nitric oxide (NO) and reduce the degree of vasodilation, which may contribute to hypertension associated with diseases such as diabetes and obesity.[104]

Both SCFAs and oxidized LDL have been associated with the microbiota-linked mechanisms for development or exacerbation of hypertension. However, the exact role played by microbial dysbiosis in mediating hypertension requires further elaborated research.

5.6 Microbial dysbiosis and congestive heart failure

Congestive heart failure (CHF) is a chronic progressive condition that affects the pumping power of the heart muscle. In this condition, due to structural and/or functional abnormalities of the heart, fluid builds up around the heart because of inefficient pumping that results in a poor quality of life for the patient as well as places an economic burden on the healthcare system.[105]

Pasinet et al. measured the presence of pathogenic bacteria and fungi (*Candida* species) and intestinal permeability and their correlations with disease severity, venous blood congestion as right atrial pressure (RAP), and/or systemic inflammation by C-reactive protein (CRP) in CHF patients.[106] This group demonstrated that CHF patients were colonized by more pathogenic bacteria and fungi (*Candida* species *Campylobacter* and *Shigella*) than control patients, and increased intestinal permeability was positively correlated with the severity of disease. Cui et al. investigated the composition and metabolic patterns of gut microbiota in CHF patients by performing metagenomic analyses of fecal samples. Results showed that the composition of gut microbiota in CHF patients significantly differed from that in controls.[107] A decrease in *F. prausnitzii* and increase in *Ruminococcus gnavus* were found to be the essential characteristics of the gut microbiota in CHF patients. They found that an imbalance of gut microbes is involved in the metabolism of protective metabolites (such as butyrate) and harmful metabolites (such as TMAO) in both fecal and plasma samples from CHF patients. Moreover, alterations in fecal and plasma metabolic patterns was correlated with gut microbiota dysbiosis in CHF.[107] However, further studies to validate the causal relationship between gut microbiota and CHF are still needed.

Heart failure-associated decreased cardiac output results in bowel wall edema and intestine ischemia, which induces alterations in gut membrane permeability and function. These changes may favor bacterial translocation, exacerbating HF pathogenesis at least partly through activation of systemic inflammation.[108]

Several strains of bacteria including Firmicutes, Proteobacteria, *Anaerococcus hydrogenalis*, *Clostridium asparagiforme*, *Clostridium hathewayi*, *Clostridium sporogenes*, *Escherichia fergusonii*, *Proteus penneri*, *Providencia rettgeri*, *Edwardsiella tarda*, and *Desulfovibrio desulfuricans* have been identified as potential producers of TMAO in vivo. Increased serum TMAO level has been reported in patients with CHF (Fig. 3) and associated with an increased mortality in the same population.[95] However, the mechanism of increased circulating TMAO levels in patients with CHF remains to be determined. Secondary BAs transformed by the gut microbiota have been reported to be increased in CHF patients (Fig. 3).[56] Gut-derived uremic toxins, such as indoxyl sulfate (IS), indole-3 acetic acid, *p*-cresyl sulfate, TMAO, and phenylacetylglutamine are uremic toxins derived from dietary protein metabolism by gut microbiota, and have been shown to be associated with CVD in patients with CKD. IS has been linked with myocardial fibrosis and ventricular remodeling.[109]

The accumulation of gut-derived uremic toxins, such as IS, may be directly responsible for the pathogenesis of CVD in CKD. These pathophysiological effects explain the association between IS concentration and the increased risk of CVD and mortality in patients with CKD. Therefore strategies designed to reduce the serum levels of IS may be considered novel therapeutic approaches for improving the outcomes of CVD-associated with CKD.[109]

CHF is a multisystem disorder in which intestinal morphology, permeability, and absorption are modified. Increased intestinal permeability and an augmented bacterial biofilm may contribute to the origin of both chronic inflammation and malnutrition. Accumulating evidence supports that gut microbial dysbiosis might play a pivotal role in the development of CHF.[110]

Thus changes in the gut microbiota of individuals with CMDs have been observed in both human and animal model systems, but the specific mechanisms are still under study.

6 Conclusion

Changes in the composition of gut microbiota dysbiosis have been linked to CMD pathologies. Specific relationships between diet and microbes, microbes and metabolites, microbes and immune functions, and microbes and/or their metabolites and CMDs are being established. Dietary treatments with fibers benefit the metabolome profile by improving the synthesis of SCFAs and decreasing the level of molecules, such as *p*-cresyl sulfate, IS, and TMAO, involved in disease states. Although there is convincing evidence that gut microbial dysbiosis correlates with CMDs, the underlying mechanisms have not yet been elucidated in depth. The gut microbiota has emerged as a potential target to modulate risk of CMDs, possibly via dietary interventions, which have been shown to considerably alter gut microbiota composition and function. Dietary interventions that affect gut microbiota composition and function are promising, novel, and cost-efficient for reducing risk of CMDs. However, further preclinical and clinical studies are needed to advance this aspect of CMDs.

References

1. Hooper LV, Gordon JI. Commensal host-bacterial relationships in the gut. *Science*. 2001;292:1115–1118.
2. Tap J, Mondot S, Levenez F, et al. Towards the human intestinal microbiota phylogenetic core. *Environ Microbiol*. 2009;11:2574–2584.

3. Qin J, Li Y, Cai Z, et al. A metagenomewide association study of gut microbiota in type 2 diabetes. *Nature*. 2012;490:55–60.
4. Jumpertz R, Le DS, Turnbaugh PJ, et al. Energy-balance studies reveal associations between gut microbes, caloric load, and nutrient absorption in humans. *Am J Clin Nutr*. 2011;94:58–65.
5. Purchiaroni F, Tortora A, Gabrielli M, et al. The role of intestinal microbiota and the immune system. *Eur Rev Med Pharmacol Sci*. 2013;17:323–333.
6. Round JL, Mazmanian SK. The gut microbiota shapes intestinal immune responses during health and disease. *Nat Rev Immunol*. 2009;9:313–323.
7. Walter J, Ley R. The human gut microbiome: ecology and recent evolutionary changes. *Annu Rev Microbiol*. 2011;65:411–429.
8. Kho ZY, Lal SK. The human gut microbiome—a potential controller of wellness and disease. *Front Microbiol*. 2018;9:1835.
9. Clement K. Acting on gut microbiota in cardiometabolic disorders: who, what, when and how. In: *Intestinal microbiota and its host: harmony or discord?* ; 2018. L'Institut Servier Scientific 2018 Colloquium.
10. Nikhra V. The novel dimensions of cardio-metabolic health: gut microbiota, dysbiosis and its fallouts. *Curr Res Diabetes Obes J*. 2019;11(1), 555805.
11. Hansen TH, Gøbel RJ, Hansen T, Pedersen O. The gut microbiome in cardio-metabolic health. *Genome Med*. 2015;7:33.
12. Corrêa TAF, Rogero MM, Hassimotto NMA, Lajolo FM. The two-way polyphenols-microbiota interactions and their effects on obesity and related metabolic diseases. *Front Nutr*. 2019;6:188.
13. Gao W. DECODE Study Group does the constellation of risk factors with and without abdominal adiposity associate with different cardiovascular mortality risk? *Int J Obes*. 2008;32(5):757–762.
14. World Health Organization. *Noncommunicable Diseases Country Profiles*. Geneva: World Health Organization; 2018. https://www.who.int/nmh/publications/ncd-profiles-2018/en/.
15. Ndisang JF, Rastogi S. Cardiometabolic diseases and related complications: current status and future perspective. *Biomed Res Int*. 2013;2013, 467682.
16. Malik VS, Willett WC, Hu FB. Global obesity: trends, risk factors and policy implications. *Nat Rev Endocrinol*. 2013;9:13–27.
17. GHR R. Diet, nutrition and cardiometabolic health. *Acta Sci Nutr Health*. 2018;2(12):26–33.
18. Kataoka M, Aimi Y, Yanagisawa R, et al. Alumediated nonallelic homologous and nonhomologous recombination in the BMPR2 gene in heritable pulmonary arterial hypertension. *Genet Med*. 2013;15(12):941–947.
19. Fiorillo C, Brisca G, Cassandrini D, et al. Subclinical myopathy in a child with neutral lipid storage disease and mutations in the PNPLA2 gene. *Biochem Biophys Res Commun*. 2013;430:241–244.
20. Heianza Y, Qi L. Impact of genes and environment on obesity and cardiovascular disease. *Endocrinology*. 2019;160:81–100.
21. Loh M, Zhou L, Ng HK, Chambers JC. Epigenetic disturbances in obesity and diabetes: epidemiological and functional insights. *Mol Metab*. 2019;27(Suppl):S33–S41.
22. Slomko H, Heo HJ, Einstein FH. Minireview: epigenetics of obesity and diabetes in humans. *Endocrinology*. 2012;153(3):1025e1030.
23. Smail HO. The epigenetics of diabetes, obesity, overweight and cardiovascular disease. *AIMS Genetics*. 2019;6(3):36–45.
24. Ndisang JF. Role of heme oxygenase in inflammation, insulin-signalling, diabetes and obesity. *Mediat Inflamm*. 2010;2010:18. 359732.
25. Micha R, Peñalvo JL, Cudhea F, Imamura F, Rehm CD, Mozaffarian D. Association between dietary factors and mortality from heart disease, stroke, and type 2 diabetes in the United States. *JAMA*. 2017;317(9):912–924.

26. Yu E, Malik VS, Hu FB. Cardiovascular disease prevention by diet modification. *J Am Coll Cardiol*. 2018;72:914–926.
27. Angelis MD, Garruti G, Minervini F, Bonfrate L, Portincasa P, Gobbetti M. The food-gut human axis: the effects of diet on gut microbiota and metabolome. *Curr Med Chem*. 2019;26(19):3567–3583.
28. Zinöcker MK, Lindseth IA. The Western diet–microbiome-host interaction and its role in metabolic disease. *Nutrients*. 2018;10(3):365.
29. Wu GD, Chen J, Hofmann C, et al. Linking long-term dietary patterns with gut microbial enterotypes. *Science*. 2011;334:105–108.
30. Reddy BS, Weisburger JH, Wynder EL. Efects of high risk and low risk diets for colon carcinogenesis on fecal microfora and steroids in man. *J Nutr*. 1975;105:878–884.
31. Cordain L, Eaton SB, Sebastian A, et al. Origins and evolution of the Western diet: health implications for the 21st century. *Am J Clin Nutr*. 2005;81:341–354.
32. Park JE, Seo JE, Lee JY, Kwon H. Distribution of seven N-nitrosamines in food. *Toxicol Res*. 2015;31:279–288.
33. Tomova A, Bukovsky I, Rembert E, et al. The effects of vegetarian and vegan diets on gut microbiota. *Front Nutr*. 2019;6:47.
34. Wang F, Zheng J, Yang B, Jiang J, Fu Y, Li D. Effects of vegetarian diets on blood lipids: a systematic review and meta-analysis of randomized controlled trials. *J Am Heart Assoc*. 2015;27, e002408.
35. Rosell M, Appleby P, Spencer E, Key T. Weight gain over 5 years in 21,966 meat-eating, fish-eating, vegetarian, and vegan men and women in *EPIC-Oxford*. *Int J Obes*. 2006;30:1389–1396.
36. Berkow SE, Barnard ND. Blood pressure regulation and vegetarian diets. *Nutr Rev*. 2005;63:1–8.
37. Khan N, Afaq F, Mukhtar H. Cancer chemoprevention through dietary antioxidants: progress and promise. *Antioxid Redox Signal*. 2008;10:475–510.
38. Jenkins DJ, Kendall CW, Marchie A, et al. Type 2 diabetes and the vegetarian diet. *Am J Clin Nutr*. 2003;78:610S–616S.
39. Lopez-Legarrea P, Fuller NR, Zulet MA, Martinez JA, Caterson ID. The influence of Mediterranean, carbohydrate and high protein diets on gut microbiota composition in the treatment of obesity and associated inflammatory state. *Asia Pac J Clin Nutr*. 2014;23:360–368.
40. DeFilippis F, Pellegrini N, Vannini L, et al. High-level adherence to a Mediterranean diet beneficially impacts the gut microbiota and associated metabolome. *Gut*. 2016;65(11):1812–1821.
41. Queipo-Ortuno MI, Boto-Ordonez M, Murri M, et al. Influence of red wine polyphenols and ethanol on the gut microbiota ecology and biochemical biomarkers. *Am J Clin Nutr*. 2012;95:1323–1334.
42. Bialonska D, Ramnani P, Kasimsetty SG, Muntha KR, Gibson GR, Ferreira D. The influence of pomegranate by-product and punicalagins on selected groups of human intestinal microbiota. *Int J Food Microbiol*. 2010;140:175–182.
43. Koloverou E, Panagiotakos DB, Pitsavos C, et al. Adherence to Mediterranean diet and 10-year incidence (2002–2012) of diabetes: correlations with inflammatory and oxidative stress biomarkers in the ATTICA cohort study. *Diabetes Metab Res Rev*. 2016;32:73–81.
44. Casas R, Sacanella E, Estruch R. The immune protective effect of the Mediterranean diet against chronic low-grade inflammatory diseases. *Endocr Metab Immune Disord Drug Targets*. 2014;14:245–254.
45. Singh RK, Chang H-W, Yan D, et al. Infuence of diet on the gut microbiome and implications for human health. *J Transl Med*. 2017;15:73.
46. Wu GD, Compher C, Chen EZ, et al. Comparative metabolomics in vegans and omnivores reveal constraints on diet-dependent gut microbiota metabolite production. *Gut*. 2016;65(1):63–72.

47. Franco-de-Moraes AC, de Almeida-Pititto B, da Rocha Fernandes G, Gomes EP, da Costa Pereira A, Ferreira SRG. Worse inflammatory profile in omnivores than in vegetarians associates with the gut microbiota composition. *Diabetol Metab Syndr.* 2017;9:62.
48. Sanz Y. Effects of a gluten-free diet on gut microbiota and immune function in healthy adult humans. *Gut Microbes.* 2010;1:135–137.
49. Bonder MJ, Tigchelaar EF, Cai X, et al. The influence of a short-term gluten-free diet on the human gut microbiome. *Genome Med.* 2016;8:45.
50. Potter MDE, Brienesse SC, Walker MM, Boyle A, Talley NJ. Effect of the gluten-free diet on cardiovascular risk factors in patients with coeliac disease: a systematic review. *J Gastroenterol Hepatol.* 2018;33:781–791.
51. Garcia-Mazcorro JF, Noratto G, Remes-Troche JM. The effect of gluten-free diet on health and the gut microbiota cannot be extrapolated from one population to others. *Nutrients.* 2018;10(10):1421.
52. Belizário JE, Faintuch J. Microbiome and gut dysbiosis. *Exp Suppl.* 2018;109:459–476.
53. Jason HY, Wu JHY, Micha R, Mozaffarian D. Dietary fats and cardiometabolic disease: mechanisms and effects on risk factors and outcomes. *Nature Rev Cardiol.* 2019;16(10):581–601.
54. Ilias Attaye I, Pinto-Sietsma SJ, Hilde Herrema H, Nieuwdorp M. A crucial role for diet in the relationship between gut microbiota and cardiometabolic disease. *Annu Rev Med.* 2020;71:149–161.
55. Billingsley HE, Carbone S, Lavie CJ. Dietary fats and chronic noncommunicable diseases. *Nutrients.* 2018;10(10):1385.
56. Tang WHW, Kitai T, Hazen SL. Gut microbiota in cardiovascular health and disease. *Circ Res.* 2017;120(7):1183–1196.
57. DeGruttola AK, Low D, Atsushi EM. Current understanding of dysbiosis in disease in human and animal models. *Inflamm Bowel Dis.* 2016;22(5):1137–1150.
58. Chan YK, Estaki M, Gibson DL. Clinical consequences of diet-induced dysbiosis. *Ann Nutr Metab.* 2013;63(Suppl 2):28–40.
59. Candela M, Biagi E, Maccaferri S, Turroni S, Brigidi P, et al. Intestinal microbiota is a plastic factor. Responding to environmental changes. *Trends Microbiol.* 2012;20:385–391.
60. Ley RE, Turnbaugh PJ, Klein S, Gordon JI. Microbial ecology: human gut microbes associated with obesity. *Nature.* 2006;444:1022–1023.
61. Mouzaki M, Comelli EM, Arendt BM, et al. Intestinal microbiota in patients with nonalcoholic fatty liver disease. *Hepatology.* 2013;58:120–127.
62. Turnbaugh PJ, Hamady M, Yatsunenko T, et al. A core gut microbiome in obese and lean twins. *Nature.* 2009;457:480–484.
63. Le Chatelier E, Nielsen T, Qin J, et al. Richness of human gut microbiome correlates with metabolic markers. *Nature.* 2013;500:541–546.
64. Duncan SH, Lobley GE, Holtrop G, et al. Human colonic microbiota associated with diet, obesity and weight loss. *Int J Obes.* 2008;32:1720–1724.
65. Schwiertz A, Taras D, Schäfer K, et al. Microbiota and SCFA in lean and overweight healthy subjects. *Obesity.* 2010;18:190–195.
66. Davis CD. The gut microbiome and its role in obesity. *Nutr Today.* 2016;51(4):167–174.
67. Suez J, Korem T, Zeevi D, et al. Artificial sweeteners induce glucose intolerance by altering the gut microbiota. *Nature.* 2014;514(7521):181–186.
68. Chassaing B, Koren O, Goodrich JK, et al. Dietary emulsifiers impact the mouse gut microbiota promoting colitis and metabolic syndrome. *Nature.* 2015;519(7541):92–96.
69. Lijuan Sun L, Ma L, Ma Y, Zhang F, Zhao C, Nie Y. Insights into the role of gut microbiota in obesity: pathogenesis, mechanisms, and therapeutic perspectives. *Protein Cell.* 2018;9(5):397–403.
70. Ramírez-Pérez O, Cruz-Ramón V, Chinchilla-López P, Méndez-Sánchez N. The Role of the Gut Microbiota in Bile Acid Metabolism. *Ann Hepatol.* 2017;16(Suppl 1):s21–s26.

71. Anhê FF, Varin TV, Barz MLE, et al. Gut microbiota dysbiosis in obesity-linked metabolic diseases and prebiotic potential of polyphenol-rich extracts. *Curr Obes Rep.* 2015;4(4):389–400.
72. Wong JMW. Gut microbiota and cardiometabolic outcomes: influence of dietary patterns and their associated components. *Am J Clin Nutr.* 2014;100:369S–377S.
73. Gurunga M, Lia Z, Youa H, et al. Role of gut microbiota in type 2 diabetes pathophysiology. *EBioMedicine.* 2020;51, 102590.
74. Cani PD, Amar J, Iglesias MA, et al. Metabolic endotoxemia initiates obesity and insulin resistance. *Diabetes.* 2007;56(7):1761–1772.
75. Boroni Moreira AP, de Cássia Gonçalves Alfenas R. The influence of endotoxemia on the molecular mechanisms of insulin resistance. *Nutr Hosp.* 2012;27(2):382–390.
76. Beer K. The gut microbiome in type 2 diabetes. *Clin Rev.* 2018;13–16.
77. Al-Obaide MAI, Singh R, Datta P, et al. Gut microbiota-dependent trimethylamine-N-oxide and serum biomarkers in patients with T2DM and advanced CKD. *J Clin Med.* 2017;6:e86.
78. Sircana A, Framarin L, Leone N, et al. Altered gut microbiota in Type 2 diabetes: just a coincidence? *Curr Diab Rep.* 2018;18(10):98.
79. Barengolts E. Gut microbiota, prebiotics, probiotics and synbiotics in management of obesity and prediabetes: review of randomized controlled trials. *Endod Prac.* 2016;22(10):1224–1234.
80. Jasirwan COM, Lesmana CRA, Hasan I, Sulaiman AS, Gani RA. The role of gut microbiota in non-alcoholic fatty liver disease: pathways of mechanisms. *Biosci Microbiota Food Health.* 2019;38(3):81–88.
81. Armstrong MJ, Adams LA, Canbay A, Syn WK. Extrahepatic complications of nonalcoholic fatty liver disease. *Hepatology.* 2014;59:1174–1197.
82. Loomba R, Seguritan V, Li W, et al. Gut microbiome-based metagenomic signature for non-invasive detection of advanced fibrosis in human nonalcoholic fatty liver disease. *Cell Metab.* 2017;25(5):1054–1062.
83. Grabherr F, Grander C, Effenberger M, Adolph TE, Tilg H. Gut dysfunction and non-alcoholic fatty liver disease. *Front Endocrinol.* 2019;10:611.
84. Sha W, da Costa KA, Fischer LM, et al. Metabolomic profiling can predict which humans will develop liver dysfunction when deprived of dietary choline. *Faseb J.* 2010;24:2962–2975.
85. Teodoro JS, Rolo AP, Duarte FV, et al. Differential alterations in mitochondrial function induced by a choline-deficient diet: understanding fatty liver disease progression. *Mitochondrion.* 2008;8:367–376.
86. Cheung O, Sanyal AJ. Recent advances in nonalcoholic fatty liver disease. *Curr Opin Gastroenterol.* 2010;26:202–208.
87. Wang Z, Zhao Y. Gut microbiota derived metabolites in cardiovascular health and disease. *Protein Cell.* 2018;9(5):416–431.
88. Drosos I, Tavridou A, Kolios G. New aspects on the metabolic role of intestinal microbiota in the development of atherosclerosis. *Metabolism.* 2015;64:476–481.
89. Gregory JC, Buffa JA, Org E, et al. Transmission of atherosclerosis susceptibility with gut microbial transplantation. *J Biol Chem.* 2015;290:5647–5660.
90. Lau K, Srivatsav V, Rizwan A, et al. Bridging the gap between gut microbial dysbiosis and cardiovascular diseases. *Nutrients.* 2017;9:E859.
91. Jie Z, Xia H, Zhong SL, et al. The gut microbiome in atherosclerotic cardiovascular disease. *Nat Commun.* 2017;8:845.
92. Ma J, Li H. The role of gut microbiota in atherosclerosis and hypertension. *Front Pharmacol.* 2018;9:1082.
93. Koeth RA, Wang Z, Levison BS, et al. Intestinal microbiota metabolism of L-carnitine, a nutrient in red meat, promotes atherosclerosis. *Nat Med.* 2013;19:576–585.

94. Rohrmann S, Linseisen J, Allenspach M, von Eckardstein A, Müller D. Plasma of trimethylamine-N-oxide are directly associated with dairy food consumption and low-grade inflammation in a German adult population. *J Nutr.* 2016;146:283–289.
95. Janeiro MH, Ramírez MJ, Milagro FI, Martínez JA, Solas M. Implication of trimethylamine N-oxide (TMAO) in disease: potential biomarker or new therapeutic target. *Nutrients.* 2018;10(10):1398.
96. Soliman GA. Dietary fiber, atherosclerosis, and cardiovascular disease. *Nutrients.* 2019;11(5):1155.
97. Vinod N. The novel dimensions of cardio-metabolic health: gut microbiota, dysbiosis and its fallouts. *Diab Obes J.* 2019;11(1), 555805.
98. Yang T, Santisteban MM, Rodriguez V, et al. Gut dysbiosis is linked to hypertension. *Hypertension.* 2015;65:1331–1340.
99. Karbach SH, Schonfelder T, Brandao I, et al. Gut microbiota promotes angiotensin ii-induced arterial hypertension and vascular dysfunction. *J Am Heart Assoc.* 2016;5(9), e003698.
100. Gomez-Arango LF, Barrett HL, McIntyre HD, et al. Increased systolic and diastolic blood pressure is associated with altered gut microbiota composition and butyrate production in early pregnancy. *Hypertension.* 2016;68(4):974–981.
101. Wang L, Manson JE, Gaziano M, Buring JE, Sesso HD. Fruit and vegetable intake and the risk of hypertension in middle-aged and older women. *Am J Hypertens.* 2012;25(2):180–189.
102. Marques FZ, Nelson E, Chu PY, et al. High-fiber diet and acetate supplementation change the gut microbiota and prevent the development of hypertension and heart failure in hypertensive mice. *Circulation.* 2017;135(10):964–977.
103. Packer CS, Rice AE, Johnson TC. Oxidized low density lipoprotein (OX-LDL) induced arterial muscle contraction signaling mechanisms. *Open Hyperten J.* 2014;6:20–26.
104. Peluso I, Morabito G, Urban L, Ioannone F, Serafini M. Oxidative stress in atherosclerosis development: the central role of LDL and oxidative burst. *Endocr Metab Immune Disord Drug Targets.* 2012;12(4):351–360.
105. Ramani GV, Uber PA, Mehra MR. Chronic heart failure: contemporary diagnosis and management. *Mayo Clin Proc.* 2010;85(2):180–195.
106. Pasini E, Aquilani R, Testa C, et al. Pathogenic gut flora in patients with chronic heart failure. *JACC Heart Failure.* 2016;4:220-227.
107. Cui X, Ye L, Li J, et al. Metagenomic and metabolomic analyses unveil dysbiosis of gut microbiota in chronic heart failure patients. *Scientific Reports.* 2018;8(1):635.
108. Branchereau M, Burcelin R, Heymes C. The gut microbiome and heart failure: a better gut for a better heart. *Rev Endocr Metab Disord.* 2019;20(4):407–414.
109. Wu CC, Hsieh MY, Hung SC, et al. Serum indoxyl sulfate associates with postangioplasty thrombosis of dialysis grafts. *J Am Soc Nephrol.* 2016;27:1254-1264.
110. Sandek A, Bauditz J, Swidsinski A, et al. Altered intestinal function in patients with chronic heart failure. *J Am Coll Cardiol.* 2007;50(16):1561–1569.

CHAPTER 11

Contribution of microbiota in obesity and obesity-related chronic diseases

Akhlaq A. Farooqui
Department of Molecular and Cellular Biochemistry, The Ohio State University, Columbus, OH, United States

1 Introduction

Obesity is a chronic and multifactorial condition caused by an imbalance of food intake and energy expenditure lasting for several years.[1] Obesity leads to an excessive accumulation of adipose tissue, which is now recognized not only as a main site of storage of excess energy derived from food intake but also as an endocrine organ.[2] Adipose tissue consists of multiple cell types, including adipocytes, monocytes/macrophages, pericytes, endothelial cells, and various stem cells. During adipose tissue expansion, adipose tissue macrophages (ATMs) represent the most abundant class of leukocytes that are involved in the regulation of several regulatory physiological processes, such as tissue remodeling and insulin sensitivity. They also produce a number of bioactive substances known as adipocytokines or adipokines (hormones, cytokines, complement and growth factors, extracellular matrix proteins, vasoactive factors, and microRNAs) that trigger chronic low-grade inflammation and interact with a range of processes in many different organs.[3,4] Converging evidence suggests that in progressive obesity, ATMs are key mediators of inflammation, insulin resistance, and impairment of adipocyte function. While macrophage recruitment from blood monocytes is a critical component of the generation of adipose tissue-mediated inflammation, several studies have revealed a role for ATM proliferation not only in the early stages of obesity but also in sustaining adipose tissue-induced inflammation.[4] Although the precise mechanisms are still unclear, dysregulated production or secretion of these adipokines is closely associated with the development of obesity and obesity-related metabolic diseases such as hypertension, depression, type 2 diabetes, cardiovascular disease (CVD), certain cancers, and neurological disorders.[5,6] Obesity is modulated by several factors such as

Gut Microbiota in Neurologic and Visceral Diseases
https://doi.org/10.1016/B978-0-12-821039-0.00003-4

long-term consumption of a high-calorie diet; genetic, epigenetic, behavioral, socio-economic, and environmental factors; lack of physical activity; and psychological factors (Fig. 1).[7–9] Among these factors, the composition of diet is known to regulate the population of intestinal microbiota in the intestine. Microbiota play a crucial role in maintaining immune and metabolic homeostasis and protecting against pathogens. The gut microbiota offers many benefits to the host through a range of physiological functions such as strengthening gut integrity or shaping the intestinal epithelium,[10] harvesting energy,[11] protecting against pathogens,[12] and regulating host immunity.[13] An animal-based diet, which has high fat, high simple carbohydrates, low fiber, and high salt, increases the abundance of bile-tolerant microorganisms (*Alistipes*, *Bilophila*, and *Bacteroides*) and decreases the levels of *Firmicutes*.[7] In contrast, a plant-based diet, which contains low fat, low complex carbohydrates, high fiber, and low salt, promotes the development of more diverse and stable microbial systems with increased population of *Roseburia*, *Eubacterium rectale*, and *Ruminococcus bromii*, and reduction in *Clostridium* and *Enterococcus* species.[14] The molecular mechanisms associated with animal-based, diet-mediated detrimental effects on human metabolism are not fully understood. However, it is suggested that consumption of an animal-based diet results in development of inflammation not only in the hypothalamus but also in the peripheral tissues, including the liver, adipose tissue, skeletal muscle, and intestine.[15] This chapter describes the contribution of gut microbiota in the pathogenesis of obesity and how obesity and microbiota composition are linked with the pathogenesis of chronic diseases such as type 2 diabetes, heart disease, obstructive sleep apnea, and neurological disorders.

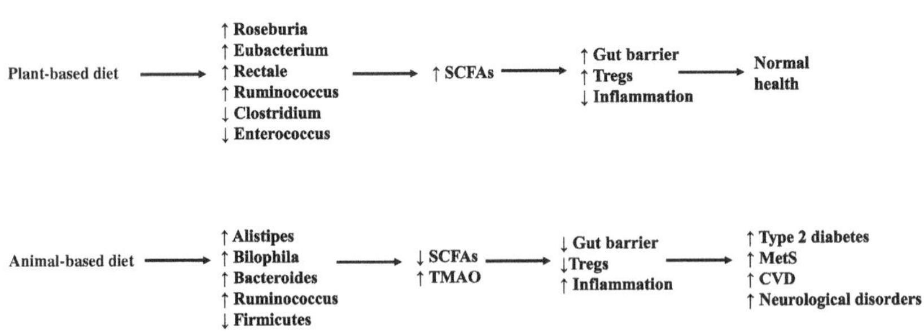

Fig. 1 Effect of plant-based and animal-based diet consumption on human gut microbiota. Short-chain fatty acids (SCFAs); trimethylamine-N-oxide (TMAO); T regulatory cells (Tregs); metabolic syndrome (MetS); cardiovascular disease (CVD).

2 Molecular mechanism of obesity

At the molecular level, obesity is causally linked to the induction of insulin resistance and inflammation.[16,17] However, little is known about how the fundamental mechanisms contributing to insulin resistance and inflammatory pathways are linked with obesity. It has been hypothesized that insulin resistance in obesity is caused by elevated levels of free fatty acids (FFAs), which originate from adipose tissue. Indeed, FFA infusion in vivo has been shown to impair the ability of insulin to suppress hepatic glucose production and to stimulate glucose uptake into skeletal muscle,[18] which in turn may lead to insulin resistance. Potential intracellular mechanisms by which FFAs contribute to insulin resistance may involve inflammatory signaling networks. For example, intracellular kinases, which are linked to inflammatory signaling and include PKC-θ, IKKα, and JNK, may play important roles in FFA-induced insulin resistance. This suggestion is based on the finding that *PKC*-θ-, *IKK*α-, and *JNK*-knockout mice are substantially protected from FFA-induced insulin resistance.[16,19,20] These pathways lead to two metabolically important effects: (1) the inhibition of insulin signaling pathways and (2) activation of chronic inflammation via the activation of Toll-like receptor 4 (TLR4) receptors. Toll-like receptors (TLRs) are a family of highly conserved receptors that recognize pathogen-associated molecular patterns and allow the host to recognize bacterial membrane/wall components and several gut-derived products. TLR-4 is the most important receptor in TLRs family because it plays a key role in innate immunity by triggering inflammatory responses. TLR-4 initiates innate immune responses via nuclear factor kappa B (NF-κB) when it is activated by its primary ligand, Gram-negative bacterial lipopolysaccharides (LPS), which results in the transcription of several genes that encode inflammatory cytokines, chemokines, and antimicrobial agents.[21] LPS crosses the intestinal barrier, from the intestinal lumen to mesenteric lymph nodes and other extra-intestinal organs and sites. The entry of LPS in systemic circulation via the portal vein and activates inflammatory pathways results in the enteric lymphatic drainage. This process not only triggers a proinflammatory state by provoking the release of cytokines, such as tumor necrosis factor-alpha (TNF-α), interleukin-6 (IL-6), and IL-1β, but also contributes to the hyperdynamic circulation and portal hypertension. Recent evidence suggests that intestinal barrier damage is due to a microbial imbalance that influences gut motility.[22] The observations of intestinal dysmotility in germ-free animals further suggest that microbiota play a crucial role in the modulation of intestinal motility.[23] Several studies on mice models and in humans have indicated that obesity is associated with a larger proportion of *Firmicutes* and relatively less *Bacteroidetes*.[24] To this end, it is reported that *Firmicutes* are significantly decreased in postgastric-bypass individuals, and *Prevotellaceae* is greatly enriched in obese individuals.[25,26] The differences in microbial composition have been explained on the basis of an increased capacity of the obesity-associated microbiome to harvest energy from the diet.[25] In contrast, Schwiertz et al.[27] reported a smaller ratio of *Firmicutes* to *Bacteroidetes* in overweight human adults compared to that in lean controls. Another study, using weight-loss diets, found no proof of the link between the proportion of *Bacteroidetes* and *Firmicutes* and human obesity.[28] Consequently, the composition of the obese microbiome is still questionable and more scientific evidence is needed to elucidate the relationship between the gut microbial composition and metabolic diseases.

3 Effects of diet and microbiota on chronic diseases

As previously stated, consumption of an animal-based diet alters gut microbiota and produces dysbiosis, a process associated with the pathogenesis of obesity and obesity-related metabolic diseases such as hypertension, depression, type 2 diabetes, CVD, certain cancers, and neurological disorders (Fig. 1).[5] In contrast, long-term consumption of plant-based food increases the level of *Prevotella* leading to stabilization of epithelial cells in the gastrointestinal tract.[29–31] Indeed, regular ingestion of plant polysaccharides is integral for maintaining a healthy balance of microbes in our lower gastrointestinal tract (Fig. 1).[32,33]

4 Obesity, microbiota composition, and type2 diabetes

Type 2 diabetes (diabetes mellitus) is a chronic metabolic lifestyle condition characterized by chronic hyperglycemia with alterations in carbohydrate, fat, and protein metabolism along with elevations in FFAs and an increase in circulating cytokines[34] due to defects in secretion and/or insulin action (Fig. 2).[35,36] Interactions among diet, gut microbiota, and host metabolism predisposes one to obesity and type 2 diabetes.[37] Gut microbiota play an important role in the development of insulin resistance, type 2 diabetes, and obesity. Many studies have indicated that the genera of *Bifidobacterium*, *Bacteroides*, *Faecalibacterium*, *Akkermansia*, and *Roseburia* are negatively associated with type 2 diabetes, while the genera of *Ruminococcus*, *Fusobacterium*,

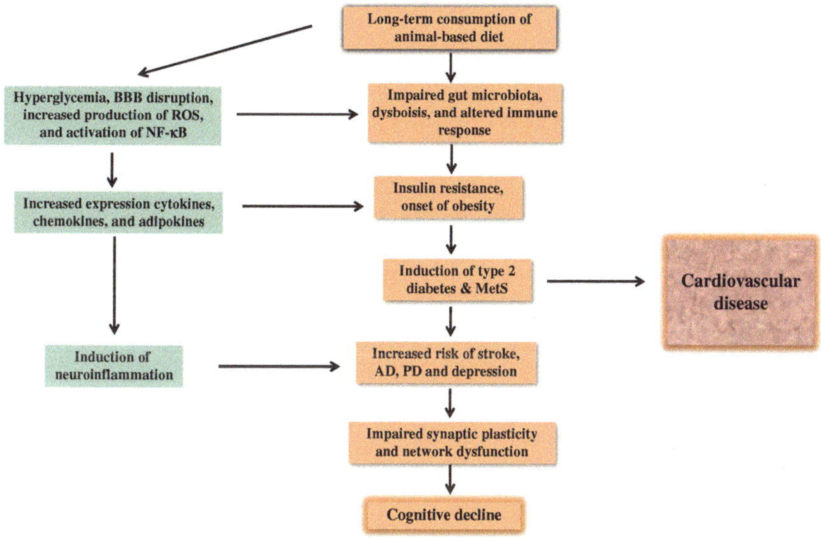

Fig. 2 Factors contributing to insulin resistance and susceptibility to cardiovascular disease, and neurological disorders in humans. Reactive oxygen species (ROS); metabolic syndrome (MetS); nuclear factor-kappaB (NF-κB).

and *Blautia* are positively associated with type 2 diabetes.[38,39] These observations indicate that the Gram-negative *Bacteroidetes* and *Proteobacteria* may contribute to the pathogenesis of type 2 diabetes through an endotoxin-induced inflammatory response as the endotoxin LPS exists in high concentrations as a main outer cell membrane component.[40] Additionally, the gut microbiome may be a new biomarker for type 2 diabetes prediction, as gut metagenome-based computational models could predict the type 2 diabetes-associated phenotype in glucose-intolerant patients.[38] Gut microbiota provide host energy by hydrolyzing the fiber (cellulose, xylan, and pectin) and partially digested starch that hosts consume daily. Meanwhile, numerous plant- and host-derived glycoconjugates including cellulose, chondroitin sulfate, hyaluronic acid, mucin, and heparin support intestinal bacterial ecosystems, which supply the host with pyruvate via glycolysis and short-chain fatty acids (SCFAs, including acetate, propionate, and butyrate). These metabolites modulate biologically active fatty acid composition, bile acid-activated metabolic pathways, and dietary choline bioavailability. Fermentation end products, such as SFCAs, can engage the epigenetic regulation of inflammatory reactions via free fatty acid receptors (FFARs) and other SFCA receptors while modulating the gut microbiota communities by selecting and promoting "pathogenic" strains (*Faecalibacterium prausnitzii*) and inhibiting beneficial strains (*Clostridium* cluster IV and *Clostridium* cluster XIVa) in T2D patients as compared to healthy control subjects.[41] Furthermore, the gut microbiota and SFCA excretion can be influenced by the type and quantity of dietary fats and carbohydrates in the population at risk of developing metabolic syndrome (MetS).[42] Collective evidence suggests that gut microbiota plays a major role in the food absorption and development of low-grade inflammation in obesity and type 2 diabetes.

5 Obesity, microbiota, and metabolic syndrome

MetS is a common and complex disorder involving abdominal obesity, dyslipidemia, endothelial dysfunction, hypertension, insulin resistance, genetic susceptibility, elevated blood pressure, hypercoagulable state, and chronic stress.[43] MetS is also accompanied by hyperglycemia, elevation in triglycerides (TAG), dyslipidemia, and low high-density lipoprotein cholesterol (HDL-C). These factors predispose the individual to increased risk of developing type 2 diabetes, MetS, and CVD (Fig. 2). In both humans and mice, the development of obesity correlates with shifts in the relative abundance of the two dominant bacterial phyla in the gut, *Bacteroidetes* and *Firmicutes*.[24] In addition, it has been shown that transfer of the gut microbiota from obese (*ob/ob*) mice to germ-free wild-type (WT) recipients leads to an increase in fat mass in the recipients, suggesting that the gut microbiota promotes obesity by increasing the capacity of the host to extract energy (calories) from ingested food.[25]

6 Obesity, microbiota, and cardiovascular disease

CVD is a chronic inflammatory atherosclerotic condition primarily associated with the large- and medium-sized arteries. Obesity and type 2 diabetes have been reported to increase a number of conventional risk factors for CVD. These factors include age, sex, genetic predisposition, diet, and lifestyle. Most type 2 diabetes and obesity patients tend to have greater levels of serum lipids, which predisposes them to atherosclerosis. Analysis of the oral, gut, and atherosclerotic plaque microbiota in atherosclerotic patients has indicated that there is an association between microbiota and atherosclerosis.[44] This association results in the development of hypertension, dyslipidemia, and endothelial dysfunction.[45] Other important risk factors for obesity and CVD are sedentary lifestyle, oxidative stress, insulin resistance, and smoking

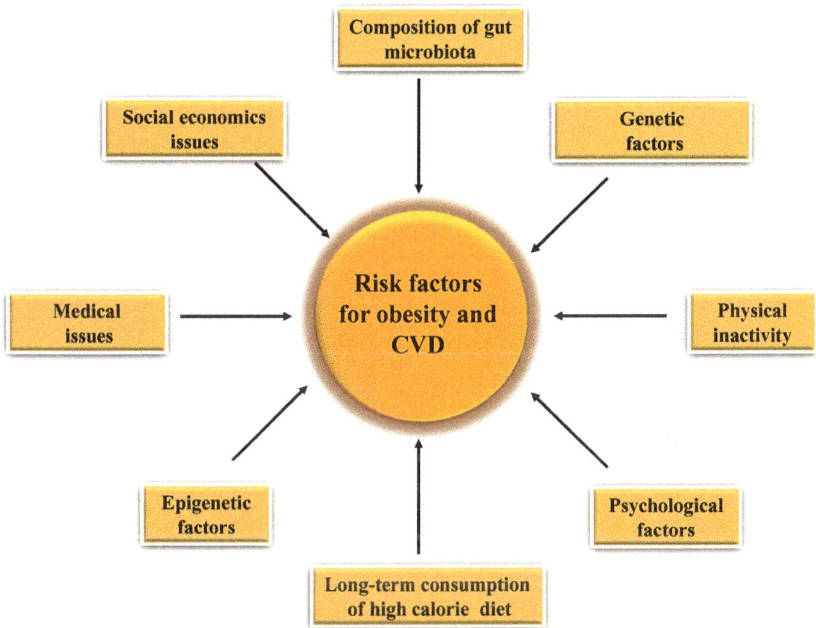

Fig. 3 Factors influencing obesity and cardiovascular disease.

(Fig. 3).[46,47] Moreover, it is widely accepted that excessive dietary intake of saturated fats and cholesterol plays a role in the onset and development of CVD through changes in plasma low-density lipoprotein-cholesterol (LDL-c).[48] The composition of the gut microbiota in people with CVD is different from those with healthy status. 16S rRNA sequencing studies have indicated that there is a significant decrease in the abundance of Coriobacteriaceae, Erysipelotrichaceae, and Ruminococcaceae at the family level and a significant decrease in the abundance of *Blautia*, *Collinsella*, unclassified Erysipelotrichaceae, and unclassified Ruminococcaceae at the genus level in the gut microbiota of CVD patients.[49] A reduction in SCFA-producing bacteria in patients with heart failure may be a notable characteristic for patients with heart CVD. However, nonlipid risk factors can also contribute to the development of CVD. Thus, dietary nutrients such as phosphatidylcholine, choline, and carnitine are utilized by gut microbes as a carbon fuel source. During this process, trimethylamine (TMA) is produced. TMA is carried via the portal circulation to the liver where it is rapidly converted by a family of enzymes, host hepatic flavin monooxygenases

(FMOs), into TMA N-oxide (TMAO).[50] Animal model studies have shown that TMAO is pro-atherogenic.[50–52] Suggested mechanisms of action of TMAO include increase of pro-inflammatory monocytes and vascular inflammation, platelet hyperreactivity,[53] cholesterol metabolism (Fig. 4),[51] and mediation of foam cell formation.[50] It has been suggested that TMAO is a link between the diet, gut microbiota, and illness. Emerging experimental and clinical evidence shows that TMAO may be involved in the etiology of hypertension, atherosclerosis, coronary artery disease, diabetes, and renal failure. On the contrary, a number of studies have shown protective functions of TMAO, such as stabilization of proteins and protection of cells from osmotic and hydrostatic stresses. Finally, it is possible that TMAO is neither a causative nor a protecting factor, but may be merely a marker of disrupted homeostasis.[54]

In contrast, recent clinical studies have revealed a paradoxically protective effect of obesity in patients with CVD, supporting the view that the biological "quality" of adipose tissue may be more important than overall adipose tissue mass or body weight.[55] As stated earlier, adipose tissue is recognized as a dynamic organ secreting a wide range of biologically active adipokines, microRNAs, gaseous messengers, and other metabolites that affect the cardiovascular system in both endocrine and paracrine ways. Despite

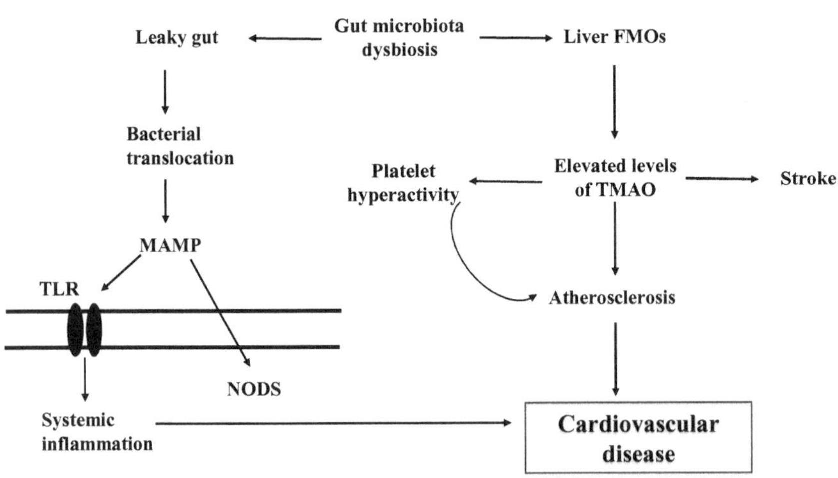

Fig. 4 Involvement of microbiota-derived trimethylamine-N-oxide in the cardiovascular disease. Flavin monooxygenases (FMOs); trimethylamine-N-oxide (TMAO); Microbe-associated molecular pattern (MAMPs); nucleotide oligomerization (NOD); Toll-like receptor (TLR).

being able to mediate normal cardiovascular function under physiological conditions, adipose tissue undergoes a phenotypic shift characterized by acquisition of pro-oxidant and pro-inflammatory properties in cases of CVD.[55] Crucially, recent evidence suggests that adipose tissue depots such as perivascular adipose tissue and epicardial adipose tissue are able to modify their phenotype in response to local signals of vascular and myocardial origin, respectively. Utilization of this unique property of certain adipose tissue depots to dynamically track cardiovascular biology may reveal novel diagnostic and prognostic tools against CVD. Better understanding of the mechanisms controlling the "quality" of adipose tissue secretome, as well as the communication links between adipose tissue and the cardiovascular system, is required for the efficient management of CVD.

7 Obesity, microbiota, and neurological disorders

The gut microbiome is known to interact with the brain through the maintenance of "gut-brain axis" homeostasis,[56] a process supported by bidirectional neurohumoral communication, production of neuroactive molecules (norepinephrine, catecholamines, serotonin (5-HT)), and regulation of the circulating levels of some cytokines. It has been demonstrated that some microbiota and their metabolites may target the brain directly via vagus nerve stimulation, or indirectly through immune-neuroendocrine mechanisms, triggering the release of pro-inflammatory cytokines that promote neuroinflammation.[57] The other mechanism of interaction between microbiome and brain is through the blood-brain barrier (BBB). The BBB is a component of the neurovascular unit (NVU). It is not just a physical barrier (due to the presence of specialized tight junctions and other changes that prevent unregulated leakage), but also acts more selectively as a transport interface (with specific transporters present on luminal and abluminal membranes), a secretory body, and a metabolic barrier (containing and releasing certain enzymes locally).[58] Microbiota and their release factors may enter into the systemic circulation from the gut. Once in the blood, the microbiota-derived factors can alter peripheral immune cells. This process may facilitate interactions with the BBB and ultimately with other elements of the NVU. Based on these studies, it is suggested that gut microbiota-derived factors can cross the BBB not only by altering BBB integrity and changing BBB transport rates, but also by inducing the release of neuroimmune substances from BBB cells.[59] The existence of a bidirectional communication between gut microbiota and brain suggests that microbes play a role in the induction of neurological disorders as well as in host perception, behavior, and emotional response. Recent studies have demonstrated that gut bacteria can produce significant amounts of amyloid and LPS, which are key players in the pathogenesis of Alzheimer's disease.[60] These observations illustrate the important role of gut microbiota in human health and suggest that manipulating them may influence disease activity. While the microbiome of a healthy individual is relatively stable, gut microbial dynamics can certainly be influenced by host lifestyle and dietary choices.[61] However, whether the observed alteration of the microbiota profile in these diverse conditions is the cause or the consequence of the disease remains to be established.

8 Conclusion

Obesity is a multifactorial condition resulting from improper balances of hormones and gene expression induced by diet. Obesity also has a strong inflammatory component that can be driven by diet-induced increases in arachidonic acid. The intestinal microbiome is intrinsically linked with overall health, including the risk of obesity, MetS, CVD, and neurological disorders. Overall, the predominant bacterial groups in the microbiome are Gram-positive *Firmicutes* and Gram-negative *Bacteroidetes*. Obesity and obesity-related metabolic disorders are characterized by specific alterations in the composition and function of the human gut microbiome. Mechanistic studies have indicated that the gastrointestinal microbiota can modulate not only energy utilization from the diet but also host genes. This raises the possibility that diet-induced alterations in the gut microbiota can promote weight loss or prevent obesity and related diseases (type 2 diabetes, MetS, and CVDs), suggesting that these conditions can be prevented and treated by targeting the microbiota through the modulation of intestinal microbiota via the consumption of live bacteria (probiotics) or digestible fiber (prebiotics).

References

1. Businaro R, Ippoliti F, Ricci S, Canitano N, Fuso A. Alzheimer's disease promotion by obesity: induced mechanisms-molecular links and perspectives. *Curr Gerontol Geriatr Res.* 2012;2012, 986823.
2. Goran MI, Alderete TL. Targeting adipose tissue inflammation to treat the underlying basis of the metabolic complications of obesity. *Nestle Nutr Workshop Ser.* 2012;73:49–60.
3. Rodríguez A, Ezquerro S, Méndez-Giménez L, Becerril S, Frühbeck G. Revisiting the adipocyte: a model for integration of cytokine signaling in the regulation of energy metabolism. *Am J Physiol Endocrinol Metab.* 2015;309:E691–E714.

4. Russo L, Lumeng CN. Properties and functions of adipose tissue macrophages in obesity. *Immunology.* 2018;155:407–417.
5. Abdelaal M, le Roux CW, Docherty NG. Morbidity and mortality associated with obesity. *Ann Transl Med.* 2017;5:161.
6. Oh SW, Shin SA, Yun YH, Yoo T, Huh BY. Cut-off point of BMI and obesity-related comorbidities and mortality in middle-aged Koreans. *Obes Res.* 2004;12:2031–2040.
7. Farooqui AA. *High Calorie Diet and the Human Brain.* Switzerland: Springer International Publishing; 2015.
8. González-Muniesa P, Mártinez-González M-A, Hu FB, et al. Obesity. *Nat Rev Dis Primers.* 2017;3:17034.
9. Hruby A, Hu FB. The epidemiology of obesity: a big picture. *Pharmacoeconomics.* 2015;33:673–689.
10. Natividad JMM, Verdu EF. Modulation of intestinal barrier by intestinal microbiota: pathological and therapeutic implications. *Pharmacol Res.* 2013;69:42–51.
11. den Besten G, van Eunen K, Groen AK, Venema K, Reijngoud DJ, Bakker BM. The role of short-chain fatty acids in the interplay between diet, gut microbiota, and host energy metabolism. *J Lipid Res.* 2013;54:2325–2340.
12. Bäumler AJ, Sperandio V. Interactions between the microbiota and pathogenic bacteria in the gut. *Nature.* 2016;535:85–93.
13. Gensollen T, Iyer SS, Kasper DL, Blumberg RS. How colonization by microbiota in early life shapes the immune system. *Science.* 2016;352:539–544.
14. Tomova A, Bukovsky I, Rembert E, et al. The effects of vegetarian and vegan diets on gut microbiota. *Front Nutr.* 2019;6:47.
15. Guillemot-Legris O, Masquelier J, Everard A, Cani PD, Alhouayek M, Muccioli GG. High-fat diet feeding differentially affects the development of inflammation in the central nervous system. *J Neuroinflammation.* 2016;13:206.
16. Hotamisligil GS. Endoplasmic reticulum stress and the inflammatory basis of metabolic disease. *Cell.* 2010;140:900–917.
17. Wellen KE, Hotamisligil GS. Inflammation, stress, and diabetes. *J Clin Invest.* 2005;115:1111–1119.
18. Boden G, She P, Mozzoli M, et al. Free fatty acids produce insulin resistance and activate the proinflammatory nuclear factor-κB pathway in rat liver. *Diabetes.* 2005;54:3458–3465.
19. Nakamura T, Furuhashi M, Li P, et al. Double-stranded RNA-dependent protein kinase links pathogen sensing with stress and metabolic homeostasis. *Cell.* 2010;140:338–348.
20. Samuel VT, Shulman GI. Mechanisms for insulin resistance: common threads and missing links. *Cell.* 2012;148:852–871.
21. Kawasaki T, Kawai T. Toll-like receptor signaling pathways. *Front Immunol.* 2014;5:461.
22. Quigley EM. Microflora modulation of motility. *J Neurogastroenterol Motil.* 2011;17:140–147.
23. Barbara G, Stanghellini V, Brandi G, Cremon C, Di Nardo G, et al. Interactions between commensal bacteria and gut sensorimotor function in health and disease. *Am J Gastroenterol.* 2005;100:2560–2568.
24. Turnbaugh PJ, Hamady M, Yatsunenko T, Cantarel BL, Duncan A, et al. A core gut microbiome in obese and lean twins. *Nature.* 2009;457:480–487.
25. Turnbaugh PJ, Ley RE, Mahowald MA, Magrini V, Mardis ER, et al. An obesity-associated gut microbiome with increased capacity for energy harvest. *Nature.* 2006;444:1027–1131.
26. Zhang HS, DiBaise JK, Zuccolo A, Kudrna D, Braidotti M, et al. Human gut microbiota in obesity and after gastric bypass. *Proc Natl Acad Sci U S A.* 2009;106:2365–2370.
27. Schwiertz A, Taras D, Schafer K, Beijer S, Bos NA, et al. Microbiota and SCFA in lean and overweight healthy subjects. *Obesity (Silver Spring).* 2010;18:190–195.

28. Duncan SH, Lobley GE, Holtrop G, Ince J, Johnstone AM, et al. Human colonic microbiota associated with diet, obesity and weight loss. *Int J Obes (Lond).* 2008;32:1720–1724.
29. David LA, Maurice CF, Carmody RN, et al. Diet rapidly and reproducibly alters the human gut microbiome. *Nature.* 2014;505:559–563.
30. El Kaoutari A, Armougom F, Gordon JI, Raoult D, Henrissat B. The abundance and variety of carbohydrate-active enzymes in the human gut microbiota. *Nat Rev Microbiol.* 2013;11:497–504.
31. Hamaker BR, Tuncil YE. A perspective on the complexity of dietary fiber structures and their potential effect on the gut microbiota. *J Mol Biol.* 2014;426:3838–3850.
32. De Filippo C, Cavalieri D, Di Paola M, et al. Impact of diet in shaping gut microbiota revealed by a comparative study in children from Europe and rural Africa. *Proc Natl Acad Sci U S A.* 2010;107:14691–14696.
33. Sonnenburg ES, Sonnenburg JL. Starving our microbial self: the deleterious consequences of a diet deficient in microbiota-accessible carbohydrates. *Cell Metab.* 2014;20:779–786.
34. Stumvoll M, Goldstein BJ, van Haeften TW. Type 2 diabetes: principles of pathogenesis and therapy. *Lancet.* 2005;365:1333–1346.
35. American Diabetes Association. Standards of medical care in diabetes—2014. *Diabetes Care.* 2014;37(Suppl. 1):S14–S80.
36. American Diabetes Association. Diagnosis and classification of diabetes mellitus. *Diabetes Care.* 2014;37(Suppl. 1):S81–S90.
37. Musso G, Gambino R, Cassader M. Interactions between gut microbiota and host metabolism predisposing to obesity and diabetes. *Annu Rev Med.* 2011;62:361–380.
38. Karlsson FH, Tremaroli V, Nookaew I, et al. Gut metagenome in European women with normal, impaired and diabetic glucose control. *Nature.* 2013;498:99–103.
39. Qin J, Li Y, Cai Z, et al. A metagenome-wide association study of gut microbiota in type 2 diabetes. *Nature.* 2012;490:55–60.
40. Larsen N, Vogensen FK, van den Berg FW, et al. Gut microbiota in human adults with type 2 diabetes differs from non-diabetic adults. *PLoS One.* 2010;5:e9085.
41. Remely M, Aumueller E, Merold C, et al. Effects of short chain fatty acid producing bacteria on epigenetic regulation of FFAR3 in type 2 diabetes and obesity. *Gene.* 2014;537:85–92.
42. Fava F, Gitau R, Griffin BA, Gibson GR, Tuohy KM, Lovegrove JA. The type and quantity of dietary fat and carbohydrate alter faecal microbiome and short-chain fatty acid excretion in a metabolic syndrome 'at-risk' population. *Int J Obes (Lond).* 2013;37:216–223.
43. Bruce KD, Byrne CD. The metabolic syndrome: common origins of a multifactorial disorder. *Postgrad Med J.* 2009;85:614–621.
44. Koren O, Spor A, Felin J, et al. Human oral, gut, and plaque microbiota in patients with atherosclerosis. *Proc Natl Acad Sci U S A.* 2011;108:4592–4598.
45. Mendelsohn AR, Larrick JW. Dietary modification of the microbiome affects risk for cardiovascular disease. *Rejuvenation Res.* 2013;16:241–244.
46. NDIC (National Diabetes Information Clearinghouse). *Diabetes, Heart Disease, and Stroke*; 2014. http://diabetes.niddk.nih.gov/dm/pubs/stroke/index.aspx.
47. Petersen MC, Shulman GI. Mechanisms of insulin action and insulin resistance. *Physiol Rev.* 2018;98:2133–2223.
48. Das M, Das DK. Resveratrol and cardiovascular health. *Mol Aspects Med.* 2010;31:503–512.
49. Luedde M, Winkler T, Heinsen FA, et al. Heart failure is associated with depletion of core intestinal microbiota. *ESC Heart Fail.* 2017;4:282–290.
50. Wang Z, Klipfell E, Bennett BJ, et al. Gut flora metabolism of phosphatidylcholine promotes CVD. *Nature.* 2011;472:57–63.

51. Koeth RA, Levison BS, Culley MK, et al. γ-Butyrobetaine is a proatherogenic intermediate in gut microbial metabolism of L-carnitine to TMAO. *Cell Metab.* 2014;20:799–812.
52. Tang WH, Wang Z, Levison BS, et al. Intestinal microbial metabolism of phosphatidylcholine and cardiovascular risk. *N Engl J Med.* 2013;368:1575–1584.
53. Zhu W, Gregory JC, Org E, et al. Gut microbial metabolite TMAO enhances platelet hyperreactivity and thrombosis risk. *Cell.* 2016;165:111–124.
54. Nowiński A, Ufnal M. Trimethylamine N-oxide: a harmful, protective or diagnostic marker in lifestyle diseases? *Nutrition.* 2018;46:7–12.
55. Elagizi A, Kachur S, Lavie CJ, et al. An overview and update on obesity and the obesity paradox in cardiovascular diseases. *Prog Cardiovasc Dis.* 2018;61:142–150.
56. Hamilton MK, Raybould HE. Bugs, guts and brains, and the regulation of food intake and body weight. *Int J Obes Suppl.* 2016;6(Suppl. 1):S8–S14.
57. Agusti A, Garcia-Pardo MP, Lopez-Almela I, et al. Interplay between the gut-brain axis, obesity and cognitive function. *Front Neurosci.* 2018;12:155.
58. Abbott NJ, Rönnbäck L, Hansson E. Astrocyte-endothelial interactions at the blood-brain barrier. *Nat Rev Neurosci.* 2006;7:41–53.
59. Logsdon MC, Vogt K, Davis DW, et al. Efficacy of an internet-based depression intervention to improve rates of treatment in adolescent mothers. *MCN Am J Matern Child Nurs.* 2018;43:324–329.
60. Pistollato F, Sumalla Cano S, Elio I, Masias Vergara M, Giampieri F, Battino M. Role of gut microbiota and nutrients in amyloid formation and pathogenesis of Alzheimer disease. *Nutr Rev.* 2016;74:624–634.
61. Wu GD, Chen J, Hoffmann C, et al. Linking long-term dietary patterns with gut microbial enterotypes. *Science.* 2011;334:105–108.

CHAPTER 12

Contribution of microbiota in the pathogenesis of ischemic stroke

Akhlaq A. Farooqui
Department of Molecular and Cellular Biochemistry, The Ohio State University, Columbus, OH, United States

1 Introduction

The human gastrointestinal tract contains approximately 10^{13}–10^{14} microbiota.[1] Humans live in an intimate symbiotic relationship with the extremely dense and diverse microbiota in the intestine. This symbiotic balance, which is established at birth, is key to the maintenance of health and well-being. The complex network of commensal microbiota exerts a number of specialized functions beneficial to the host, including absorption of nutrients, synthesis of essential organic compounds, protection against pathogens, and contribution to the development of the intestinal immune system.[2,3] In addition, the gut microbiota may also function like an endocrine organ that produces bioactive metabolites, including trimethylamine/trimethylamine *N*-oxide, and bile acids, which are also involved in host health and disease via numerous pathways.[4] Perturbations of the gut microbiota ecology is known as "dysbiosis," a process that has been implicated in a number of visceral diseases (obesity, malnutrition, diabetes, inflammatory bowel disease, and heart disease) and neurological disorders (stroke, Alzheimer's disease, Parkinson's disease, and depression).[2,5,6] The causes of these diseases are complex and include contributions from genetic, geographic, and habitual factors. In addition, intestinal dysbiosis is also involved in a range of chronic inflammatory disorders of the gastrointestinal tract, including Crohn's disease (CD), ulcerative colitis (UC),[7] and celiac disease (CeD).[8,9]

Stroke is a metabolic trauma caused either by a decrease in cerebral blood flow (CBF) due to the formation of a clot (ischemic stroke) or disruption of CBF to the brain due to the rupture of a blood vessel resulting in spillage of blood inside the brain or around the brain (hemorrhagic stroke).[10,11] Both types of strokes result in considerable brain damage. Stroke-mediated damage to the brain occurs in two primary zones: the ischemic core and the ischemic penumbra. Failure to meet energy demands

Gut Microbiota in Neurologic and Visceral Diseases
https://doi.org/10.1016/B978-0-12-821039-0.00007-1

is the primary cause of neurodegeneration within the ischemic core, where neurons die by necrosis. The ischemic core is characterized by the presence of reactive astrocytes. These astrocytes are capable of tissue reorganization and formation of glial scars.[12] In ischemic penumbra, neurons remain viable and neurodegeneration occurs slowly via apoptotic cell death. As stated, stroke-mediated injury is caused by a decrease in CBF, which reduces the availability of oxygen and glucose in the brain. This process not only impairs energy synthesis in the brain but also results in neuronal and glial depolarization leading to metabolic and ionic imbalance (increase in Ca^{2+}), and excitotoxicity (overstimulation of glutamate receptors)[10,13] leading to neurodegeneration in the ischemic core due to necrosis. The ischemic penumbra is sandwiched between the poorly perfused ischemic core and normally perfused healthy tissue. The penumbra remains viable for hours due to collateral blood flow, presenting a window of opportunity to re-perfuse and salvage this region. Ischemic stroke-mediated brain injury is accompanied by the death of neural cells through a cascade of events including failure of energy and sodium-potassium pump, depolarization of neural membranes, generation of free radicals, disruption of blood-brain barrier (BBB), and induction of apoptosis.[14]

2 Molecular aspects of ischemic stroke

At the molecular level, stroke-mediated injury involves the overstimulation of NMDA receptors, rapid Ca^{2+}-influx, and stimulation of phospholipases A_2, C, and D (PLA_2, PLC, and PLD), mitogen-activated protein kinases (MAPKs) such as extracellular signal-regulated kinase (ERK), p38, and c-Jun N-terminal kinase (JNK), death-associated protein kinase 1 (DAPK1), nitric oxide synthases (NOS), and endonucleases (Fig. 1).[13] Among these enzymes, stimulation of phospholipases contributes to the accumulation of free fatty acids and free radical generation. DAPK1 is a Ca^{2+}/calmodulin (CaM)-dependent serine/threonine protein kinase (160 kDa). The phosphorylation of DAPK1 may be responsible for certain forms of apoptotic cell death, including Fas, tumor necrosis factor (TNF)-α, ceramide, caspase, and p53-mediated apoptosis.[15] DAPK1activity is increased in injured neurons due to stroke.[16] A specific peptide inhibitor of DAPK1 significantly attenuates brain injury after ischemic stroke[17] (Fig. 2). In stroke-mediated brain injury, activated DAPK1 directly interacts with glutamate receptor (GluN2B) and phosphorylates at Ser1303, thereby increasing neuronal cell death by enhancing Ca^{2+} influx.[16,18] The massive influx of Ca^{2+} not only

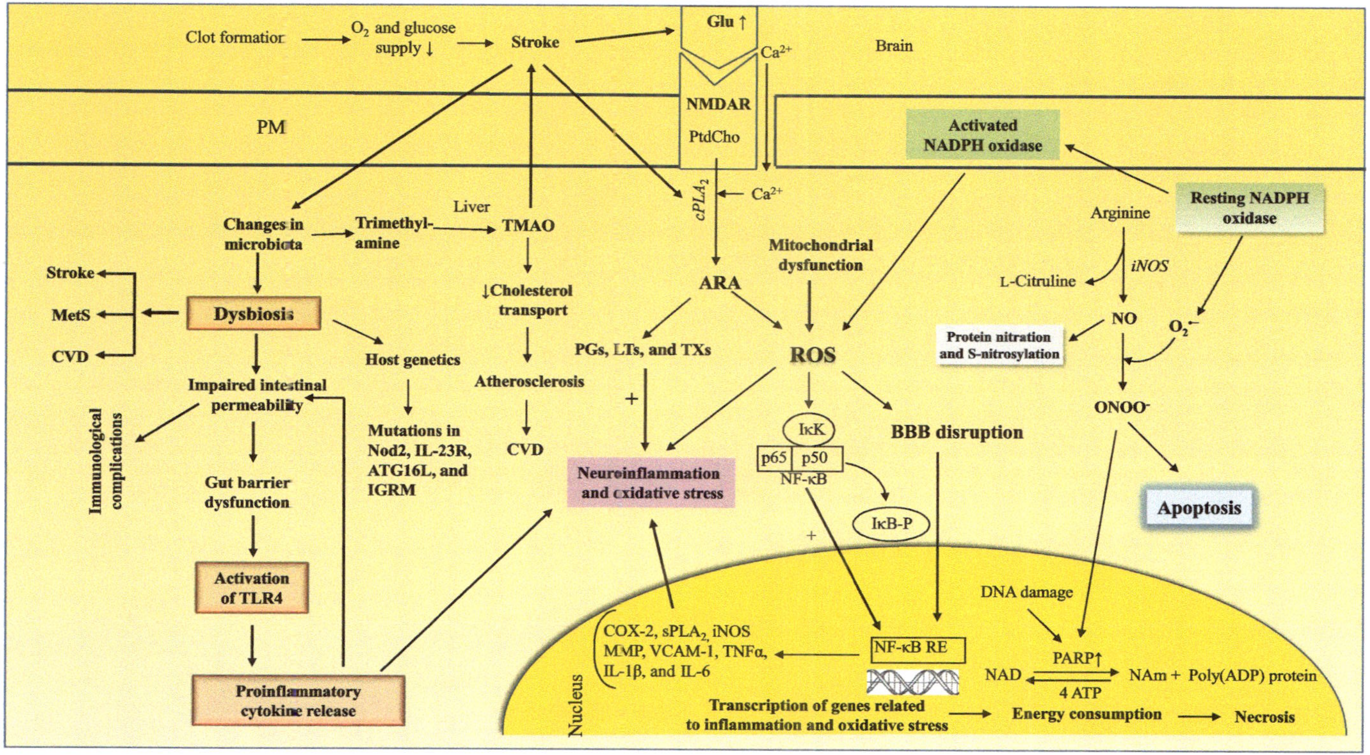

Fig. 1 Molecular mechanisms contributing to the pathogenesis of stroke. Plasma membrane (PM); *N*-methyl-D-aspartate receptor (NMDA-R); glutamate (Glu); phosphatidylcholine (PtdCho); Cytosolic phospholipase A_2 ($cPLA_2$); secretory phospholipase A_2 ($sPLA_2$); cyclo-oxygenase (COX); arachidonic acid (ARA); reactive oxygen species (ROS); nuclear factor kappaB (NF-κB); nuclear factor kappaB response element (NF-κB-RE); inhibitory subunit of NF-κB (IκB); inducible nitric oxide synthase (iNOS); peroxynitrite ($ONOO^-$); prostaglandins (PGs); leukotrienes (LTs); thromboxanes (TXs); Superoxide ($^{\cdot}O_2$); trimethylamine oxide (TMAO); matrix metalloproteinases (MMPs); vascular cell adhesion molecule-*1* (VCAM-1); poly(ADP-ribose) polymerase (PARP); nicotinamide (Nam); nicotinamide adenine dinucleotide (NAD); blood-brain barrier (BBB); cardiovascular disease (CVD); metabolic syndrome (MetS); positive sign (+) represents upregulation.

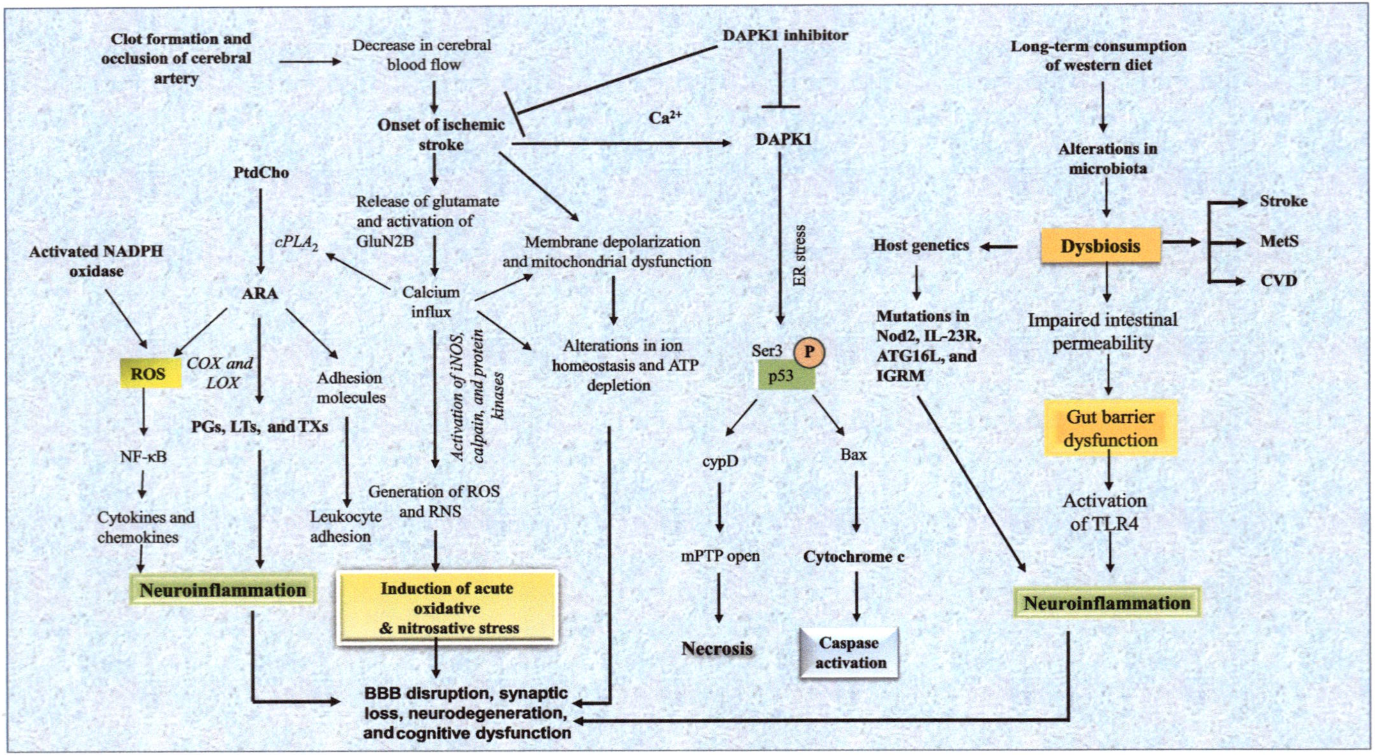

Fig. 2 Prevention of stroke-mediated injury by DAPK1 inhibitors. Phosphatidylcholine (PtdCho); Cyclooxygenase (COX); lipoxygenase (LOX); poly [ADP-ribose] polymerase-1(PARP-1); reactive oxygen species (ROS); reactive nitrogen species (RNS); phospholipase A_2 (PLA_2); inducible nitric oxide synthase (iNOS); nitric oxide (NO); death-associated protein kinase 1 (DAPK1); arachidonic acid (ARA); nuclear factor kappaB (NF-κB); inducible nitric oxide synthase (iNOS); peroxynitrite ($ONOO^-$); prostaglandins (PGs); leukotrienes (LTs); thromboxanes (TXs); Superoxide ($^{\bullet}O_2$); mitochondrial permeability transition pore (mPTP); BCL2 associated X (Bax); endoplasmic reticulum (ER); superoxide anion ($O_2^{\bullet -}$); peroxynitrite ($ONOO^-$); guanosine-5′-triphosphate (GTP); cyclic guanosine monophosphate (cGMP); guanylate cyclase (GC); Cyclophilin D (CypD); mitochondrial permeability transition pore (mPTP); apoptotic protein (bax); cardiovascular disease (CVD); metabolic syndrome (MetS).

activates the neuronal Ca^{2+}-dependent enzymes, but also results in depolarization of mitochondrial membrane and the opening of the mitochondrial permeability transition pore (mPTP), causing mitochondrial permeability transition along with increase in reactive oxygen species (ROS), and neuronal cell death.[13] NOS generates NO, which reacts with ROS (e.g., superoxide) to form the highly reactive peroxynitrite, hydrogen peroxide, hypochloride ions, and hydroxyl radicals. Inflammatory iNOS has been implicated in the pathogenesis of stroke, in which elevated NO mediates axonal degeneration and activates cyclooxygenases to provoke neuroinflammation. NO also instigates a downregulated secretion of brain-derived neurotrophic factor, which is essential for neuronal survival, development and differentiation, synaptogenesis, and learning and memory. The gut-brain axis denotes communication between the enteric nervous system (ENS) of the gastrointestinal tract and the central nervous system (CNS) of the brain, and the modes of communication include the vagus nerve, passive diffusion, and carrier by oxyhemoglobin. Ischemic injury to the brain also involves neuroinflammation, a process orchestrated by microglia and astrocytes. Microglia, the resident immune cells in the brain, are essential for modulating neurogenesis, influencing synaptic remodeling, and regulating neuroinflammation by surveying the brain microenvironment. Microglial dysfunction has been implicated in the onset and progression of stroke and neurodegenerative diseases. Microglia not only respond to local signals within the brain but also receive input from the periphery, including the gastrointestinal tract. Recent preclinical findings suggest that the gut microbiome plays a pivotal role in regulating microglial maturation and function. Induction of neuroinflammation involves the stimulation of NF-κB. This transcription factor migrates from cytoplasm to the nucleus, where it binds to the NF-κB response element (NF-κB-RE) and promotes the expression and release of pro-inflammatory cytokines (TNF-α; IL-6, IL-1β).[19,20] During this subacute phase of stroke, pro-inflammatory cytokines and chemokines not only activate PLA_2, COX-2, and 5-LOX, but also activate matrix metalloproteinases (MMPs). The activation of these enzymes not only worsens BBB disruption and promotes brain edema, but also stimulates neurodegeneration.[20,21]

3 Contribution of gut microbiota in poststroke injury

In healthy individuals, the composition of the gut microbiota remains relatively stable, dominated mainly by a few phyla, for example, Firmicutes,

Bacteroidetes, Proteobacteria, Actinobacteria, and Verrucomicrobia.[22] Firmicutes (F) and Bacteroidetes (B) account for more than 90% of the bacterial population in the colon,[23] while Actinobacteria and Proteobacteria (which includes Enterobacteriaceae) are regularly present but are scarce (less than 1%–5%).[24,25] The Bacteroidetes/Firmicutes phylum ratio remains remarkably stable over time.[25] These organisms inhabit different ecological niches on mucosal surfaces and in the gut lumen, forming complex biochemical interaction networks between themselves and their host.[26] However, there is considerable variation in bacterial diversity between individuals that is caused by differences in the host genome and by lifestyle factors such as diet, drug use, and environmental exposure.[27] Alterations in the relative abundance of each microbiota species in the gut microbial community not only influence immune performance and physiological homeostasis, but also healthy aging.[28–30] As stated, gut microbiota is in a symbiotic relationship with the host intestine. For instance, structural components of the microbiota, such as lipopolysaccharide (LPS) and peptidoglycan, can directly interact with host intestinal cells via Toll-like receptors (TLRs).[31] Among its myriad of physiological functions, the most significant function of the gut microbiota is supporting the digestion of food components to allow metabolites to interact with our human biology.[32]

The gastrointestinal tract is a major immune organ equipped with the largest pool of immune cells representing more than 70% of the entire immune system and the largest population of macrophages in the human body. There is a continuous bidirectional communication between the brain and the gut through the microbiota gut-brain axis. This bidirectional communication occurs not only through the gut–brain axis and vagus nerve but also through the immune system, neuroendocrine pathways, and microbiota-derived metabolites.[33,34] Stroke often leads to gut dysmotility, gut microbiota dysbiosis, "leaky" gut, gut hemorrhage, and even gut-origin sepsis, which is often associated with poor prognosis leading to abnormalities in gut functions (nausea, satiety, and pain) and neurotransmission. Emerging evidence suggests that gut inflammatory and immune response plays a key role in the pathophysiology of stroke, hypertension, diabetes, and obesity.[35] Studies on the acute middle cerebral artery occlusion model of stroke indicate that stroke-mediated brain injury involves reduction in bacterial biomass after 24 hours in the ileum and colon, promoting gut microbiota dysbiosis (Fig. 2) (reduction in number of Bacteroidetes). Ischemic brain tissue produces damage-associated molecular patterns to initiate innate and adaptive immune response both locally and systemically through specialized pattern-recognition receptors (e.g., TLRs). After stroke, innate immune cells including neutrophils, microglia or macrophages, mast cells, innate lymphocytes (IL-17 secreting γδ T-cell), and natural killer T cells respond within hours, followed by the adaptive immune response through activation of T and B lymphocytes. Subpopulations of T cells can help or worsen ischemic brain injury. Pro-inflammatory Th1, Th17, and γδ T cells are often associated with increased inflammatory damage, whereas regulatory T cells are known to suppress postischemic inflammation by increasing the secretion of antiinflammatory cytokine IL-10.[36] Converging evidence suggests that these changes in turn affect stroke outcomes via immune-mediated mechanisms.[37–40] Stroke-mediated injury also promotes the migration of intestinal lymphocytes to the ischemic brain. Therapeutic transplantation of fecal microbiota normalizes brain lesion-induced dysbiosis and improves stroke outcome.[39] These results support a novel mechanism in which the gut microbiome is a target of stroke-induced systemic alterations and an effector with substantial impact on stroke outcome.

Gut microbiota release many metabolites including short-chain fatty acids (acetic, propionic, and butyric acids), choline, phosphatidylcholine, L-carnitine, betaine, and trimethylamine, as well as minor bioactive molecules including tryptophan, indole, hydrogen peroxide, γ-butyrobetaine, and crotono-betaine. These bioactive products enter the circulation and modulate circulatory system homeostasis.[41,42] Among these bacterial metabolites, choline is metabolized to trimethylamine (TMA), which is transported to the liver via the portal vein where it is oxidized by liver flavin-containing monooxygenase-3 to trimethylamine-*N*-oxide (TMAO), a quaternary amine compound with the molecular weight of 75.1 Da. It is a colorless, odorless, solid soluble in water and is usually in the form of a dihydrate.[42] TMAO has a lot of important physiological and biochemical functions in the stability of protein structure as well as osmotic regulation, resistance to ionic instability, and resistance to water pressure.[43]

Elevated plasma TMAO levels are biomarkers for several visceral and neurological diseases such as type 2 diabetes, metabolic syndrome (MetS), cardiovascular disease, and stroke (Fig. 3).[42,44] Long-term consumption of a Western diet not only contributes to hypertension and atherosclerosis but also leads to elevated circulating TMAO levels, subsequently contributing to cardiac fibrosis and dysfunction. Multiple studies indicate that TMAO can affect heart failure (HF) susceptibility and development, partially via increased myocardial fibrosis, ventricular remodeling, and the accompanying cardiac dysfunction. The underlying mechanism by which TMAO

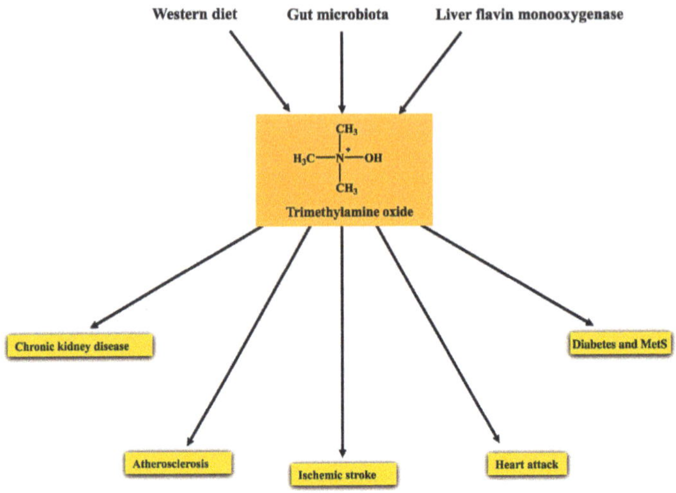

Fig. 3 Synthesis and contribution of TMAO in visceral and neurological disorders.

aggravates myocardial fibrosis is not fully understood. TMAO prolongs the effect of angiotensin, eventually leading to adverse cardiac remodeling.[45] Another mechanistic link is that TMAO promotes myocardial fibrosis via activation of NLRP3 inflammasome-associated TGF-β/Smad3 signaling. In addition, the finding that NLRP3 silencing can abolish the increases in the levels of IL-1β, TGF-β, and other profibrotic markers induced by TMAO in cardiac fibroblasts suggests that TMAO can be a new potential target for retarding the development of myocardial remodeling and the progression of myocardial functional impairment.[46] TMAO is mainly produced by *Prevotella* species, linking them to a greater risk of atheroma formation.[41] TMAO is not only known to modulate platelet hyperresponsiveness and clot formation rate but also activate arterial endothelial cell activation in vivo.[47] In vivo, TMAO is known to increase scavenger receptor, CD36, and SR-A1 expression, leading to more uptake of modified low-density lipoprotein (LDL) for macrophages to form foam cells.[42] On the other hand, TMAO decreases expression of two key enzymes, CYP7A1 and CYP27A1, essential for bile acid biosynthesis and multiple bile acid transporters (OATP1, OATP4, MRP2, and NTCP) in the liver, which decreases the bile acid pool, resulting in decreased reverse cholesterol efflux.[41] Moreover, TMAO increases endoplasmic reticulum calcium release in platelet cells, consequently leading to platelet aggregation and thrombosis.[48] Decreasing the consumption of TMA precursors in diet and inhibiting TMAO production protect from the harmful effects of microbiota. All these processes are associated with cardiovascular disease (CVD), a condition that is an important risk factor for stroke.

Stroke-mediated injury in rats significantly decreases gastrointestinal motility and mediates intestinal mucosal injury involving necrosis and shedding of the intestinal epithelium 6 hours after stroke, which worsens 24 hours after stroke.[49,50] Intestinal lymphocytes infiltrate the brain and aggravate ischemic injury.[37] Furthermore, an active interplay of commensal bacteria and the gut-associated lymphatic system (GALT) modulate systemic immunity, affecting outcomes after brain lesions. In addition, BBB disruption following stroke injury allows infiltration of immune cells into the brain. Concurrent with changes in BBB integrity, microbiota may bias the immune system by inducing, instructing, and shifting immune cell populations (such as regulatory T cells or IL-17 producing Th17 cells).[51,52]

Aging is another parameter that is not only a risk factor for stroke but that also affects the gut microbiome (Fig. 4).[53] While gut microbiota of older individuals seems to be relatively stable over time, its composition differs substantially from that of younger people and is very heterogeneous.[54] Older and younger people respond differently to stroke not only in terms of brain damage but also in terms of microbiota composition. In a mouse model of stroke, antibiotics are known to alter the intestinal microbiota composition and reduce ischemic lesion volume and neurological deficits.[37]

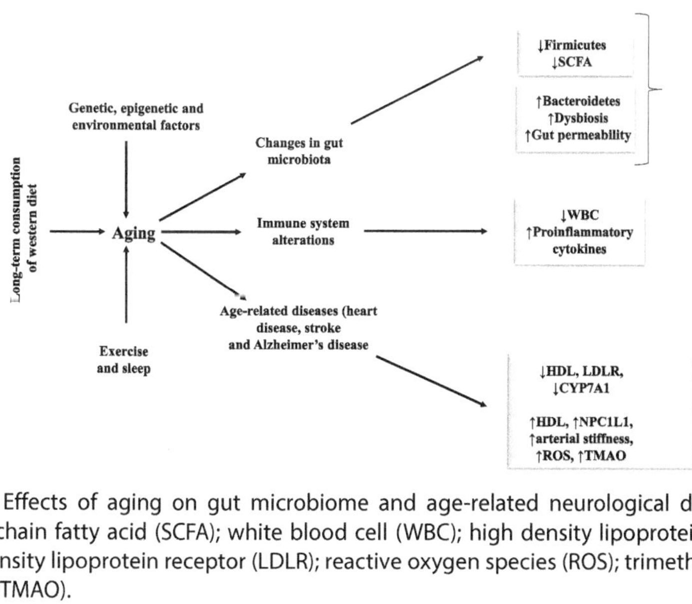

Fig. 4 Effects of aging on gut microbiome and age-related neurological disorders. Short-chain fatty acid (SCFA); white blood cell (WBC); high density lipoprotein (HDL); low density lipoprotein receptor (LDLR); reactive oxygen species (ROS); trimethylamine oxide (TMAO).

Systemic inflammation can increase gut permeability.[55] In mice subjected to stroke, increased gut permeability facilitated bacterial translocation, leading to infection, sepsis, and increased mortality.[56] While stroke increases gut permeability and bacterial translocation in both young and aged mice, aged animals are found to be particularly more susceptible to sepsis and immune dysfunction. Bacterial translocation from the gut to mesenteric lymph nodes and peripheral organs such as the spleen, liver, and lungs triggers adaptive and innate immune responses.[56] Poststroke infections are common complications of stroke[57] and stroke patients are often treated with a potent combination of antibiotics,[58] which may lead to a dramatic alteration in the composition of gut microbiota,[59] adding a further level of complexity. Converging evidence suggests that gut microbiota play an important role in shaping the outcome after acute cerebral ischemia. Gut dysbiosis-mediated inflammatory responses are also known to affect cardiac function after stroke.[60,61] However, more studies are needed on this important topic.

4 Conclusion

Stroke is an acute metabolic brain trauma caused by the formation of a clot in the blood vessel, which decreases blood flow to a specific region in the brain. It results in neuronal death (neurodegeneration) due to the overstimulation of glutamate receptors, a massive influx of Ca^{2+}, prolonged decrease in ATP, stimulation of Ca^{2+}-dependent enzymes, induction of mitochondrial dysfunction, increase in oxidative and nitrosative stress, hyperexcitability, and onset of acute neuroinflammation due to the activation of microglia and astrocytes around the injury site. Contribution of NO in nitrosative stress involves its reactions with ROS (e.g., superoxide), forming the highly reactive peroxynitrite. Among various NOS enzymes, inflammatory iNOS has been implicated in the pathogenesis of stroke. During stroke injury, elevation in NO levels contributes to axonal degeneration and activation of cyclooxygenases/lipoxygenases, provoking neuroinflammation. NO also instigates a downregulated secretion of brain-derived neurotrophic factor, which is essential for neuronal survival, development and differentiation, synaptogenesis, and learning and memory. In addition, stroke-mediated injury also disrupts the BBB and alters gut microbiota composition, inducing gut dysbiosis and increasing systemic levels of TMAO. This pro-atherogenic and pro-thrombotic metabolite is produced by gut microbiota from dietary trimethylamine (TMA)-containing nutrients such as choline or carnitine. High levels of TMAO predict incident cardiovascular and cerebrovascular diseases (heart disease and stroke). Emerging evidence suggests that gut inflammatory and immune responses play a key role in the pathophysiology of stroke. Ischemic brain tissue also produces damage-associated molecular patterns to initiate innate and adaptive immune response both locally and systemically through specialized pattern-recognition receptors (e.g., TLRs).

References

1. Clarke G, Sandhu KV, Griffin BT, Dinan TG, Cryan JF, Hyland NP. Gut reactions: breaking down xenobiotic-microbiome interactions. *Pharmacol Rev.* 2019;71:198–224.
2. Lozupone CA, Stombaugh JI, Gordon JI, Jansoon JK, Knight R. Diversity, stability and resilience of the human gut microbiota. *Nature.* 2012;489:242–249.
3. Brestoff JR, Artis D. Commensal bacteria at the interface of host metabolism and the immune system. *Nat Immunol.* 2013;14:676–684.
4. Xu H, Wang X, Feng W, et al. The gut microbiota and its interactions with cardiovascular disease. *J Microbial Biotechnol.* 2020;13:637–656.
5. McLean MH, Dieguez Jr D, Miller LM, Young HA. Does the microbiota play a role in the pathogenesis of autoimmune disease? *Gut.* 2015;64:332–341.
6. Farooqui AA. *Insulin Resistance as a risk factor in Visceral and Neurological Disorders.* San Diego, CA: Academic Press, An Imprint of Elsevier; 2020.
7. Knights D, Lassen KG, Xavier RJ. Advances in inflammatory bowel disease pathogenesis: linking host genetics and the microbiome. *Gut.* 2013;62:1505–1515.
8. de Sousa Moraes LF, Grzeskowiak LM, de Sales Teixeira TF, Peluzio MCG. Intestinal microbiota and probiotics in celiac disease. *Clin Microbiol Rev.* 2014;27:482–489.
9. M'Koma AE. Inflammatory bowel disease: an expanding global health problem. *Clin Med Insights Gastroenterol.* 2013;14:33–47.
10. Farooqui AA. *Neurochemical Aspects of Neurotraumatic and Neurodegenerative Diseases.* New York: Springer; 2010.
11. Keep RF, Hua Y, Xi G. Intracerebral haemorrhage: mechanisms of injury and therapeutic targets. *Lancet Neurol.* 2012;11:720–731.
12. Sofroniew M. Astrocyte failure as a cause of CNS dysfunction. *Mol Psychiatry.* 2000;5:230–232.
13. Farooqui AA. *Ischemic and Traumatic Brain and Spinal Cord Injuries.* San Diego, CA: Academic Press, An Imprint of Elsevier; 2018.
14. Kanekar SG, Zacharia T, Roller R. Imaging of stroke: part 2, pathophysiology at the molecular and cellular levels and corresponding imaging changes. *AJR Am J Roentgenol.* 2012;198:63–74.
15. Raveh T, Droguett G, Horwitz MS, DePinho RA, Kimchi A. DAP kinase activates a p19ARF/p53-mediated apoptotic checkpoint to suppress oncogenic transformation. *Nat Cell Biol.* 2001;3:1–7.
16. Wang S, Shi X, Li H, et al. DAPK1 signaling pathways in stroke: from mechanisms to therapies. *Mol Neurobiol.* 2017;54:4716–4722.
17. Velentza AV, Wainwright MS, Zasadzki M, et al. An aminopyridazine-based inhibitor of a pro-apoptotic protein kinase attenuates hypoxia-ischemia induced acute brain injury. *Bioorg Med Chem Lett.* 2003;13:3465–3470.
18. Kim N, Chen D, Zhou XZ, Lee TH. Death associated protein kinase 1 phosphorylation in neuronal cell death and neurodegenerative diseases. *Int J Mol Sci.* 2019;20:3131.
19. Liesz A, Zhou W, Mracsko E, et al. Inhibition of lymphocyte trafficking shields the brain against deleterious neuroinflammation after stroke. *Brain.* 2011;134:704–720.

20. Yilmaz G, Granger DN. Leukocyte recruitment and ischemic brain injury. *Neuromol Med.* 2010;12:193–204.
21. Jin R, Yang G, Li G. Inflammatory mechanisms in ischemic stroke: role of inflammatory cells. *J Leukoc Biol.* 2010;87:779–789.
22. Human Microbiome Project Consortium. Structure, function and diversity of the healthy human microbiome. *Nature.* 2012;486:207.
23. Ley RE, Hamady M, Lozupone C, et al. Evolution of mammals and their gut microbes. *Science.* 2008;320:1647–1651.
24. Chwalba A, Otto-Buczkowska E. Participation of the microbiome in the pathogenesis of diabetes mellitus. *Clin Diabetol.* 2017;6:178–181.
25. Faith JJ, Guruge JL, Charbonneau M, et al. The long-term stability of the human gut microbiota. *Science.* 2013;341:1237439.
26. Shkoporov AN, Hill C. Bacteriophages of the human gut: the "known unknown" of the microbiome. *Cell Host Microbe.* 2019;25:195–209.
27. Tamburini S, Shen N, Wu HC, Clemente JC. The microbiome in early life: implications for health outcomes. *Nat Med.* 2016;22:713–722.
28. Candela M, Turroni S, Biagi E, Carbonero F, Rampelli S, et al. Inflammation and colorectal cancer, when microbiota-host mutualism breaks. *World J Gastroenterol.* 2014;20:908–922.
29. Miele L, Giorgio V, Alberelli MA, De Candia E, Gasbarrini A, Grieco A. Impact of gut microbiota on obesity, diabetes, and cardiovascular disease risk. *Curr Cardiol Rep.* 2015;17:120.
30. Ottman N, Smidt H, de Vos WM, Belzer C. The function of our microbiota: who is out there and what do they do? *Front Cell Infect Microbiol.* 2012;2:104.
31. Larsson E, Tremaroli V, Lee YS, et al. Analysis of gut microbial regulation of host gene expression along the length of the gut and regulation of gut microbial ecology through MyD88. *Gut.* 2012;61:1124–1131.
32. Cani PD, Plovier H, Van Hul M, et al. Endocannabinoids–at the crossroads between the gut microbiota and host metabolism. *Nat Rev Endocrinol.* 2016;12:133–143.
33. Cenit MC, Sanz Y, Codoñer-Franch P. Influence of gut microbiota on neuropsychiatric disorders. *World J Gastroenterol.* 2017;23:5486–5498.
34. Quigley EMM. Microbiota-brain-gut axis and neurodegenerative diseases. *Curr Neurol Neurosci Rep.* 2017;17:94.
35. Winek K, Dirnagl U, Meisel A. Role of the gut microbiota in ischemic stroke. *Neurol Int Open.* 2017;1:E287–E293.
36. Arya AK, Hu B. Brain–gut axis after stroke. *Brain Circ.* 2018;4:165–173.
37. Benakis C, Brea D, Caballero S, et al. Commensal microbiota affects ischemic stroke outcome by regulating intestinal γδt cells. *Nat Med.* 2016;22:516.
38. Benakis C, Martin-Gallausiaux C, Trezzi J-P, Liesz A, Wilmes P. The microbiome-gut-brain axis in acute and chronic brain diseases. *Curr Opin Neurobiol.* 2020;61:1–9.
39. Singh V, Roth S, Llovera G, et al. Microbiota dysbiosis controls the neuroinflammatory response after stroke. *J Neurosci.* 2016;36:7428–7440.
40. Stanley D, Mason LJ, Mackin KE, et al. Translocation and dissemination of commensal bacteria in post-stroke infection. *Nat Med.* 2016;22:1277–1284.
41. Koeth RA, Wang Z, Levison BS, et al. Intestinal microbiota metabolism of L-carnitine, a nutrient in red meat, promotes atherosclerosis. *Nat Med.* 2013;19:576–585.
42. Wang Z, Klipfell E, Bennett BJ, et al. Gut flora metabolism of phosphatidylcholine promotes cardiovascular disease. *Nature.* 2011;472:57–63.
43. Yancey PH. Organic osmolytes as compatible, metabolic and counteracting cytoprotectants in high osmolarity and other stresses. *J Exp Biol.* 2005;208:2819–2830.
44. Tang WH, Wang Z, Levison BS, et al. Intestinal microbial metabolism of phosphatidylcholine and cardiovascular risk. *N Eng J Med.* 2013;368:1575–1584.

45. Ufnal M, Jazwiec R, Dadlez M, Drapala A, Sikora M, Skrzypecki J. Trimethylamine-N-oxide: a carnitine-derived metabolite that prolongs the hypertensive effect of angiotensin II in rats. *Can J Cardiol.* 2014;30:1700–1705.
46. Li X, Geng J, Zhao J, et al. Trimethylamine N-oxide exacerbates cardiac fibrosis via activating the NLRP3 inflammasome. *Front Physiol.* 2019;10:866.
47. Seldin MM, Meng Y, Qi H, et al. Trimethylamine N-oxide promotes vascular inflammation through signaling of mitogen-activated protein kinase and nuclear factor-kappaB. *J Am Heart Assoc.* 2016;5:e002767.
48. Zhu W, Gregory JC, Org E, et al. Gut microbial metabolite TMAO enhances platelet hyperreactivity and thrombosis risk. *Cell.* 2016;165:111–124.
49. Arpaia N, Campbell C, Fan X, et al. Metabolites produced by commensal bacteria promote peripheral regulatory T-cell generation. *Nature.* 2013;504:451–455.
50. Liu X, Almast J, Ekholm S. Lesions masquerading as acute stroke. *J Magn Reson Imaging.* 2013;37:15–34.
51. Gibbons DL, Spencer J. Mouse and human intestinal immunity: same ballpark, different players; different rules, same score. *Mucosal Immunol.* 2011;4:148–157.
52. Round JL, Mazmanian SK. The gut microbiota shapes intestinal immune responses during health and disease. *Nat Rev Immunol.* 2009;9:313–323.
53. O'Toole PW, Claesson MJ. Gut microbiota: changes throughout the lifespan from infancy to elderly. *Int Dairy J.* 2010;20:281–291.
54. Claesson MJ, Cusack S, O'Sullivan O, et al. Composition, variability, and temporal stability of the intestinal microbiota of the elderly. *Proc Natl Acad Sci U S A.* 2011;108(Suppl. 1):4586–4591.
55. Kelly JR, Kennedy PJ, Cryan JF, et al. Breaking down the barriers: the gut microbiome, intestinal permeability and stress-related psychiatric disorders. *Front Cell Neurosci.* 2015;9:392.
56. Crapser J, Ritzel R, Verma R, et al. Ischemic stroke induces gut permeability and enhances bacterial translocation leading to sepsis in aged mice. *Aging.* 2016;8:1049–1060.
57. Westendorp WF, Nederkoorn PJ, Vermeij JD, et al. Post-stroke infection: a systematic review and meta-analysis. *BMC Neurol.* 2011;11:110.
58. Harms H, Hoffmann S, Malzahn U, et al. Decision-making in the diagnosis and treatment of stroke-associated pneumonia. *J Neurol Neurosurg Psychiatry.* 2012;83:1225–1230.
59. Dethlefsen L, Huse S, Sogin ML, et al. The pervasive effects of an antibiotic on the human gut microbiota, as revealed by deep 16s rRNA sequencing. *PLoS Biol.* 2008;6:e280.
60. Adrie C, Parlato M, Salmi L, et al. Bacterial translocation and plasma cytokines during transcatheter and open-heart aortic valve implantation. *Shock.* 2015;43:62–67.
61. Nagatomo Y, Tang WH. Intersections between microbiome and heart failure: revisiting the gut hypothesis. *J Card Fail.* 2015;21:973–980.

CHAPTER 13

Bioactive lipids and gut microbiota interact to regulate health and disease

Undurti N. Das[a,b]
[a]UND Life Sciences, Battle Ground, WA, United States
[b]BioScience Research Centre and Department of Medicine, GVP Medical College and Hospital, Visakhapatnam, India

1 Introduction

Several studies suggest that gut microbiota have a major role in both health and disease (local and systemic). For instance, it has been suggested that intestinal dysbiosis has a role in insulin resistance and the development and progression of obesity, type 2 diabetes mellitus, hypertension, and cardiovascular diseases.[1–7] These results imply that manipulation of the gut microbiota may have the potential to prevent metabolic syndrome and its associated components. Though the exact mechanism(s) by which gut microbiota produce their beneficial actions is not clear, it has been suggested that it could be a result of their ability to secrete small-chain fatty acids (SCFAs) such as acetate, butyrate, and propionate.[8–10] These SCFAs produced by the beneficial gut microbiota seem to possess antiinflammatory actions; reduce splenic effector memory T cells and splenic T_H17 cells; decrease cardiac immune cell infiltration; abrogate regulatory T-cell-dependent angiotensin-II-induced hypertension and other actions; decrease gut permeability and thus prevent endotoxemia; act via G protein-coupled receptors or histone deacetylase to act on the brain via direct humoral actions, and indirect hormonal and immune pathways to regulate the gut-brain-immune axis; regulate cognitive function and emotions; and regulate and communicate via vagal action.[11–28] Some, if not all, the beneficial actions of gut microbiota can be reproduced by SCFAs, implying a significant role for these lipids.

Nevertheless, it needs to be noted that some of the gut microbiota may also have harmful actions.[5,29–31] It is reasonable to propose that under normal physiological conditions there will be a balance between useful and harmful gut microbiota, especially between the invading microorganisms and/or harmful gut microbiota and the host tissue/cells for their survival,

Gut Microbiota in Neurologic and Visceral Diseases
https://doi.org/10.1016/B978-0-12-821039-0.00011-3

growth, and dietary needs. These dietary needs include carbohydrates, lipids, protein, and several minerals, vitamins, and trace elements. This competition suggests that the capability to absorb and utilize these dietary nutrients determines the health of the gut microbiota and plays a role in the prevention of various diseases.

2 Essential fatty acids and their clinical significance

Essential fatty acids (EFAs), linoleic acid (LA, 18:2 n-6), and alpha-linolenic acid (ALA, 18:3 n-3) are needed for normal health. Since EFAs are widely distributed in our diet, EFA deficiency is extremely rare in humans. EFA deficiency is common in those who are on long-term parenteral nutrition therapy, provided the intravenous solutions do not contain adequate amounts of EFAs. EFAs are converted to their respective long-chain metabolites. Dietary LA is converted to gamma-linolenic acid (GLA, 18:3 n-6), dihomo-gamma-linolenic acid (DGLA, 20:3 n-6), and arachidonic acid (AA, 20:4 n-6) by the action of delta-6- and delta-5-desaturases as shown in Fig. 1. Similarly, ALA is converted to eicosapentaenoic acid (EPA, 20:5 n-3) and docosahexaenoic acid (DHA, 22:6 n-3) by the same set of desaturases as shown in Fig. 1. DGLA is the precursor of the 1 series of prostaglandins (PGs), whereas AA is the precursor of 2-series PGs, thromboxanes (TXs), and 4-series leukotrienes (LTs). EPA is the precursor of 3-series PGs and TXs and 5-series LTs. The 2- and 3-series PGs and TXs, and the 4- and 5-series LTs, are proinflammatory in nature, whereas the 3-series PGs and TXs and 5-series LTs are much less potent pro-inflammatory compounds compared to the 2-series PGs and TXs and 4-series LTs. AA is also the precursor of lipoxin A4 (LXA4), a potent antiinflammatory compound. Similarly, EPA is the precursor of resolvins that are also antiinflammatory in nature. EPA can be converted to DHA and under some specific conditions DHA can be retroconverted to EPA. DHA is the precursor of resolvins, protectins, and maresins that are also potent antiinflammatory compounds. Thus, both AA and EPA are the precursors of both pro- and antiinflammatory compounds. This implies that under normal physiological conditions, a delicate balance is maintained between pro- and antiinflammatory products formed from AA and EPA/DHA.[32–35] It is evident from several studies that EFAs and their metabolites have many actions under physiological and pathological conditions attesting to the fact that they could have important clinical implications.[36–56] Thus, the various actions of EFAs and their metabolites seem to be involved in the pathobiology of

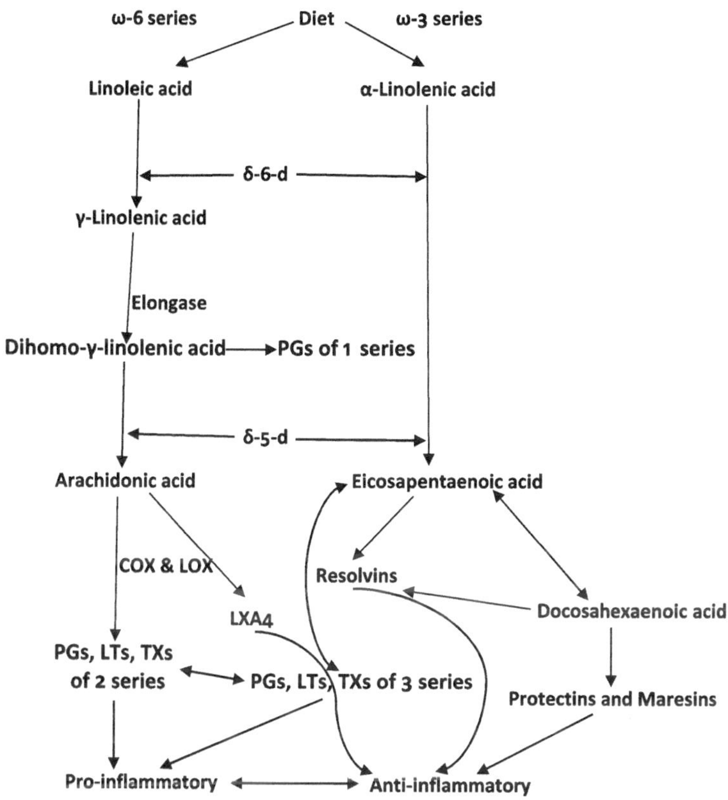

Fig. 1 Scheme showing the metabolism of essential fatty acids and their metabolites. COX = Cyclo-oxygenase enzyme; LOX = Lipoxygenase enzymes; δ-6-d and δ-5-d = Delat-6 and delta-5-desaturases; LXA4 = Lipoxin A4; PGs = Prostaglandins; TXs = Thromboxanes; LTs = Leukotrienes.

inflammatory, degenerative diseases including cancer. Such wide variety of action of EFAs and their metabolites is understandable since they have important regulatory action on inflammation, resolution of inflammation, and immunoregulatory functions.[57,58] It may be noted here that EFAs/GLA/DGLA/AA/EPA/DHA, LXA4, resolvins, protectins, maresins, PGs, TXs, and LTs are all termed as bioactive lipids (BALs) in the present discussion. Thus, for the sake of simplicity and for the present discussion, GLA/DGLA/AA/EPA/DHA, LXA4, resolvins, protectins, and maresins are referred to as beneficial BALs, whereas PGs of the 2-series and 3-series, TXs of the 2- and 3-series, and LTs of the 4- and 5-series are referred to as harmful or pathological BALs.

3 EFA deficiency suppresses autoimmune diseases: But how and why?

EFAs are needed for skin integrity and to prevent water loss from skin. EFA deficiency causes water loss, renal and hepatic dysfunctions, immunosuppression, splenomegaly, and growth retardation.[59–66] Despite the fact that EFA deficiency is harmful, it is surprising to note that EFA-deficient animals are resistant to the development of autoimmune diseases such as type 1 diabetes mellitus and glomerulonephritis.[67–74] These results are interesting suggesting that EFA deficiency is not always harmful. However, the exact mechanism of action of this beneficial action of EFA deficiency is not clear.

It was reported that EFA deficiency reduced the frequency of diabetes in DR-BB rats, an effect that was found to be associated with depletion of n-6 fatty acids, especially AA. This indicates that AA and/or its metabolite(s) may play a key role in the pathobiology of diabetes mellitus and associated inflammatory events. Subsequent studies revealed that EFA deficiency may affect macrophage migration into the glomerulus glomerulonephritis or to β cells in the context of type 1 diabetes mellitus. This functional defect in macrophage migration and chemotactic response in EFA-deficient animals was found to be due to the failure of generation of lipid mediators, probably LTs, by the inflamed glomeruli that elaborates this chemoattractant for monocytes. Thus, it is likely that EFA deficiency leads to diminished LTB4 synthesis, a pro-inflammatory molecule, but not of antiinflammatory metabolites from AA (such as LXA4) (see Figs. 2 and 3). It is possible that EFA deficiency reduces the ability of glomeruli to synthesize LTB4 by depleting them of macrophages[68–75] with no decrease in the synthesis and release of LXA4, a potent antiinflammatory molecule. This results in shifting the balance between various metabolites of AA and other PUFAs toward anti-inflammatory events, which results in prevention of autoimmune diseases. Since EFA deficiency is not compatible with life, it is not a preferred method of inducing remission or prevention of autoimmune diseases. Hence, it is preferable to design methods to enhance the formation of LXA4, which is known to inhibit the formation and action of pro-inflammatory LTB4. This could be achieved by supplementation of AA, as shown by our recent studies.[47,48,76–83] AA brings about its beneficial actions by augmenting the production of LXA4, a potent antiinflammatory molecule that has the ability to enhance generation of M2 macrophages, aid in wound healing, suppress the production of pro-inflammatory IL-6, TNF-α, IL-17, IL-1β, and IL-2, and increase the synthesis and action of antiinflammatory IL-4 and IL-10.[50,51,57,58] Such antiinflammatory and immunomodulatory action

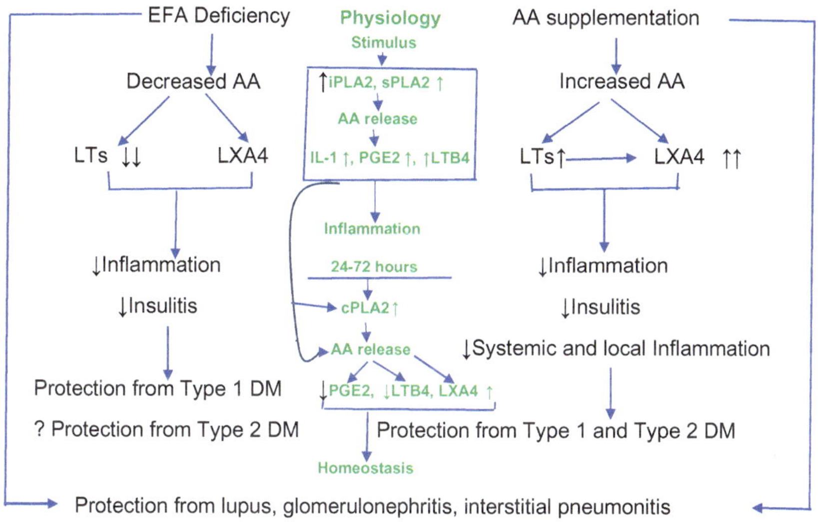

Fig. 2 Scheme showing potential relationship between EFA deficiency and AA supplementation and their effect on the development of type 1 and type 2 DM and other inflammatory conditions. Under physiological conditions, in response to a stimulus, activation of iPLA2 and sPLA2 occurs leading to release of AA from the cell membrane lipid pool. The released AA is converted to PGE2 and LTB4 (and other pro-inflammatory eicosanoids). Simultaneously, there is release of IL-1 and other pro-inflammatory cytokines from macrophages and other cells. Once PGE2 and LTB4 reach their peak, they stimulate cPLA2, leading to induction of a second wave of AA release from which LXA4 is formed and PGE2 and LTB4 formed from the first wave of AA release are diminished. IL-1 stimulates the formation of LXA4 and antiinflammatory IL-4 and IL-10. Furthermore, LTB4 and other LTs may be converted to LXA4 (see Fig. 3). Accumulation of mead acid due to EFA deficiency leads to formation and release of PGE2-like and LTC3 and LTD3 by the target tissues that have chemo-attractive properties with little or no change in LXA4 formation. In contrast to this, AA supplementation in disease states like glomerulonephritis, lupus, and type 1 and type 2 diabetes mellitus leads to the increased formation of LXA4 and little or no change in the formation and release of PGE2 and LTB4/LTD4. Thus the balance is tilted toward antiinflammatory events that lead to suppression of autoimmune diseases including type 1 diabetes mellitus, glomerulonephritis, and lupus. In a similar fashion, EPA and DHA and their antiinflammatory metabolites resolvins, protectins, and maresins may also protect against the development of autoimmune diseases, but may be less potent than LXA4. Thus, it appears that both EFA deficiency states and AA supplementation may have similar consequences, namely, suppression of inflammation. Since EFA deficiency is not compatible with normal life, adequate supplementation of AA is recommended.

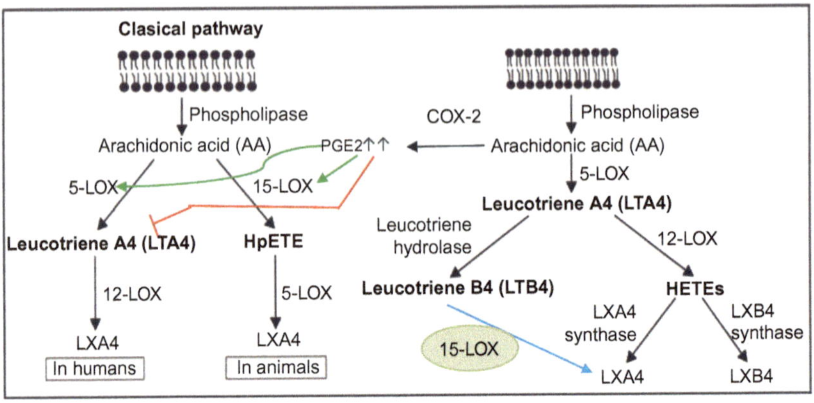

Fig. 3 Scheme showing the metabolism of AA and how pro-inflammatory LTA4 can be converted to antiinflammatory LXA4. Similar conversion of pro-inflammatory PGs, TXs, and LTs to antiinflammatory resolvins, protectins, and maresins may occur.

of AA, LXA4, and other BALs may have relevance to several other conditions in which inflammation and immune dysregulation play a major role, especially in diseases such as multiple sclerosis, rheumatoid arthritis, lupus, inflammatory bowel disease (IBD), hypertension, coronary heart disease, insulin resistance, nonalcoholic fatty liver disease (NAFLD), uveitis, and cancer.[50,57,58] Thus, it is anticipated that any intervention or strategy that acts on AA/LXA4 and other BAL pathways will play a significant role in all these diseases. In this context, it is noteworthy that gut microbiota is also believed to have a role in the aforementioned diseases and so it is worth studying the potential relationship between BALs (especially AA/LXA4) and gut microbiota.

4 Gut microbiota

Several studies have emphasized the potential role of gut microbiota and their metabolites in the pathobiology of several diseases including but not limited to obesity, diabetes mellitus, hypertension, metabolic syndrome, insulin resistance, NAFLD, lupus, IBD, and cancer.[1–31] It has been shown that gut microbiota produces SCFAs such as acetate, butyrate, and palmitate that are capable of modulating inflammatory events and immune response and thus bring about beneficial actions.[84–103] These SCFAs have several actions on the gut and gut-associated events including vagus fibers and through the vagus nerve on hypothalamic neurotransmitters. Some of the important actions of SCFAs include: (1) modulation of gene(s) expression concerned with inflammation and immune response; (2) activation of p42/p44 mitogen-activated protein kinase and inhibition of histone deacetylases; (3) promotion of sympathetic outflow via GPR41 and suppression of SNS activity by antagonizing GPR41; (4) exertion of antitumor action via receptor GPR43, which is frequently lost in colon cancer cells; (5) upregulation of p21 and decreasing the levels of cyclin D3 and cyclin-dependent kinases (CDKs) 1 and 2 and thus preventing colon cancer, especially by propionate; (6) induction of peroxisome proliferator-activated receptor α (PPARα) expression in a time-dependent and concentration-dependent manner on intestinal epithelial cells by propionate; (7) enhancement of β-cell function by potentiating glucose-stimulated insulin release and maintaining β-cell mass through inhibition of apoptosis; (8) augment the intestinal barrier by inhibiting NLRP3 inflammasome and autophagy; (9) blocking lipid raft redox signaling to produce $O_2^{\bullet-}$; (10) suppression of gluconeogenesis by binding to hepatic GPR43 to elicit CaMKKβ-dependent activation of AMPK through intracellular Ca^{2+} increase leading to suppression of gluconeogenesis; (11) increase melatonin receptor-1 and other neurotransmitters (such as serotonin) receptor expression and synthesis by intestinal epithelial cells; (12) stimulate IL-22 production that is essential for gut homeostasis and β-cell survival; and (13) enhance the synthesis and secretion of cholecystokinin (CCK), ghrelin, leptin, peptide tyrosine tyrosine (PYY), glucagon-like peptide-1 (GLP-1), and 5-hydroxytryptamine (5-HT; serotonin) by the enterochromaffin cells and colonic epithelial cells of the gut.[84–103] Thus, SCFAs have pleiotropic actions on several aspects of the gut that are ultimately responsible for their beneficial actions.

5 Potential interaction between EFAs, their metabolites and gut microbiota

In view of the several overlapping beneficial actions shown by EFAs and their metabolites and gut microbiota, it is reasonable to expect that there could occur an interaction between these two factors. For instance, it is known that dietary LA and ALA are utilized by the intestinal epithelial cells and T cells/NK cells to form GLA/DGLA/AA/EPA/DHA for their own benefit, especially to inactivate harmful microbes. This is supported by the fact that GLA/DGLA/AA/EPA/DHA have antibacterial, antifungal, and antiviral actions.[104–112] These results imply that the ability of intestinal epithelial cells to synthesize GLA/DGLA/AA/EPA/DHA could be one mechanism by which they can inhibit and recover from microbial infections.

Similar action is also shown by T cells, NK cells, and other immunocytes located in the gut. It is well documented that T cells/NK cells can release GLA/DGLA/AA/EPA/DHA (or similar lipids) to inactivate not only invading microbes but also tumor cells.[57,113] In addition, Intestinal epithelial cells and local T cells/NK cells could use GLA/DGLA/AA/EPA/DHA to form their beneficial metabolites such as PGE1, PGI2, NO, CO, H2S, LXA4, resolvins, protectins, and maresins to maintain gut epithelial barrier function, maintain gut integrity, protect the gut from various infections, and maintain adequate blood supply (since PGE1, PGI2, and LXA4 are vasodilators, prevent platelet aggregation, suppress inappropriate leukocytes and macrophage activation, and inhibit excess production of pro-inflammatory cytokines) and nutrition to the gut. This implies that dietary LA and ALA are appropriately converted to their respective metabolites by desaturases, COX, and LOX enzymes by gut epithelial cells.[114,115] Thus, the activities of desaturases and COX and LOX enzymes will determine the ability of the gut to interact with the gut microbiome, as well as the gut's inflammatory and immune response to various endogenous and exogenous insults.

In this contrast, it needs to be noted that the gut is also constantly exposed to various pathogenic microbes that activate NF-kappaB and induce pro- and antiinflammatory events.[116] Thus, manipulating the NF-κB pathway may result in either enhanced degree of inflammation or recovery form inflammation by triggering antiinflammatory events. Similarly, some bacteria augment COX-2 expression in macrophages infected with *Escherichia coli*, whereas commensal bacteria fail do so, suggesting that pathological gut bacteria enhance the expression of COX-2 to augment their production of PGE2, an immunosuppressor.[117] Similar events may also occur in the intestinal epithelial cells upon exposure to pathological bacteria/viruses/fungi. The excess PGE2 produced by the resident macrophages and other immunocytes and intestinal epithelial cells will ultimately result in inflammatory events. This data implies that the immunosuppression observed in those with intestinal tuberculosis could be due to excess production of PGE2 locally.[118] It is noteworthy that COX-2 induction by cytomegalovirus (CMV) leads to an increase in the production of PGE2 in the host cells that significantly increased CMV proliferation.[119] These results emphasize the importance of gut microbiota in the regulation of production of pro-inflammatory and antiinflammatory cytokines and BALs and maintenance of balance between them to ensure gut homeostasis.[120,121] One mechanism by which gut microbiota (useful bacteria) suppress inappropriate inflammation is by producing SCFAs that have antiinflammatory actions. However, it is equally possible that gut microbiota act on the intestinal epithelial cells to induce the production of antiinflammatory LXA4, resolvins, protectins, and maresins from GLA, DGLA, AA, EPA, and DHA.[122–128]

Based on this evidence, it is proposed that beneficial gut microbes enhance the production of GLA/DGL/AA/EPA/DHA that will be utilized for the synthesis and release of LXA4, resolvins, protectins, and maresins not only to suppress inappropriate inflammation but also to enhance wound healing and prevent colon cancer. In contrast to this, harmful gut microbes utilize GLA/DGLA/AA/EPA/DHA to form pro-inflammatory PGE2 and LTs and to suppress the production of LXA4, resolvins, protectins, and maresins. Furthermore, gut microbiota influence the production of cytokine-useful bacteria to enhance the production of antiinflammatory cytokines, whereas harmful bacteria augment pro-inflammatory cytokines IL-6 and TNF-α. Similar actions are also shown by SCFAs produced by gut microbiota. There is evidence to suggest that SCFAs suppress the production of PGE2 and LTB4,[129,130] thus tilting the balance toward antiinflammatory status by indirectly promoting LXA4 (including resolvins, protectins, and maresins) formation since under physiological conditions a balance is maintained between pro- and antiinflammatory BALs (bioactive lipids).

Based on the preceding discussion, it is evident that BALs (especially AA/EPA/DHA) can influence the gut microbiota and vice versa, and SCFAs produced by gut microbiota have antiinflammatory actions, at least, in part, by influencing the production of PGE2/LTB4. Gut microbiota are known to play a significant role in many diseases and similarly BALs also have a major role in all these diseases. Hence, it is likely that a close interaction exists between gut microbiota and BALs that eventually influences health and disease.

6 BALs influence gut microbiota

In this context, it is noteworthy that BALs, especially AA, EPA, and DHA, augment the survival and proliferation of useful gut microbiota[23,131–142] that can inhibit the production of pro-inflammatory cytokines and thus bring about their beneficial actions. It has not been studied, but it is likely that metabolites of various fatty acids such as PGE1/PGE2/PGI2/LTs/LXA4/resolvins/protectins/maresin may also have significant action on gut microbiota and an increase in the elaboration of SCFAs by the gut microbiota. It is not known whether gut microbiota can metabolize PUFAs to SCFAs. However, this is a distinct possibility[142] that needs to be investigated in

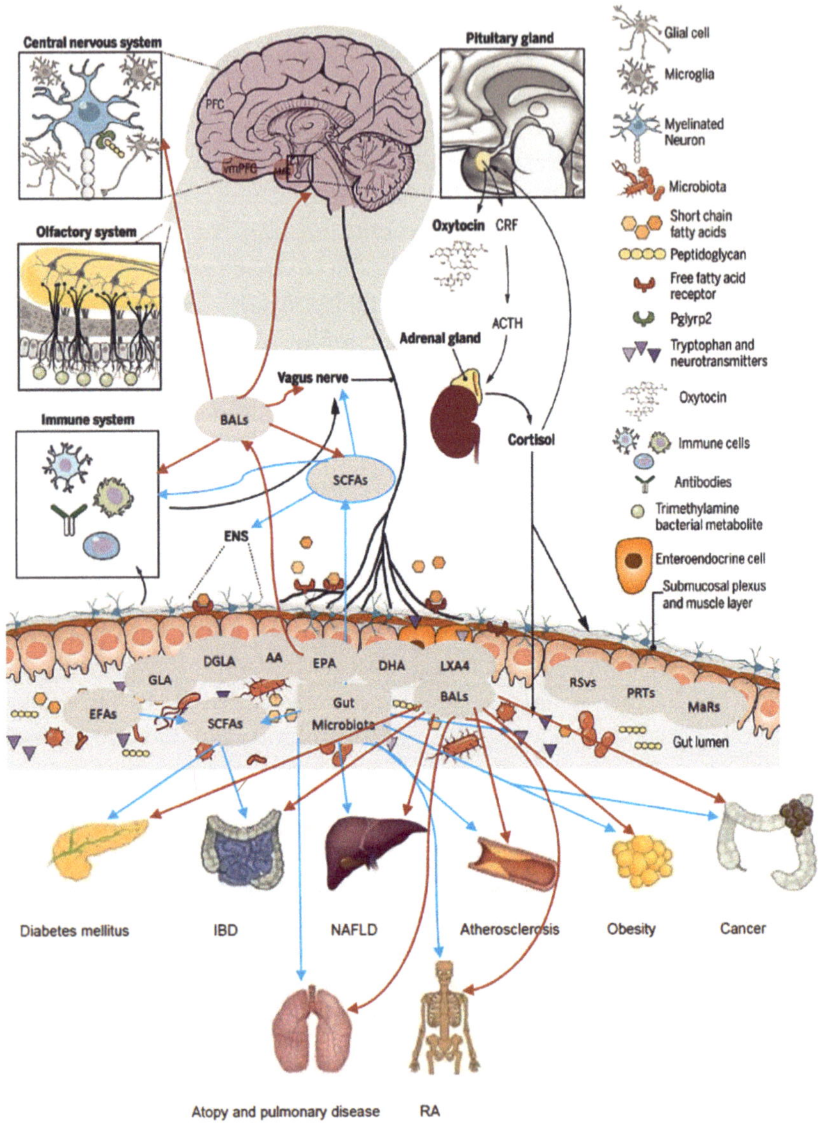

Fig. 4 Scheme showing potential relationship among gut microbiota, their SCFAs, BALs, and various cells/tissues in the body and diseases. SCFAs = Short-chain fatty acids, BALs = Bioactive lipids, EFAs = Essential fatty acids, ENS = Enteric nervous system, GLA = Gamma-linolenic acid, DGLA = Dihomo-GLA, AA = Arachidonic acid, EPA = Eicosapentaenoic acid, DHA = Docosahexaenoic acid, LXA4 = Lipoxin A4, RSvs = Resolvins, PRTs = Protectins, MaRs = Maresins, IBD = Inflammatory bowel disease, RA = Rheumatoid arthritis, NAFLD = Nonalcoholic fatty liver disease.

depth. What is known is that SCFAs can alter the metabolism of BALs,[143] suggesting that PUFAs influence SCFAs formation in a similar fashion. More importantly, it is possible that dietary EFAs may be converted to SCFAs by gut microbiota.

It is noteworthy that human breast milk is a rich source of AA/EPA/DHA/LA/ALA and it is well documented that breast milk is not only the source of beneficial microbiota but is also essential for their survival, proliferation, and function.[144–147] Thus, there seems to occur a close interaction(s) among human breast milk, BALs, and gut microbiota and their ability to produce SCFAs.

7 Conclusion

It is evident from the preceding discussion that both gut microbiota and BALs have similar beneficial actions. In addition, several lipids such as AA/EPA/DHA and their metabolites seem to possess the ability to influence the survival, proliferation, and function(s) of gut microbiota. The ability of gut microbiota to alter the metabolism of EFAs,[143] and the potential of reciprocal influence of EFAs, GLA/DGLA/AA/EPA/DHA, and their metabolites on the synthesis of SCFAs by gut microbiota, is of immense clinical interest. Gut microbiota and their metabolites, especially SCFAs and BALs, have similar beneficial actions on several diseases (see Fig. 4) implying that gut microbiota, SCFAs, and BALs interact and modulate the pathobiology of various diseases. This suggests that a combination of BALs and gut microbiota and their SCFAs could have potential clinical application in a variety of diseases. Future studies examining this potential application are needed in order to develop newer therapeutic approaches to several human diseases.

References

1. Vinjé S, Stroes E, Nieuwdorp M, Hazen SL. The gut microbiome as novel cardio-metabolic target: the time has come! *Eur Heart J.* 2014;35:883–887.
2. Van Olden CC, Van de Laar AW, Meijnikman AS, et al. A systems biology approach to understand gut microbiota and host metabolism in morbid obesity: design of the BARIA Longitudinal Cohort Study [published online ahead of print, 2020 Jul 8]. *J Intern Med.* 2020. https://doi.org/10.1111/joim.13157.
3. Lau E, Neves JS, Ferreira-Magalhães M, Carvalho D, Freitas P. Probiotic ingestion, obesity, and metabolic-related disorders: results from NHANES, 1999-2014. *Nutrients.* 2019;11:1482.
4. Palmu J, Salosensaari A, Havulinna AS, et al. Association between the gut microbiota and blood pressure in a population cohort of 6953 individuals [published online ahead of print, 2020 Jul 21]. *J Am Heart Assoc.* 2020;e016641. https://doi.org/10.1161/JAHA.120.016641.

5. O'Connor KM, Lucking EF, Cryan JF, O'Halloran KD. Bugs, breathing and blood pressure: microbiota-gut-brain axis signalling in cardiorespiratory control in health and disease [published online ahead of print, 2020 Jul 11]. *J Physiol.* 2020. https://doi.org/10.1113/JP280279.
6. Sun T, Zhang Y, Yin J, et al. Association of gut microbiota-dependent metabolite trimethylamine n-oxide with first ischemic stroke [published online ahead of print, 2020 Jul 9]. *J Atheroscler Thromb.* 2020. https://doi.org/10.5551/jat.55962.
7. Yan F, Li N, Shi J, et al. Lactobacillus acidophilus alleviates type 2 diabetes by regulating hepatic glucose, lipid metabolism and gut microbiota in mice. *Food Funct.* 2019;10(9):5804–5815. https://doi.org/10.1039/c9fo01062a.
8. Wang L, Zhu Q, Lu A, et al. Sodium butyrate suppresses angiotensin II-induced hypertension by inhibition of renal (pro)renin receptor and intrarenal renin-angiotensin system. *J Hypertens.* 2017;35(9):1899–1908. https://doi.org/10.1097/HJH.0000000000001378.
9. Bartolomaeus H, Balogh A, Yakoub M, et al. Short-chain fatty acid propionate protects from hypertensive cardiovascular damage. *Circulation.* 2019;139(11):1407–1421. https://doi.org/10.1161/CIRCULATIONAHA.118.036652.
10. Dalile B, Van Oudenhove L, Vervliet B, et al. The role of short-chain fatty acids in microbiota–gut–brain communication. *Nat Rev Gastroenterol Hepatol.* 2019;16:461–478. https://doi.org/10.1038/s41575-019-0157-3.
11. Cerquetti MC, Hovsepian E, Sarnacki SH, Goren NB. *Salmonella enterica* serovar enteritidis dam mutant induces low NOS-2 and COX-2 expression in macrophages via attenuation of MAPK and NF-kappaB pathways. *Microbes Infect.* 2008;10:1431–1439.
12. Ahmed S, Busetti A, Fotiadou P, et al. *In vitro* characterization of gut microbiota-derived bacterial strains with neuroprotective properties. *Front Cell Neurosci.* 2019;13:402.
13. Guo Y, Xie JP, Deng K, et al. Prophylactic effects of *Bifidobacterium adolescentis* on anxiety and depression-like phenotypes after chronic stress: a role of the gut microbiota-inflammation axis. *Front Behav Neurosci.* 2019;13:126.
14. Lim SM, Kim DH. Bifidobacterium adolescentis IM38 ameliorates high-fat diet-induced colitis in mice by inhibiting NF-κB activation and lipopolysaccharide production by gut microbiota. *Nutr Res.* 2017;41:86–96.
15. Lim SM, Jeong JJ, Woo KH, Han MJ, Kim DH. *Lactobacillus sakei* OK67 ameliorates high-fat diet-induced blood glucose intolerance and obesity in mice by inhibiting gut microbiota lipopolysaccharide production and inducing colon tight junction protein expression. *Nutr Res.* 2016;36:337–348.
16. Hansen CHF, Larsen CS, Petersson HO, et al. Targeting gut microbiota and barrier function with prebiotics to alleviate autoimmune manifestations in NOD mice. *Diabetologia.* 2019;62:1689–1700.
17. Albuquerque RCMF, Brandão ABP, De Abreu ICME, et al. *Saccharomyces boulardii* Tht 500101 changes gut microbiota and ameliorates hyperglycaemia, dyslipidaemia, and liver inflammation in streptozotocin-diabetic mice. *Benefic Microbes.* 2019;10:901–912.
18. Everard A, Matamoros S, Geurts L, Delzenne NM, Cani PD. *Saccharomyces boulardii* administration changes gut microbiota and reduces hepatic steatosis, low-grade inflammation, and fat mass in obese and type 2 diabetic db/db mice. *mBio.* 2014;5:e01011–e01014.
19. Zhang WQ, Zhao TT, Gui DK, et al. Sodium butyrate improves liver glycogen metabolism in type 2 diabetes mellitus. *J Agric Food Chem.* 2019;67:7694–7705.
20. Noce A, Marrone G, Di Daniele F, et al. Impact of gut microbiota composition on onset and progression of chronic non-communicable diseases. *Nutrients.* 2019;11:1073.
21. Wang F, Zhu H, Hu M, et al. Perilla oil supplementation improves hypertriglyceridemia and gut dysbiosis in diabetic KKAy mice. *Mol Nutr Food Res.* 2018;62:e1800299.

22. Tian Y, Wang H, Yuan F, et al. Perilla oil has similar protective effects of fish oil on high-fat diet-induced nonalcoholic fatty liver disease and gut dysbiosis. *Biomed Res Int.* 2016;2016:9462571.
23. Wang J, He Y, Yu D, Jin L, Gong X, Zhang B. Perilla oil regulates intestinal microbiota and alleviates insulin resistance through the PI3K/AKT signaling pathway in type-2 diabetic KKAy mice. *Food Chem Toxicol.* 2020;135:110965.
24. Xu YH, Gao CL, Guo HL, et al. Sodium butyrate supplementation ameliorates diabetic inflammation in db/db mice. *J Endocrinol.* 2018;238:231–244.
25. Sata Y, Marques FZ, Kaye DM. The emerging role of gut dysbiosis in cardio-metabolic risk factors for heart failure. *Curr Hypertens Rep.* 2020;22:38.
26. Tang WH, Kitai T, Hazen SL. Gut microbiota in cardiovascular health and disease. *Circ Res.* 2017;120:1183–1196.
27. Zeng C, Tan H. Gut microbiota and heart, vascular injury. *Adv Exp Med Biol.* 2020;1238:107–141.
28. Bonaz B, Bazin T, Pellissier S. The vagus nerve at the interface of the microbiota-gut-brain axis. *Front Neurosci.* 2018;12:49.
29. Adachi K, Sugiyama T, Yamaguchi Y, et al. Gut microbiota disorders cause type 2 diabetes mellitus and homeostatic disturbances in gut-related metabolism in Japanese subjects. *J Clin Biochem Nutr.* 2019;64:231–238.
30. Zhu Y, Shui X, Liang Z, et al. Gut microbiota metabolites as integral mediators in cardiovascular diseases. *Int J Mol Med.* 2020;46:936–948.
31. Li J, Yang X, Zhou X, Cai J. The role and mechanism of intestinal flora in blood pressure regulation and hypertension development [published online ahead of print, 2020 May 14]. *Antioxid Redox Signal.* 2020. https://doi.org/10.1089/ars.2020.8104.
32. Das UN. Essential fatty acids: biochemistry, physiology and pathology. *Biotechnol J.* 2006;1:420–439.
33. Das UN. Biological significance of essential fatty acids. *J Assoc Physicians India.* 2006;54:309–319.
34. Das UN. Essential fatty acids—a review. *Curr Pharm Biotechnol.* 2006;7:467–482.
35. Dobryniewski J, Szajda SD, Waszkiewicz N, Zwierz K. Biology of essential fatty acids (EFA). *Przegl Lek.* 2007;64:91–99.
36. Das UN. Essential fatty acids and their metabolites could function as endogenous HMG-CoA reductase and ACE enzyme inhibitors, anti-arrhythmic, anti-hypertensive, anti-atherosclerotic, anti-inflammatory, cytoprotective, and cardioprotective molecules. *Lipids Health Dis.* 2008;7:37.
37. Das UN. Do polyunsaturated fatty acids behave like an endogenous "polypill"? *Med Hypotheses.* 2008;70:430–434.
38. Das UN. Long-chain polyunsaturated fatty acids interact with nitric oxide, superoxide anion, and transforming growth factor-beta to prevent human essential hypertension. *Eur J Clin Nutr.* 2004;58:195–203.
39. Das UN. Can perinatal supplementation of long-chain polyunsaturated fatty acids prevents schizophrenia in adult life? *Med Sci Monit.* 2004;10:HY33–HY37.
40. Das UN. Perinatal supplementation of long-chain polyunsaturated fatty acids, immune response and adult diseases. *Med Sci Monit.* 2004;10:HY19–HY25.
41. Das UN. Beneficial effect(s) of n-3 fatty acids in cardiovascular diseases: but, why and how? *Prostaglandins Leukot Essent Fat Acids.* 2000;63:351–362.
42. Das UN. Can perinatal supplementation of long-chain polyunsaturated fatty acids prevent atopy, bronchial asthma and other inflammatory conditions? *Med Sci Monit.* 2006;12:RA99–RA111.
43. Das UN. Folic acid and polyunsaturated fatty acids improve cognitive function and prevent depression, dementia, and Alzheimer's disease—but how and why? *Prostaglandins Leukot Essent Fat Acids.* 2008;78:11–19.

44. Hari AD, Vegi NG, Das UN. Arachidonic and eicosapentaenoic acids induce oxidative stress to suppress proliferation of human glioma cells. *Arch Med Sci*. 2020;16:974–983.
45. Bathina S, Gundala NKV, Rhenghachar P, et al. Resolvin D1 ameliorates nicotinamide-streptozotocin-induced type 2 diabetes mellitus by its anti-inflammatory action and modulating PI3K/Akt/mTOR pathway in the brain. *Arch Med Res*. 2020. https://doi.org/10.1016/j.arcmed.2020.05.002. S0188-4409(19)30572-7.
46. Gundala NKV, Naidu VGM, Das UN. Arachidonic acid and lipoxinA4 attenuate streptozotocin-induced cytotoxicity to RIN5 F cells in vitro and type 1 and type 2 diabetes mellitus in vivo. *Nutrition*. 2017;35:61–80.
47. Gundala NKV, Naidu VGM, Das UN. Amelioration of streptozotocin-induced type 2 diabetes mellitus in Wistar rats by arachidonic acid. *Biochem Biophys Res Commun*. 2018;496:105–113.
48. Das UN. Arachidonic acid and lipoxin A4 as possible endogenous anti-diabetic molecules. *Prostaglandins Leukot Essent Fat Acids*. 2013;88:201–210.
49. Yin Y, Chen F, Wang W, Wang H, Zhang X. Resolvin D1 inhibits inflammatory response in STZ-induced diabetic retinopathy rats: possible involvement of NLRP3 inflammasome and NF-κB signaling pathway. *Mol Vis*. 2017;23:242–250.
50. Das UN. Bioactive lipids in age-related disorders. *Adv Exp Med Biol*. 2020;1260:33–83.
51. Das UN. Molecular pathobiology of scleritis and its therapeutic implications. *Int J Ophthalmol*. 2020;13:163–175.
52. Das UN. Cytokines, angiogenic, and antiangiogenic factors and bioactive lipids in preeclampsia. *Nutrition*. 2015;31:1083–1095.
53. Das UN. Bioactive lipids as modulators of immune check point inhibitors. *Med Hypotheses*. 2020;135:109473.
54. Das UN. Bioactive lipids in intervertebral disc degeneration and its therapeutic implications. *Biosci Rep*. 2019;39. https://doi.org/10.1042/BSR20192117, BSR20192117.
55. Das UN. Bioactive lipids as mediators of the beneficial actions of statins. *J Cardiovasc Pharmacol*. 2019;74:4–8.
56. Das UN. Beneficial role of bioactive lipids in the pathobiology, prevention, and management of HBV, HCV and alcoholic hepatitis, NAFLD, and liver cirrhosis: a review. *J Adv Res*. 2018;17:17–29.
57. Das UN. *Molecular Biochemical Aspects of Cancer*. New York: Humana Press; 2020.
58. Das UN. *Molecular Basis of Health and Disease*. New York: Springer; 2011.
59. Soares AF, Santiago RC, Aléssio ML, Descomps B, de Castro-Chaves C. Biochemical, functional, and histochemical effects of essential fatty acid deficiency in rat kidney. *Lipids*. 2005;40:1125–1133.
60. Korotkova M, Ohlsson C, Gabrielsson B, Hanson LA, Strandvik B. Perinatal essential fatty acid deficiency influences body weight and bone parameters in adult male rats. *Biochim Biophys Acta*. 2005;1686:248–254.
61. Phinney SD, Clarke SD, Odin RS, Moldawer LL, Blackburn GL, Bistrian BR. Thermogenesis secondary to transdermal water loss causes growth retardation in essential fatty acid-deficient rats. *Metabolism*. 1993;42:1022–1026.
62. Thomas IK, Erickson KL. Dietary fatty acid modulation of murine T-cell responses in vivo. *J Nutr*. 1985;115:1528–1534.
63. Erickson KL, Adams DA, McNeill CJ. Dietary lipid modulation of immune responsiveness. *Lipids*. 1983;18:468–474.
64. McKenna MC, Campagnoni AT. Effect of pre- and postnatal essential fatty acid deficiency on brain development and myelination. *J Nutr*. 1979;109:1195–1204.
65. Parnham MJ, Vincent JE, Zijlstra FJ, Bonta IL. The use of essential fatty acid deficient rats to study pathophysiological roles of prostaglandins. Comparison of prostaglandin production with some parameters of deficiency. *Lipids*. 1979;14:407–412.

66. Henry CJ, Ghusain-Choueiri A, Payne PR. Protein utilization, growth and survival in essential-fatty-acid-deficient rats. *Br J Nutr.* 1996;75:237–248.
67. Dvorak B, Stepankova R. Effects of dietary essential fatty acid deficiency on the development of the rat thymus and immune system. *Prostaglandins Leukot Essent Fat Acids.* 1992;46:183–190.
68. Lefkowith J, Schreiner G, Cormier J, et al. Prevention of diabetes in the BB rat by essential fatty acid deficiency. Relationship between physiological and biochemical changes. *J Exp Med.* 1990;171:729–743.
69. Wright Jr JR, Lefkowith JB, Schreiner G, Lacy PE. Essential fatty acid deficiency prevents multiple low-dose streptozotocin-induced diabetes in CD-1 mice. *Proc Natl Acad Sci U S A.* 1988;85:6137–6141.
70. Fraser RB, Rowden G, Colp P, Wright Jr JR. Immunophenotyping of insulitis in control and essential fatty acid deficient mice treated with multiple low-dose streptozotocin. *Diabetologia.* 1997;40:1263–1268.
71. Wright Jr JR, Haliburton B, Russell H, Henry M, Fraser R, Cook HW. The anti-diabetogenic effect of essential fatty acid deficiency in multiple low-dose streptozotocin-treated mice persists if essential fatty acid repletion occurs outside of a brief window of susceptibility. *Diabetologia.* 1991;34:709–714.
72. Hurd ER, Gilliam JN. Beneficial effect of an essential fatty acid deficient diet in NZB/NZW F1 mice. *J Invest Dermatol.* 1981;77:381–384.
73. Lefkowith JB, Schreiner G. Essential fatty acid deficiency depletes rat glomeruli of resident macrophages and inhibits angiotensin II-induced eicosanoid synthesis. *J Clin Invest.* 1987;80:947–956.
74. Schreiner GF, Rovin B, Lefkowith JB. The antiinflammatory effects of essential fatty acid deficiency in experimental glomerulonephritis. The modulation of macrophage migration and eicosanoid metabolism. *J Immunol.* 1989;143:3192–3199.
75. Rovin BH, Lefkowith JB, Schreiner GF. Mechanisms underlying the anti-inflammatory effects of essential fatty acid deficiency in experimental glomerulonephritis. Inhibited release of a monocyte chemoattractant by glomeruli. *J Immunol.* 1990;145:1238–1245.
76. Krishna Mohan I, Das UN. Prevention of chemically-induced diabetes mellitus in experimental animals by polyunsaturated fatty acids. *Nutrition.* 2001;17:126–151.
77. Suresh Y, Das UN. Protective action of arachidonic acid against alloxan-induced cytotoxicity and diabetes mellitus. *Prostaglandins Leukot Essent Fat Acids.* 2001;64:37–52.
78. Suresh Y, Das UN. Long-chain polyunsaturated fatty acids and chemically-induced diabetes mellitus: effect of ω-3 fatty acids. *Nutrition.* 2003;19:213–228.
79. Suresh Y, Das UN. Differential effect of saturated, monounsaturated, and polyunsaturated fatty acids on alloxan-induced diabetes mellitus. *Prostaglandins Leukot Essent Fat Acids.* 2006;74:199–213.
80. Suresh Y, Das UN. Long-chain polyunsaturated fatty acids and chemically induced diabetes mellitus: effect of omega-6 fatty acids. *Nutrition.* 2003;19:93–114.
81. Naveen KVG, Naidu VGM, Das UN. Arachidonic acid and lipoxin A4 attenuate alloxan-induced cytotoxicity to RIN5F cells in vitro and type 1 diabetes mellitus in vivo. *Biofactors.* 2017;43:251–271.
82. Naveen KVG, Naidu VGM, Das UN. Arachidonic acid and lipoxin A4 attenuate streptozotocin-induced cytotoxicity to RIN5F cells in vitro and type 1 and type 2 diabetes mellitus in vivo. *Nutrition.* 2017;35:61–80.
83. Naveen GV, Das UN. Arachidonic acid rich ARASCO oil has anti-inflammatory and anti-diabetic actions against high fat diet-induced type 2 diabetes mellitus in Wistar rats. *Nutrition.* 2019;66:203–218.
84. Jayasimhan A, Mariño E. Dietary SCFAs, IL-22, and GFAP: the three musketeers in the gut-neuro-immune network in type 1 diabetes. *Front Immunol.* 2019;10:2429.

85. van de Wouw M, Boehme M, Lyte JM, et al. Short-chain fatty acids: microbial metabolites that alleviate stress-induced brain-gut axis alterations. *J Physiol.* 2018;596:4923–4944.
86. Huang J, Pearson JA, Peng J, et al. Gut microbial metabolites alter IgA immunity in type 1 diabetes. *JCI Insight.* 2020;5:e135718.
87. Rizzetto L, Fava F, Tuohy KM, Selmi C. Connecting the immune system, systemic chronic inflammation and the gut microbiome: the role of sex. *J Autoimmun.* 2018;92:12–34.
88. Yu K, Zhang Y, Chen H, Zhu W. Hepatic metabolomic and transcriptomic responses induced by cecal infusion of sodium propionate in a fistula pig model. *J Agric Food Chem.* 2019;67:13073–13081.
89. Zhang Y, Yu K, Chen H, Su Y, Zhu W. Caecal infusion of the short-chain fatty acid propionate affects the microbiota and expression of inflammatory cytokines in the colon in a fistula pig model. *J Microbial Biotechnol.* 2018;11:859–868.
90. Yoshida H, Ishii M, Akagawa M. Propionate suppresses hepatic gluconeogenesis via GPR43/AMPK signaling pathway. *Arch Biochem Biophys.* 2019;672:108057.
91. Wang B, Zhang L, Zhu SW, Zhang JD, Duan LP. Short chain fatty acids contribute to gut microbiota-induced promotion of colonic melatonin receptor expression. *J Biol Regul Homeost Agents.* 2019;33:763–771.
92. Bolognini D, Barki N, Butcher AJ, et al. Chemogenetics defines receptor-mediated functions of short chain free fatty acids. *Nat Chem Biol.* 2019;15:489–498.
93. Akiba Y, Inoue T, Kaji I, et al. Short-chain fatty acid sensing in rat duodenum. *J Physiol.* 2015;593:585–599.
94. Feng Y, Wang Y, Wang P, Huang Y, Wang F. Short-chain fatty acids manifest stimulative and protective effects on intestinal barrier function through the inhibition of NLRP3 inflammasome and autophagy. *Cell Physiol Biochem.* 2018;49:190–205.
95. Yuan X, Wang L, Bhat OM, Lohner H, Li PL. Differential effects of short chain fatty acids on endothelial Nlrp3 inflammasome activation and neointima formation: antioxidant action of butyrate. *Redox Biol.* 2018;16:21–31.
96. Corrêa RO, Vieira A, Sernaglia EM, et al. Bacterial short-chain fatty acid metabolites modulate the inflammatory response against infectious bacteria. *Cell Microbiol.* 2017;19. https://doi.org/10.1111/cmi.12720.
97. Pingitore A, Chambers ES, Hill T, et al. The diet-derived short chain fatty acid propionate improves beta-cell function in humans and stimulates insulin secretion from human islets in vitro. *Diabetes Obes Metab.* 2017;19:257–265.
98. Higashimura Y, Naito Y, Takagi T, Uchiyama K, Mizushima K, Yoshikawa T. Propionate promotes fatty acid oxidation through the up-regulation of peroxisome proliferator-activated receptor α in intestinal epithelial cells. *J Nutr Sci Vitaminol (Tokyo).* 2015;61:511–515.
99. Montalban-Arques A, De Schryver P, Bossier P, et al. Selective manipulation of the gut microbiota improves immune status in vertebrates. *Front Immunol.* 2015;6:512.
100. Kimura I, Inoue D, Maeda T, et al. Short-chain fatty acids and ketones directly regulate sympathetic nervous system via G protein-coupled receptor 41 (GPR41). *Proc Natl Acad Sci U S A.* 2011;108:8030–8035.
101. Tang Y, Chen Y, Jiang H, Robbins GT, Nie D. G-protein-coupled receptor for short-chain fatty acids suppresses colon cancer. *Int J Cancer.* 2011;128:847–856.
102. Alvaro A, Solà R, Rosales R, et al. Gene expression analysis of a human enterocyte cell line reveals downregulation of cholesterol biosynthesis in response to short-chain fatty acids. *IUBMB Life.* 2008;60:757–764.
103. Jansen MS, Nagel SC, Miranda PJ, Lobenhofer EK, Afshari CA, McDonnell DP. Short-chain fatty acids enhance nuclear receptor activity through mitogen-activated protein kinase activation and histone deacetylase inhibition. *Proc Natl Acad Sci U S A.* 2004;101:7199–7204.

104. Heczko PB, Lutticken R, Hryniewicz W, Neugebauer M, Pulverer G. Susceptibility of *Staphylococcus aureus* and group A, B, C, and G streptococci to free fatty acids. *J Clin Microbiol*. 1979;9:333–335.
105. Wyss O, Ludwig BJ, Joiner RR. The fungistatic and fungicidal action of fatty acids and related compounds. *Arch Biochem*. 1945;7:415–424.
106. Kohn A, Gitelman J, Inbar M. Unsaturated free fatty acids inactivate animal enveloped viruses. *Arch Virol*. 1980;66:301–306.
107. Das UN. Anti-biotic-like action of essential fatty acids. *Can Med Assoc J*. 1985;132:1350.
108. Das UN. Can essential fatty acid deficiency predispose to AIDS? *Can Med Assoc J*. 1985;132:900–902.
109. Das UN. Arachidonic acid and other unsaturated fatty acids and some of their metabolites function as endogenous antimicrobial molecules: a review. *J Adv Res*. 2018;11:57–66.
110. Stock CC, Francis Jr T. The inactivation of the virus of epidemic influenza by soaps. *J Exp Med*. 1940;71:661–681.
111. Sands J, Auperin D, Snipes W. Extreme sensitivity of enveloped viruses, including herpes simplex, to long chain unsaturated monoglycerides and alcohols. *Antimicrob Agents Chemother*. 1979;15:67–73.
112. Gimenez AP, Wu YZ, Paya M, Delclaux C, Touqui L, Goossens PL. High bactericidal efficiency of type iia phospholipase A2 against *Bacillus anthracis* and inhibition of its secretion by the lethal toxin. *J Immunol*. 2004;173:521–530.
113. Baranov V, Nagaeva O, Hammarstrom S, et al. Lipids are a constitutive component of cytolytic granules. *Histochem Cell Biol*. 2000;14:167–171.
114. Chen Q, Nilsson A. Desaturation and chain elongation of n - 3 and n - 6 polyunsaturated fatty acids in the human CaCo-2 cell line. *Biochim Biophys Acta*. 1993;1166:193–201.
115. Huang YS, Liu JW, Koba K, Anderson SN. N-3 and n-6 fatty acid metabolism in undifferentiated and differentiated human intestine cell line (Caco-2). *Mol Cell Biochem*. 1995;151:121–130.
116. Lu J, Dong B, Chen A, et al. *Escherichia coli* promotes DSS-induced murine colitis recovery through activation of the TLR4/NF-κB signaling pathway. *Mol Med Rep*. 2019;19:2021–2028.
117. Raisch J, Rolhion N, Dubois A, Darfeuille-Michaud A. Bringer M-A (2015) intracellular colon cancer-associated *Escherichia coli* promote protumoral activities of human macrophages by inducing sustained COX-2 expression. *Lab Invest*. 2015;95:296–307.
118. Pandey RK, Dahiya Y, Sodhi A. *Mycobacterium indicus pranii* downregulates MMP-9 and iNOS through COX-2 dependent and TNF-α independent pathway in mouse peritoneal macrophages in vitro. *Microbes Infect*. 2012;14:348–356.
119. Hooks JJ, Chin SM, Srinivasan K, et al. Human cytomegalovirus induced cyclooxygenase 2 in human retinal pigment epithelial cells augments viral replication through a prostaglandin pathway. *Microbes Infect*. 2006;8:2236–2244.
120. Lim S-M, Jang HM, Jang S-E, Han MJ, Kim D-H. *Lactobacillus fermentum* IM12 attenuates inflammation in mice by inhibiting NF-κB-STAT3 signalling pathway. *Benefic Microbes*. 2017;8:407–419.
121. Liu Y-W, Ong W-K, Su Y-W, Hsu C-C, Cheng T-H, Tsai Y-C. Anti-inflammatory effects of *Lactobacillus brevis* K65 on RAW 264.7 cells and in mice with dextran sulphate sodium-induced ulcerative colitis. *Benefic Microbes*. 2016;7:387–396.
122. Kure I, Nishiumi S, Nishitani Y, et al. Lipoxin A(4) reduces lipopolysaccharide-induced inflammation in macrophages and intestinal epithelial cells through inhibition of nuclear factor-kappaB activation. *J Pharmacol Exp Ther*. 2010;332:541–548.
123. Quiros M, Feier D, Birkl D, et al. Resolvin E1 is a pro-repair molecule that promotes intestinal epithelial wound healing. *Proc Natl Acad Sci U S A*. 2020;117:9477–9482.

124. Zhong X, Lee HN, Surh YJ. RvD1 inhibits TNFα-induced c-Myc expression in normal intestinal epithelial cells and destabilizes hyper-expressed c-Myc in colon cancer cells. *Biochem Biophys Res Commun*. 2018;496:316–323.
125. Gobbetti T, Dalli J, Colas RA, et al. Protectin $D1_{n\text{-}3\ DPA}$ and resolvin $D5_{n\text{-}3\ DPA}$ are effectors of intestinal protection. *Proc Natl Acad Sci U S A*. 2017;114:3963–3968.
126. Campbell EL, MacManus CF, Kominsky DJ, et al. Resolvin E1-induced intestinal alkaline phosphatase promotes resolution of inflammation through LPS detoxification. *Proc Natl Acad Sci U S A*. 2010;107:14298–14303.
127. Campbell EL, Louis NA, Tomassetti SE, et al. Resolvin E1 promotes mucosal surface clearance of neutrophils: a new paradigm for inflammatory resolution. *FASEB J*. 2007;21:3162–3170.
128. Kucharzik T, Gewirtz AT, Merlin D, Madara JL, Williams IR. Lateral membrane LXA4 receptors mediate LXA4's anti-inflammatory actions on intestinal epithelium. *Am J Physiol Cell Physiol*. 2003;284:C888–C896.
129. Cox MA, Jackson J, Stanton M, et al. Short-chain fatty acids act as antiinflammatory mediators by regulating prostaglandin E(2) and cytokines. *World J Gastroenterol*. 2009;15:5549–5557.
130. Yamada K, Mori M, Matsuo N, Shoji K, Ueyama T, Sugano M. Effects of fatty acids on accumulation and secretion of histamine in RBL-2H3 cells and leukotriene release from peritoneal exudate cells isolated from Wistar rats. *J Nutr Sci Vitaminol (Tokyo)*. 1996;42:301–311.
131. Hakimian JK, Dong TS, Barahona JA, et al. Dietary supplementation with omega-3 polyunsaturated fatty acids reduces opioid-seeking behaviors and alters the gut microbiome. *Nutrients*. 2019;11:1900.
132. Liu F, Smith AD, Solano-Aguilar G, et al. Mechanistic insights into the attenuation of intestinal inflammation and modulation of the gut microbiome by krill oil using in vitro and in vivo models. *Microbiome*. 2020;8:83.
133. Zhang XM, Medrano RF, Wang M, et al. Corn oil supplementation enhances hydrogen use for biohydrogenation, inhibits methanogenesis, and alters fermentation pathways and the microbial community in the rumen of goats. *J Anim Sci*. 2019;97:4999–5008.
134. Zackular JP, Kirk L, Trindade BC, Skaar EP, Aronoff DM. Misoprostol protects mice against severe *Clostridium difficile* infection and promotes recovery of the gut microbiota after antibiotic perturbation. *Anaerobe*. 2019;58:89–94.
135. Ilag LL. Are long-chain polyunsaturated fatty acids the link between the immune system and the microbiome towards modulating cancer? *Medicines (Basel)*. 2018;5:102.
136. Zhuang P, Shou Q, Lu Y, et al. Arachidonic acid sex-dependently affects obesity through linking gut microbiota-driven inflammation to hypothalamus-adipose-liver axis. *Biochim Biophys Acta Mol Basis Dis*. 2017;1863:2715–2726.
137. Yu H-N, Zhu J, Pan WS, Shen SR, Shan W-G, Das UN. Effects of fish oil with high content of n-3 polyunsaturated fatty acids on mouse gut microbiota. *Arch Med Res*. 2014;45:195–202.
138. Miyamoto J, Igarashi M, Watanabe K, et al. Gut microbiota confers host resistance to obesity by metabolizing dietary polyunsaturated fatty acids. *Nat Commun*. 2019;10:4007.
139. Watson H, Mitra S, Croden FC, et al. A randomised trial of the effect of omega-3 polyunsaturated fatty acid supplements on the human intestinal microbiota. *Gut*. 2018;67:1974–1983.
140. Djuric Z, Bassis CM, Plegue MA, et al. Increases in colonic bacterial diversity after ω-3 fatty acid supplementation predict decreased colonic prostaglandin E2 concentrations in healthy adults. *J Nutr*. 2019;149:1170–1179. https://doi.org/10.1093/jn/nxy255.
141. Cao W, Wang C, Chin Y, et al. DHA-phospholipids (DHA-PL) and EPA-phospholipids (EPA-PL) prevent intestinal dysfunction induced by chronic stress. *Food Funct*. 2019;10:277–288.

142. Das UN. Essential fatty acids as possible enhancers of the beneficial actions of probiotics. *Nutrition*. 2002;18:786–789.
143. Kindt A, Liebisch G, Clavel T, et al. The gut microbiota promotes hepatic fatty acid desaturation and elongation in mice. *Nat Commun*. 2018;9:3760.
144. Wang Z, Neupane A, Vo R, White J, Wang X, Marzano SL. Comparing gut microbiome in mothers' own breast milk- and formula-fed moderate-late preterm infants. *Front Microbiol*. 2020;11:891.
145. Lyons KE, Ryan CA, Dempsey EM, Ross RP, Stanton C. Breast milk, a source of beneficial microbes and associated benefits for infant health. *Nutrients*. 2020;12:1039.
146. Trinchese G, Cavaliere G, Canani RB, et al. Human, donkey and cow milk differently affects energy efficiency and inflammatory state by modulating mitochondrial function and gut microbiota. *J Nutr Biochem*. 2015;26:1136–1146.
147. Collado MC, Cernada M, Neu J, Pérez-Martínez G, Gormaz M, Vento M. Factors influencing gastrointestinal tract and microbiota immune interaction in preterm infants. *Pediatr Res*. 2015;77:726–731.

CHAPTER 14

Contribution of gut microbiota and multiple organ failure in the pathogenesis of COVID-19 infection

Akhlaq A. Farooqui
Department of Molecular and Cellular Biochemistry, The Ohio State University, Columbus, OH, United States

1 Introduction

The coronavirus disease 2019 (COVID-19) is caused by a RNA virus. This virus belongs to the subfamily Orthocoronavirinae in the family of Coronaviridae. It is characterized by club-like spikes that project from its surface.[1] The virus, which originated in Wuhan, China, has spread rapidly throughout the world leading to the World Health Organization (WHO) declaring COVID-19 infection a global pandemic.[2] COVID-19 is a single-stranded enveloped RNA virus that is different from other known coronaviruses, including the common cold (229E, OC43, NL63, and HKU1), but is similar to severe acute respiratory syndrome (ARDS) (SARS-CoV-2 or SARS; 2002–03) and Middle East respiratory syndrome (MERS; 2012-ongoing) outbreaks.[1]

COVID-19 enters human cells through its Spike (S) protein, a type-I surface glycoprotein that binds to the angiotensin I converting enzyme 2 (ACE2), a zinc dipeptidyl carboxydipeptidase that is homologous to human angiotensin 1 converting enzyme. Thus, this enzyme acts as a receptor for COVID-19.[3] ACE2 is not only expressed in the respiratory tract, lungs, liver, and heart but also in the digestive system and kidneys. It negatively regulates the Renin-Angiotensin System (RAS) and plays an important role in neurohumoral regulation of the cardiovascular system. The binding of COVID-19 with the ACE2 receptor results not only in the activation of ACE2 signaling pathways but also in enhancement of its pathogenicity, leading to acute myocardial, lung,

Gut Microbiota in Neurologic and Visceral Diseases
https://doi.org/10.1016/B978-0-12-821039-0.00014-9

and digestive tract injuries.[4,5] The expression of ACE2 on various organs may explain the multiorgan dysfunction that has been described in COVID-19 cases.[6,7]

ACE2 receptors are abundantly distributed in the epithelia of the lungs, especially the alveolar type II (ATII) cells. Once the virus binds to the receptor, it spreads widely upon entering the blood circulation. Multiple tissues and organs can be potential targets for COVID-19 infection.[8]

COVID-19 infection spreads primarily via respiratory droplets during close face-to-face contact (via coughs and sneezes). Infection can spread by asymptomatic, presymptomatic, and symptomatic carriers. Symptoms of COVID-19 usually appear 10–14 days after viral exposure and typically include fever, dry cough, shortness of breath, fatigue, sputum production, sore throat, headache, myalgia or arthralgia, chills, nausea or vomiting, nasal congestion, diarrhea, hemoptysis, conjunctival congestion, and pneumonia (Table 1).[2,9] Radiographic and laboratory abnormalities, such as lymphopenia and elevated lactate dehydrogenase, are also common, but nonspecific. Diagnosis is made by detection of COVID-19 via reverse transcription polymerase chain reaction testing, although false-negative test results may occur in up to 20%–67% of patients. Severe cases of COVID-19 often display respiratory, hepatic, gastrointestinal, and neurological complications that ultimately culminate in hospitalization and death.

Table 1 Symptoms of COVID-19 infection in human patients.

Organ involved	Minor symptoms	Reference
Body	Fever	1,2,9
Lungs	Dry cough	1,2,9
Muscle	Myalgia (muscle pain)	1,2,9
Muscle	Fatigue	1,2,9
Digestive tract	Diarrhea	1,2,9
Severe symptoms		
Respiratory tract	Respiratory failure	1,2,9
Heart	Heart failure and sudden cardiac arrest	1,2,9
Lungs	Lung edema	1,2,9
Lungs	Proteinaceous exudate with globules	1,2,9
Lungs	Patchy inflammatory cellular infiltration	1,2,9
Lungs	Moderate formation of hyaline membranes	1,2,9

2 COVID-19 infection associated changes in respiratory tract and lungs

COVID-19 infection is accompanied by respiratory changes in the nose, sinuses, and throat and include nasal congestion, runny nose, sore throat, sneezing, achy muscles, headache, cough, diarrhea, fever, shortness of breath, and tiredness. COVID-19 invades lung epithelial cells and alveolar macrophages to produce viral nucleic acid, which stimulates the infected cells to release cytokines and chemokines, activating macrophages, dendritic cells, and others.[10] Furthermore, COVID-19 entry into the lungs has been reported to cause pneumonia, especially in older patients because they have reduced lung capacity, require longer recovery time, and have weakened immune systems due to their age.[9] Pneumonia that is caused by coronavirus may be more severe, affect many parts of the lungs, and cause shock, organ damage, abnormal blood clotting, ARDS, and deteriorating health.[11] Computed tomography (CT) has been used as an important complement for the diagnosis of COVID-19 pneumonia in the current pandemic.[12]

More than one-third of COVID-19 patients show dangerously high levels of blood clotting.[13] Furthermore, 36% of hospitalized COVID-19 patients in Wuhan, China, show elevated blood levels of D-dimer. This dimeric fragment of fibrin is most widely used as a diagnostic blood marker of coagulopathy. In one study from Wuhan, levels of D-dimer greater than 1 mg L^{-1} have been reported, indicating an 18-fold increase in risk of mortality.[14] Studies on COVID-19 patients from New York City have indicated that survival of COVID-19 patients is markedly increased when patients are mechanically ventilated and receive systemic anticoagulant therapy.[15] Thrombosis and disseminated intravascular clotting are common in COVID-19 infections.[13] Levels of coagulation factors are decreased during coagulation. Among coagulation factors, levels of anticoagulant protein S (PROS1) play an important role. This protein is present in the blood at a concentration of ~ 300 nM (~ 60% bound to the complement factor C4BP), where it normally acts in concert with activated protein C to degrade factor Va and factor VIIIa, thereby terminating the coagulation reaction.[16] It is an essential inhibitory factor in the coagulation cascade, as congenital *PROS1* deficiencies in people can lead to profound coagulopathies.[17] In later stages of COVID-19 infection, the integrity of the epithelial-endothelial barrier is compromised. In addition to epithelial cells, the COVID-19 virus infects pulmonary capillary endothelial cells, accentuating the inflammatory response and triggering an influx of monocytes and neutrophils.[18] CT studies have shown the accumulation of mononuclear cells and macrophages in lung infiltrates of autopsies along with the development of edema in the alveolar spaces.[18] Collective evidence suggests that COVID-19 infection contributes to endothelial barrier disruption, dysfunctional alveolar-capillary oxygen transmission, and impaired oxygen diffusion in lungs.

3 COVID-19 associated changes in gut microbiota and digestive tract

Several studies have indicated that COVID-19 infection produces significant alterations in the fecal microbiome and digestive tract compared with controls.[19] COVID-19 infection is characterized by the enrichment of opportunistic pathogens and depletion of beneficial commensals at the time of hospitalization and at all time points during hospitalization.[20–22] Depletion of gut microbiota and gut dysbiosis persists even after clearance of COVID-19 (determined from throat swabs) and resolution of respiratory symptoms. The baseline abundance of *Coprobacillus*, *Clostridium ramosum*, and *Clostridium hathewayi* correlates with COVID-19 severity, and there is an inverse correlation between abundance of *Faecalibacterium prausnitzii* (an antiinflammatory bacterium) and disease severity.[21,22] Another study indicated that patients with COVID-19 have an altered gut microbiome with depletion of beneficial commensals (*Eubacterium ventriosum*, *F. prausnitzii*, *Roseburia*, and Lachnospiraceae taxa) and enrichment of opportunistic pathogens (*C. hathewayi*, *Actinomyces viscosus*, and *Bacteroides nordii*) during hospitalization.[21] In addition, over the course of hospitalization, the number of *Bacteroides dorei*, *Bacteroides thetaiotaomicron*, *Bacteroides massiliensis*, and *Bacteroides ovatus* downregulates the expression of ACE2 in murine gut and inversely correlates with COVID-19 load in fecal samples from patients with COVID-19.[21,22] At the molecular level, these bacteria may produce dysbiosis by releasing immunogenic endotoxins called lipopolysaccharides (LPS) (Fig. 1). High levels of circulating LPS are known to suppress the expression of tight junction proteins, leading to an increase in intestinal permeability and subsequently the translocation of LPS from the gut into the blood (Fig. 2).[23,24] LPS is known to activate Toll-like receptors (TLR3, TLR4, TLR7, TLR8, and TLR9) and their downstream targets.[25,26] Studies on TLR and LDL receptors in double knockout mice indicate that a deficiency of TLRs reduces atherosclerosis without affecting inflammation.[27] Moreover, clinical investigations have revealed that upregulation of TLRs not only contributes to inflammation but also promotes the development of atherosclerosis.[28]

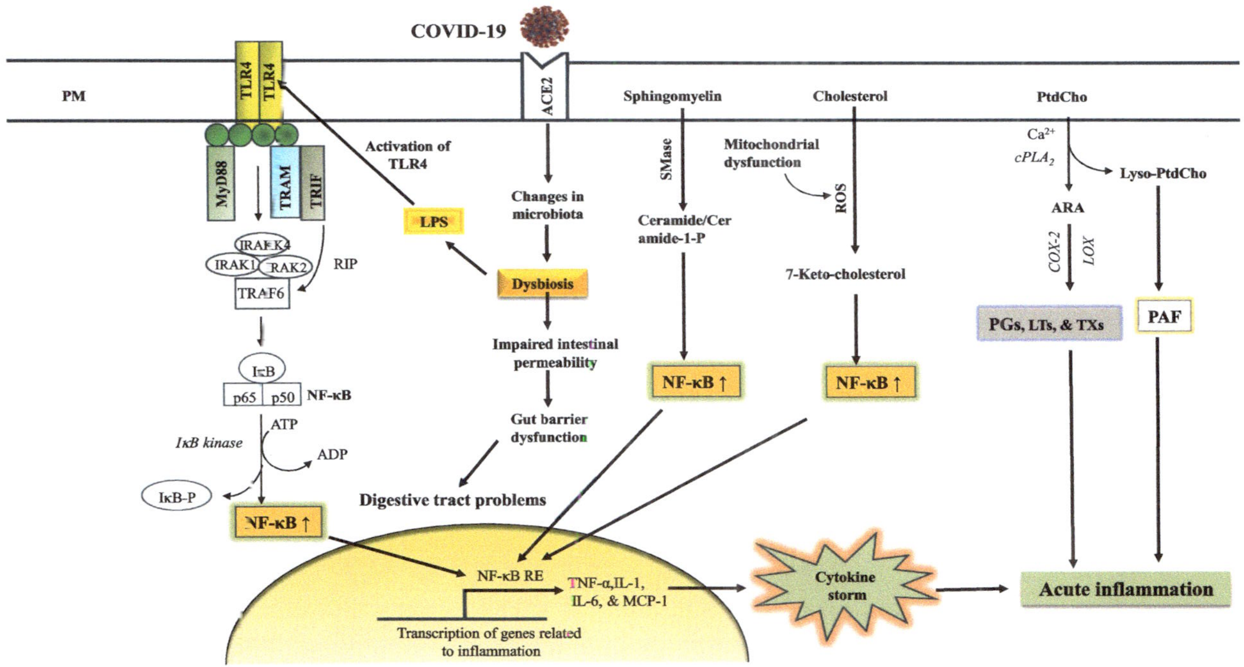

Fig. 1 Hypothetical diagram showing the induction of cytokine storm in COVID-19 infection. Plasma membrane (PM); phosphatidylcholine (PtdCho); arachidonic acid (ARA); lysophosphatidylcholine (lyso-PtdCho); platelet activating factor (PAF); cytosolic phospholipase A_2 ($cPLA_2$); cyclooxygenase (COX); lipoxygenase (LOX); reactive oxygen species (ROS); nuclear factor-kappa B (NF-κB); nuclear factor-kappa B response element (NF-κB-RE); tumor necrosis factor-alpha (TNF-α); interleukin-1beta (IL-1β); interleukin-6 (IL-6); monocyte chemotactic protein-1 (MCP1); Toll-like receptors 4 (TLR4); adaptor protein (MyD88); IL-1R-associated kinase (IRAK). tumor necrosis factor receptor-associated factor adaptor protein 6 (TRAF6); NF-κB-inducing kinase (NIK); IκB kinase (IKK); TIR-domain-containing adapter-inducing interferon-β (TRIF); interferon-beta (IFN-β); interferon regulatory transcription factor-3 (IRF-3).

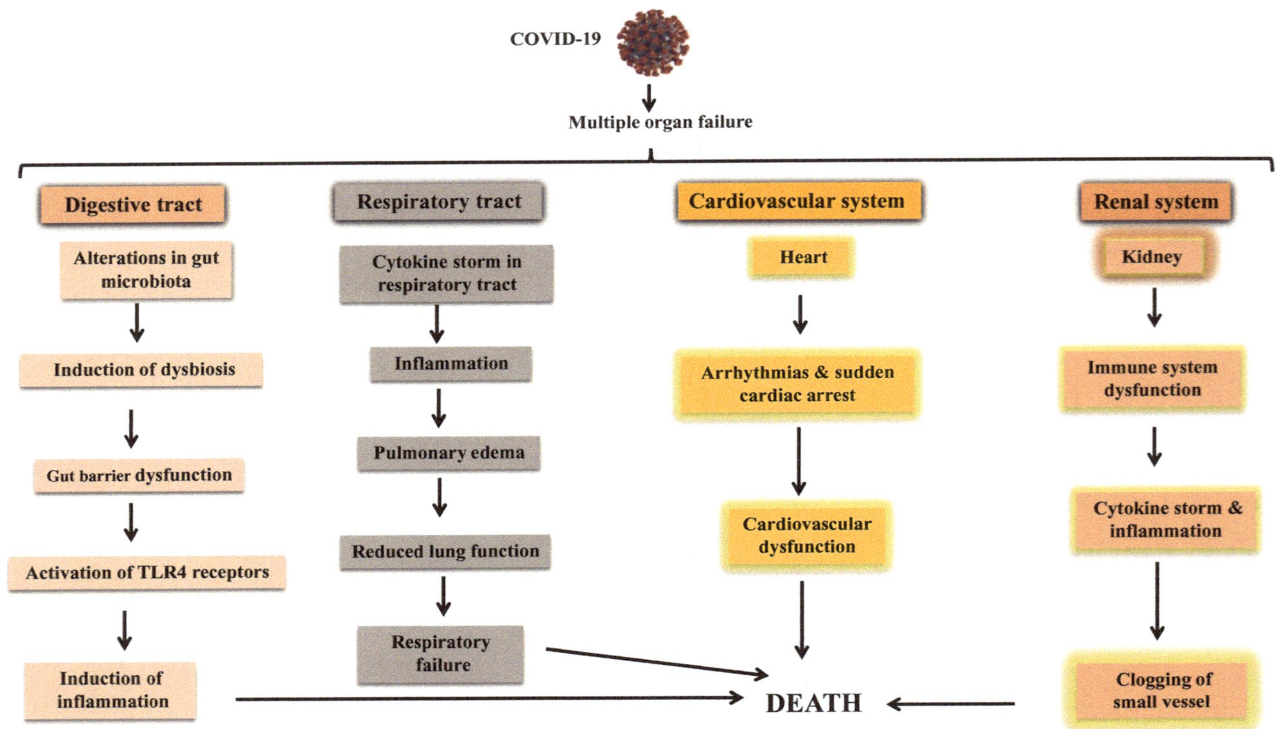

Fig. 2 Involvement of multiple organ failure in COVID-19-mediated death.

4 COVID-19 infection associated changes in cardiovascular system

COVID-19 patients show high prevalence of cardiovascular disease, and more than 7% of patients experience myocardial injury from the infection (22% of critically ill patients).[29] This abnormal cardiovascular dysfunction may be due to myocarditis, possible acute coronary syndrome, cardiac arrhythmias, heart failure, and cardiogenic shock.[30,31] Elevated troponin levels have also been observed in many patients with COVID-19, with significant differences between those who died and those who survived.[32] In a metaanalysis of four studies, levels of cardiac troponin I levels were shown to be much greater in those with severe disease compared to those with nonsevere COVID-19 disease.[32] Interestingly, the median troponin I (TnI) among survivors did not change, while it increased exponentially in nonsurvivors.[33] Along with increase in troponins, other inflammatory biomarkers such as D-dimers, interleukin-6, and so on also show a significant increase, reflecting severe pan-inflammatory response.[34] Another group of patients presented with predominant cardiac symptoms mimicking viral myocarditis or acute coronary syndrome. Two cases recently reported highlight the possible direct effect of COVID-19 on the cardiovascular system. One patient presented with typical chest pain and ST elevation in electrocardiogram (ECG), left ventricular (LV) dysfunction, and positive troponins, but nonobstructive coronary arteries.[35] Another patient presented with severe myocarditis and severe LV dysfunction along with respiratory infection.[36] Both the cases responded well to immunoglobulins and steroids with complete normalization of all parameters. Collective evidence suggests that COVID-19 infection markedly affects cardiac function.

5 COVID-19 associated changes in kidney

Multiple mechanisms may be involved in COVID-19-mediated damage to the kidneys. First, COVID-19 may exert direct cytopathic effects on kidney tissue. This proposal is supported by the detection of polymerase chain reaction fragments of coronavirus in blood and urine in both patients with the 2003 SARS virus[37] and those with COVID-19. To this end, it is well known that like other coronaviruses, COVID-19 uses ACE2 as a cell entry receptor[11]. Recent human tissue RNA-sequencing data have demonstrated that ACE2 expression in urinary organs (kidney) is nearly 100-fold greater than in respiratory organs (lungs).[38,39] Therefore, kidney disease may be caused by coronavirus entering kidney cells through an ACE2-dependent pathway.

Second, deposition of immune complexes of viral antigen or virus-induced specific immunological effector mechanisms (specific T-cell lymphocyte or antibody) may be another mechanism of kidney damage. Microscopic studies of kidney specimens from SARS patients indicate the presence of normal glomerular structure and absence of electron-dense deposits.[40] Thus, these findings do not support the presence of an active immune-mediated glomerulonephritis. Third, a virus-induced cytokine storm may exert indirect damage to renal tissue resulting in hypoxia, shock, and rhabdomyolysis. In fact, some patients with the 2009 H1N1 virus show mild-to-moderate elevations of serum creatine kinase.[41] Similarly, in one study, 138 patients with COVID-19 had increased levels of creatine kinase.[29]

6 Role of cytokine storm and multiple organ failure in COVID-19 associated death

Converging evidence suggests that complications of COVID-19 infection include impairment in function of the heart, brain, lungs, digestive tract, kidneys, and coagulation system (Fig. 2). As stated above, COVID-19 binds with ACE2 receptors of host cells resulting in membrane fusion and the release of viral RNA. This receptor has pathogen-associated molecular patterns (PAMPs), which can be detected by pattern recognition receptors (PRRs). In the host, toll-like receptors (TLR3, TLR4, TLR7, TLR8, and TLR9) are coupled with onset of inflammation.[42,43] Like other coronaviruses, the COVID-19 viral RNA receptor not only promotes the expression of the retinoic-acid inducible gene I (RIG-I)[44] and cytosolic receptor melanoma differentiation-associated gene 5 (MDA5), but also nucleotidyltransferase cyclic GMP-AMP synthase. These are responsible for the recognition of viral RNA and DNA in the cytoplasm (cGAS).[45] This complex signaling recruits adaptor proteins such as TIR-domain-containing adaptor protein (IFN-β (TRIF), mitochondrial antiviral-signaling protein (MAVS), and stimulator of interferon genes protein (STING))[46,47] to trigger and control downstream cascades. This signaling pathway involves adaptor molecule MyD88 and promotes the activation of nuclear factor-κB (NF-κB). The activation of this transcription factor ultimately leads to the expression of interferon regulatory factor 3 (IRF3) and the type I Interferons (IFN-α/β) along with a series of pro-inflammatory cytokines.[48] Hence, virus-cell interactions produce a diverse set of immune mediators against the invading virus.[49] COVID-19 infection results in overexpression of several plasma cytokines and chemokines, including IL-1, IL-2, IL-4, IL-7, IL-10, IL-12,

IL-13, IL-17, GCSF, macrophage colony-stimulating factor (MCSF), IP-10, MCP-1, MIP-1α, hepatocyte growth factor (HGF), IFN-γ, and TNF-α.[1,50] A study on COVID-19 pneumonia corpses indicates that COVID-19 infection produces an inflammatory response in the lower airway, leading to lung injury.[51] Collectively these studies indicate that COVID-19 particles first invade the respiratory mucosa and then infect other cells such as glial and neuronal cells, inducing a series of immune responses and the production of cytokines and chemokines resulting in a cytokine storm in the body, which may be responsible for the critical condition of COVID-19 patients. Collective evidence suggests that COVID-19 infection facilitates the overexpression and overproduction of inflammatory cytokines and chemokines with a wide range of biological activity in a variety of body tissues and cells. These cytokines drive a positive feedback on other cells (macrophages, neutrophils, and T cells from circulation) and continue to recruit them to the sites of inflammation, begetting the exponential expansion of inflammation and organ damage. This process results in a cytokine storm, which is a crucial cause of ARDS, a systemic inflammatory response, and multiple organ failure involving damage of the vascular barrier, capillaries, and alveoli, and ultimately death. Lung injury is one consequence of the cytokine storm that can progress into acute lung injury or its more severe form ARDS.[52] Low oxygen saturation levels during ARDS is a major cause of mortality in COVID-19. Although the exact mechanism of ARDS in COVID-19 patients is not fully understood, the excessive production of pro-inflammatory cytokines is considered one of the major contributing factors.[1,53] Cytokine storm not only produces damage in multiple tissues (lungs, heart, kidneys, and intestine) at multiple sites during COVID-19 infection[54] but also mediates ARDS aggravation and widespread tissue damage resulting in multiple organ failure and death. Targeting cytokines during the management of COVID-19 patients may improve survival rates and reduce mortality. Based on this information, it can be implied that death in COVID-19 patients may be due to cytokine storm and multiple organ failure.[29] Currently, there is no treatment or vaccine for COVID-19 infection. Apart from the efforts to develop a vaccine, several therapeutic options are currently being evaluated, including drug repurposing of distinct antiviral, immunomodulatory or antiinflammatory agents, as well as monoclonal antibodies targeting COVID-19 entry into human cells. To add further to the complexity of COVID-19 infection, most inflammatory mediators that contribute to cytokine storm demonstrate pleotropic downstream effects and are frequently interdependent in their biological activity. The cross talk among these mediators and the pathways is neither linear nor uniform. Furthermore, although their quantified levels may suggest severity of responses, they do not necessarily imply pathogenesis. This complex interplay illustrates the limitations of interfering in the acute inflammatory response based on single mediators and at indiscriminate time points.

7 Conclusion

COVID-19 infection is accompanied by multiple organ function failure. It has been suggested that digestive tract, respiratory tract, renal, and cardiovascular system dysfunctions not only lead to hyperinflammatory responses in the body but are also linked with the plausible direct effects of severe ARDS on body-wide organs via ACE2. It is proposed that complications of COVID-19 infection are linked with ARDS, heart failure, renal failure, liver damage, and shock resulting in multiorgan failure-mediated death. During COVID-19 infection, the cytokine storm is readily followed by the immune system "attacking" the body, which in turn may cause ARDS and multiple organ failure, the final result being death, at least in the most severe cases of COVID-19 infection. Acknowledging the comorbidities and potential organ injuries throughout the course of COVID-19 is therefore crucial in the clinical management of patients.

References

1. Chen N, Zhou M, Dong X, et al. Epidemiological and clinical characteristics of 99 cases of 2019 novel coronavirus pneumonia in Wuhan, China: a descriptive study. *Lancet*. 2020;395:507–513.
2. World Health Organization. *Report of the WHO-China Joint Mission on Coronavirus Disease 2019 (COVID-19)*; February 28, 2020. https://www.who.int/publications-detail/reportof-the-who-china-joint-mission-on-coronavirus-disease-2019-(covid-19).
3. Hoffmann M, Kleine-Weber H, Krüger N, Müller M, Drosten C, Stefan Pöhlmann S. *The novel coronavirus 2019 (2019-nCoV) uses the SARS-coronavirus receptor ACE2 and the cellular protease TMPRSS2 for entry into target cells*. BioRxIV; 2020. https://doi.org/10.1101/2020.01.31.929042.
4. Li B. Prevalence and impact of cardiovascular metabolic diseases on COVID-19 in China. *Clin Res Cardiol*. 2020;109:531–538.
5. Xiong T-Y, Redwood S, Prendergast B, Chen M. Coronaviruses and the cardiovascular system: acute and long-term implications. *Eur Heart J*. 2020;41:1798–1800.
6. Tikellis C, Thomas MC. Angiotensin-converting enzyme 2 (ACE2) is a key modulator of the renin angiotensin system in health and disease. *Int J Pept*. 2012;2012:256294.
7. Zhang H, Penninger JM, Li Y. Angiotensin-converting enzyme 2 (ACE2) as a SARS-CoV-2 receptor: molecular mechanisms and potential therapeutic target. *Intensive Care Med*. 2020;46:586–590.
8. Fan C, Li K, Ding Y, Lu WL, Wang J. *ACE2 Expression in kidney and testis may cause kidney and testis damage after 2019-nCoV infection*. medRxiv; 2020. https://doi.org/10.1101/2020.02.12.20022418.

9. Tang N, Li D, Wang X, Sun Z. Abnormal coagulation parameters are associated with poor prognosis in patients with novel coronavirus pneumonia. *J Thromb Haemost.* 2020;18:844–847.
10. Fehr AR, Perlman S. Coronaviruses: an overview of their replication and pathogenesis. *Methods Mol Biol.* 2015;1282:1–23.
11. Zhou P, Yang XL, Wang XG. A pneumonia outbreak associated with a new coronavirus of probable bat origin. *Nature.* 2020;579:270–273.
12. Hani C, Trieu NH, Saab I, et al. COVID-19 pneumonia: a review of typical CT findings and differential diagnosis. *Diagn Interv Imaging.* 2020;101:263–268.
13. Levi M, Thachil J, Iba T, Levy JH. Coagulation abnormalities and thrombosis in patients with COVID-19. *Lancet Haematol.* 2020;7:e438–e440.
14. Zhou F, Yu T, Du R, et al. Clinical course and risk factors for mortality of adult in-patients with COVID-19 in Wuhan, China: a retrospective cohort study. *Lancet.* 2020;395:1054–1062.
15. Paranjpe I, Fuster V, Lala A, et al. Association of treatment dose anticoagulation with in-hospital survival among hospitalized patients with COVID-19. *J Am Coll Cardiol.* 2020;76:122–124.
16. Castoldi E, Hackeng TM. Regulation of coagulation by protein S. *Curr Opin Hematol.* 2008;15:529–536.
17. Burstyn-Cohen T, Heeb MJ, Lemke G. Lack of protein S in mice causes embryonic lethal coagulopathy and vascular dysgenesis. *J Clin Invest.* 2009;119:2942–2953.
18. Xu Z, Shi L, Wang Y, et al. Pathological findings of COVID-19 associated with acute respiratory distress syndrome. *Lancet Respir Med.* 2020;8:420–422.
19. Mao R, Qiu Y, He JS, et al. Manifestations and prognosis of gastrointestinal and liver involvement in patients with COVID-19: a systematic review and meta-analysis. *Lancet Gastroenterol Hepatol.* 2020;5:667–678.
20. Su S, Shen J, Zhu L, et al. Involvement of digestive system in COVID-19: manifestations, pathology, management and challenges. *Therap Adv Gastroenterol.* 2020;13. 1756284820934626.
21. Zuo T, Zhang F, Lui GCY, et al. Alterations in gut microbiota of patients with COVID-19 during time of hospitalization. *Gastroenterology.* 2020;159:944–955.e8. S0016-5085(20)34701-4706.
22. Zuo T, Liu Q, Zhang F, et al. Depicting SARS-CoV-2 faecal viral activity in association with gut microbiota composition in patients with COVID-19. *Gut.* 2020;72:1–19. gutjnl-2020-322294.
23. Cândido FG, Valente XF, Grześkowiak LM, Moreira AP, Rocha DM Alfenas R. Impact of dietary fat on gut microbiota and low-grade systemic inflammation: mechanisms and clinical implications on obesity. *Int J Food Sci Nutr.* 2017;69:125–143.
24. Harris K, Kassis A, Major G, Chou CJ. Is the gut microbiota a new factor contributing to obesity and its metabolic disorders? *J Obes.* 2012;2012, 879151.
25. Chacon MR, Lozano-Bartolome J, Portero-Otin M, et al. The gut mycobiome composition is linked to carotid atherosclerosis. *Benefic Microbes.* 2017;9:1–14.
26. Lundin A, Bok CM, Aronsson L, et al. Gut flora, toll-like receptors and nuclear receptors: a tripartite communication that tunes innate immunity in large intestine. *Cell Microbiol.* 2008;10:1093–1103.
27. Ding Y, Subramanian S, Montes VN, Goodspeed L, Wang S, et al. Toll-like receptor 4 deficiency decreases atherosclerosis but does not protect against inflammation in obese LDL receptor-deficient mice. *Arterioscler Thromb Vasc Biol.* 2012;32:1596–1604.
28. Edfeldt K, Swedenborg J, Hansson GK, Yan ZQ. Expression of toll-like receptors in human atherosclerotic lesions: a possible pathway for plaque activation. *Circulation.* 2002;105:1158–1161.
29. Wang D, Hu B, Hu C, et al. Clinical characteristics of 138 hospitalized patients with 2019 novel coronavirus–infected pneumonia in Wuhan, China. *JAMA.* 2020;323:1061–1069.

30. Murthy S, Gomersall CD, Fowler RA. Care for critically ill patients with COVID-19. *JAMA*. 2020;323:1499–1500. (online ahead of print); PMID: 32159735.
31. Ruan Q, Yang K, Wang W, Jiang L, Song J. Clinical predictors of mortality due to COVID-19 based on an analysis of data of 150 patients from Wuhan, China. *Intensive Care Med*. 2020;3:1–3.
32. Lippi G, Lavie CJ, Sanchis-Gomar F. Cardiac troponin I in patients with coronavirus disease (COVID-19): evidence from a meta-analysis. *Prog Cardiovasc Dis*. 2020;63:390–391.
33. Zheng Y-YS, Ma Y-T, Zhang J-Y, Xie X. COVID-19 and the cardiovascular system. *Nat Rev Cardiol*. 2020;5:1–2.
34. Oudit GY, Kassiri Z, Jiang C. SARS coronavirus modulation of myocardial ACE2 expression and inflammation in patients with SARS. *Eur J Clin Invest*. 2009;39:618–625.
35. Hu H, Ma F, Wei X, Fang Y. Coronavirus fulminant myocarditis saved with glucocorticoid and human immunoglobulin. *Eur Heart J*. 2020;ehaa190.
36. Zeng JH, Liu YX, Yuan J, Wang FX, Wu WB, et al. First case of COVID-19 infection with fulminant myocarditis complication: case report and insights. *Infection*. 2020;48:773–777. preprints. 2020030180.
37. Peiris JSM, Chu CM, Cheng VCC. Clinical progression and viral load in a community outbreak of coronavirus-associated SARS pneumonia: a prospective study. *Lancet*. 2003;361:1767–1772.
38. Li Z, Wu M, Guo J. *Caution on kidney dysfunctions of 2019-nCoV patients*; 2020. MedRxiv preprint. Available at: (Accessed 14 March 2020).
39. Li YC, Bai WZ, Hashikawa T. The neuroinvasive potential of SARS-CoV2 may play a role in the respiratory failure of COVID-19 patients. *J Med Virol*. 2020;92:552–555.
40. Chu KH, Tsang WK, Tang CS. Acute renal impairment in coronavirus-associated severe acute respiratory syndrome. *Kidney Int*. 2005;67:698–705.
41. Kumar A, Zarychanski R, Pinto R. Critically ill patients with 2009 influenza A(H1N1) infection in Canada. *JAMA*. 2009;302:1872–1879.
42. Alexopoulou L, Holt AC, Medzhitov R, Flavell RA. Recognition of double-stranded RNA and activation of NF-kappaB by toll-like receptor 3. *Nature*. 2001;413:732–738.
43. Wu J, Chen ZJ. Innate immune sensing and signaling of cytosolic nucleic acids. *Annu Rev Immunol*. 2014;32:461–488.
44. Yoo JS, Kato H, Fujita T. Sensing viral invasion by RIG-I like receptors. *Curr Opin Microbiol*. 2014;20:131–138.
45. Wu J, Sun L, Chen X, et al. Cyclic GMP-AMP is an endogenous second messenger in innate immune signaling by cytosolic DNA. *Science*. 2013;339:826–830.
46. Ishikawa H, Barber GN. STING is an endoplasmic reticulum adaptor that facilitates innate immune signalling. *Nature*. 2008;455:674–678.
47. Seth RB, Sun L, Ea CK, Chen ZJ. Identification and characterization of MAVS, a mitochondrial antiviral signaling protein that activates NF-kappaB and IRF 3. *Cell*. 2005;122:669–682.
48. Kawai T, Akira S. The role of pattern-recognition receptors in innate immunity: update on toll-like receptors. *Nat Immunol*. 2010;11:373–384.
49. Takeuchi O, Akira S. Innate immunity to virus infection. *Immunol Rev*. 2009;227:75–86.
50. Huang C, Wang Y, Li X. Clinical features of patients infected with 2019 novel coronavirus in Wuhan, China. *Lancet*. 2020;395:497–506.
51. Liu Q, Wang R, Qu G, et al. General anatomy report of novel coronavirus pneumonia death corpse. *J Forensic Med*. 2020;36:19–21.
52. Shimizu M. Clinical features of cytokine storm syndrome. In: Cron R, Behrens E, eds. *Cytokine Storm Syndrome*. Cham: Springer; 2019:31–42.
53. Lai C, Shih T, Ko W, Tang H, Hsueh P-R. Severe acute respiratory syndrome coronavirus 2 (SARS-CoV-2) and coronavirus disease-2019 (COVID-19): the epidemic and the challenges. *Int J Antimicrob Agents*. 2020;55:105924.
54. Ragab D, Eldin HS, Taeimah M, Khatlab R, Salem R. The COVID-19 cytokine storm; what we know so far. *Front Immunol*. 2020;11:1446.

CHAPTER 15

Contribution of gut microbiota in the pathogenesis of migraine headache

Tahira Farooqui
Department of Entomology, The Ohio State University, Columbus, OH, United States

1 Introduction

Migraine is a severe and disabling brain condition, listed as the sixth most disabling disorder globally by the World Health Organization, and the most disabling of all neurological disorders.[1] It is a common, multifactorial, and recurring condition that causes throbbing headache, photophobia, phonophobia, nausea, and disability.[2] Of these symptoms, nausea is one of the most common, which worsens with activity. Migraine frequently occurs on one side of the head; however, it may also occur anywhere on the head, neck, and face, or all over. Migraine attacks can last for hours to days, and the pain can be so severe that it interferes with a person's daily activities.

More than 56% of migraine patients have episodic migraine (EM), whereas the remaining migraine patients have chronic migraine (CM). EM is characterized by 0–14 headache days per month, and CM is characterized by 15 or more headache days per month.[3]

According to the *International Classification of Headache Disorders, third edition*, chronic migraine is defined as "headaches on at least 15 days per month for at least 3 months, with the features of migraine on at least 8 days per month." However, diagnostic criteria distinguishing EM from CM continue to evolve.[4] Persons with EM can remit, not change, or progress to high-frequency EM or CM over time. However, CM is associated with a substantially greater personal and societal burden, more frequent comorbidities, and possibly with persistent and progressive brain abnormalities.[4, 5]

Migraine affects more than 17% of women and 5%–8% of men. Migraine susceptibility is multifactorial. Risk factors for migraine headache include:

(1) sex: migraine is more prevalent in females than males with a ratio of 3:1
(2) age: migraine is common among individuals aged 15–45 years, with most people experiencing their first migraine during adolescence

Gut Microbiota in Neurologic and Visceral Diseases
https://doi.org/10.1016/B978-0-12-821039-0.00015-0

(3) genetic factors: about 70%–80% of patients with migraine have a family history
(4) hormonal changes: due to changes in female hormone levels, headaches begin just before or shortly after onset of menstruation
(5) lifestyle: this includes diet, rest/sleep, drug usage, and physical activity[6–9]

The physiopathology of migraine is complex and still not fully understood; however, several factors are among the most important causes of migraine.[6, 10–17] These factors include magnesium deficiency; production of inflammatory agents around the nerves and cerebrospinal fluid; low serotonin levels; increased calcitonin gene-related peptide (CGRP), matrix metalloproteinase 9 (MMP-9), homocysteine, and nitric oxide (NO) levels; mitochondrial dysfunction; abnormal levels of vitamin D; altered transport of ions across the cell membrane; decreased levels of metabolic enzymes; and difference in gut microbial community composition when comparing a diseased group with healthy controls (dysbiosis).

2 Theories underlying migraine headache

At present, there are four major theories with regard to migraine mechanism.

(I) Vascular theory is based on uncontrolled excitation of the trigeminal nerve due to cerebral vasoconstriction during aura, followed by inflammatory neuropeptide release (CGRP, SP, NK-A) causing vasodilation, resulting in migraine headache.
(II) Neurological theory in which excessive neuronal excitability (abnormal neurological firing) in the cerebral cortex via cortico-trigeminal pathways increases release of extracellular Glu, Ca^{2+}, Na^{+}, and K^{+} ions initiating depolarization, causing cortical spreading depression (CSD).
(III) Activation of trigeminal sensory fibers leads to painful neurogenic inflammation within the meningeal vasculature (dura matter) mediated by inflammatory neuropeptide release (CGRP, SP, NK-A), suggesting that CGRP release is associated with neurogenic dural vasodilation that may be important in the generation of migraine pain.
(IV) Gut-brain axis (GBA) theory, in which alterations in gut microbiota composition may affect the gut-brain axis and cause inflammation, which may lead to migraine (Fig. 1).

I discuss the validity of these theories in the sections that follow.

2.1 Vascular theory due to cerebral vasodilation

The vascular theory of migraine, first proposed by Willis and later articulated by Wolff, suggests that migraine headache occurs due to dilation of cranial vessels.[18, 19] It posits that dilation of extracerebral vessels occurs during a migraine attack, and stimulation of cranial blood vessels provokes an ipsilateral headache. Therefore, vasoconstrictor drugs (ergot derivatives) alleviate the pain, whereas vasodilators like nitrates may induce an attack. The headache phase of migraine is associated with cerebral vasodilation and inflammation, presumably mediated by the release of vasoactive substances and neuropeptides such as CGRP (Fig. 2). In serum, elevated CGRP levels have been detected during migraine headache. As such, a common strategy to treat migraine is to inhibit the release of neuropeptides or antagonize neuropeptide receptors.[11] Serotonin is thought to be the underlying neurotransmitter involved in migraine, based on a lower than normal level of serotonin (5-HT) observed in migraine patients (Fig. 2).[12] However, between attacks, some migraine patients have reported with higher 5-HT levels. Pharmacological studies of 5-HT_4 receptor binding suggest an inverse relationship with brain 5-HT levels, indicating higher brain 5-HT levels in migraine patients compared to controls.[12]

The cornerstone of pharmacological treatment is acute therapy. The serotonin (5-hydroxytryptamine [5-HT]) receptor subtype $1_B/1_D$ agonists (triptans) are used as first-line acute therapy for patients who experience moderate-to-severe migraine attacks.[20] Triptans' mechanism of action is based on vasoconstriction; they inhibit the release of neuropeptides involved in migraine and act as vasoconstrictors (Fig. 2). Triptans work best if used within

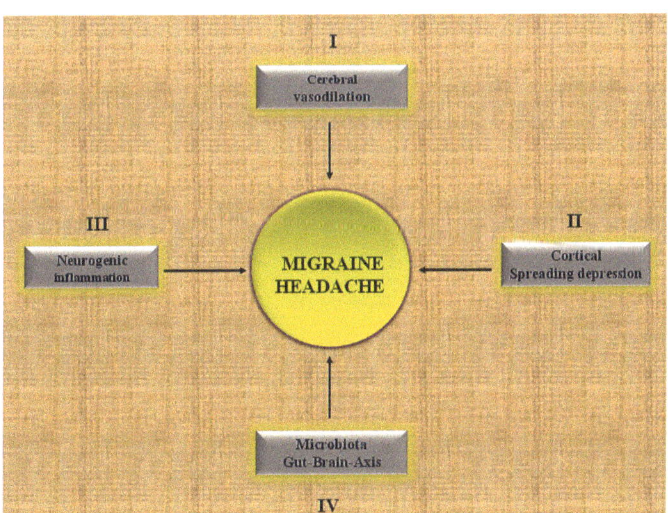

Fig. 1 Theories of migraine headache.

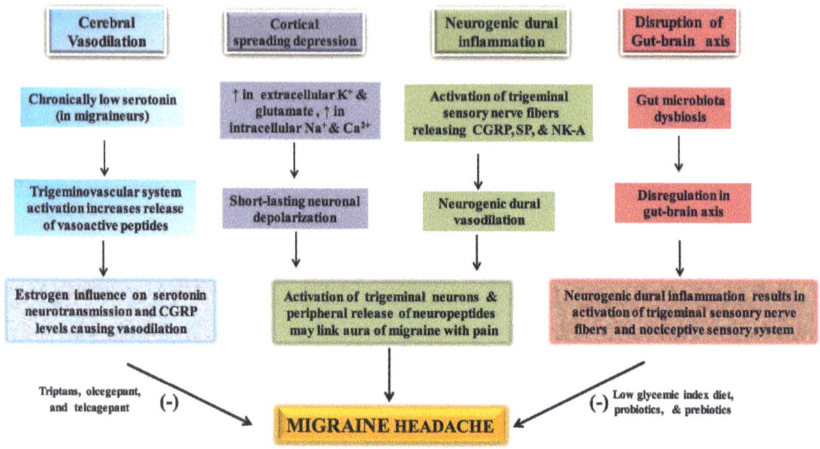

Fig. 2 Mechanisms underlying four migraine theories. Calcitonin gene-related peptide (CGRP), Substance P (SP), Neurokinin-A (NK-A), Triptans (serotonin receptor subtype $1_B/1_D$ agonists), and CGRP receptor antagonists (olcegepant and telcagepant).

the first 2h of the attack.[21] This coincides with the vasodilatory period following CGRP administration.[22] However, a large percentage of patients are not satisfied with this acute treatment, due to either lack of response or side effects. One group of researchers infused a vasoactive intestinal peptide (VIP) in migraineurs, and reported a marked dilation of cranial arteries without triggering any migraine attacks, providing no further evidence in support of vascular theory of migraine.[23] Moreover, another study from the same group used pituitary adenylate cyclase-activating peptide-38 (PACAP38), a strong vasodilator found in trigeminal sensory and parasympathetic perivascular nerve fibers. Infusion of PACAP38 caused headache and vasodilatation in both healthy subjects and migraine patients. In migraine sufferers, it caused delayed migraine-like attacks, suggesting that possible mechanisms of PACAP38 induction of headache and migraine-like attacks are peripheral sensitization of trigeminal sensory neurons, mast cell degranulation secondarily leading to activation of peripheral sensory trigeminal fibers, and facilitation of pain by sensitization of central second-order trigeminal neurons.[24] Thus induction of migraine headache may not be due to dilation, but it may be mediated by pituitary adenylate cyclase-activating polypeptide type I (PAC1) receptor site activation in response to PACAP138. Therefore, it is currently believed that vascular change is neither necessary, sufficient, nor needed in migraine; it may be an epiphenomenon of the neural substrates that are activated by the underlying pathophysiology of the disorder.[25]

2.2 Cortical spreading depression and migraine

CSD is an experimental phenomenon well characterized in animal models as well as shown in human brains. CSD, which refers to a slowly propagated wave of depolarization followed by suppression of brain activity, was first described by Leẹ̆o in 1944.[26] It is a remarkably complex event that involves dramatic changes in neural and vascular function. In ~30% of migraine cases, an aura phase may occur during some attacks and likely correlates with a CSD-like event; a slowly propagating wave of neuronal and glial cell depolarization and hyperpolarization.[27] These visual disturbances appear to progress across the visual field at the same rate as spreading depression, implicating a strong association between CSD and aura.[27–32]

In addition to the cortex, CSD is well characterized in the hippocampus, cerebellum, and retina, among other regions.[33] CSD is initiated by local elevations in extracellular potassium, which chronically depolarizes neurons. The initial accumulation of extracellular potassium occurs because of repeated depolarization and repolarization of hyperexcitable neurons in the cerebral cortex, and accumulation of extracellular K then further depolarizes the cells from which it was released. This large efflux of K is associated with a major disruption of cell membrane ionic gradients, influx of sodium and calcium, and release of glutamate (Fig. 2).[33] The excessive excitement of nerve cells in the cerebral cortex CSD activates trigeminal vascular sensitization, causing reduction in brain electrical activity, and decreases blood flow, leading to migraine.

Neuronal-glial cell communication is also important in CSD propagation, which is mediated by connexin-containing gap junctions and nonjunctional connexin-containing hemichannels. Accordingly, connexin-containing gap junctions may be viable targets for migraine prophylaxis by inhibiting neuronal-glial communication and thereby interfering with CSD propagation and possibly the development of peripheral sensitization.[34]

There are many arguments against this theory. CSD is difficult to evoke in humans, and electroencephalography (EEG) readings are not flattened during migraine as opposed to EEG readings during CSD. In contrast to CSD, migraine can occur bilaterally, and is not accompanied by a disrupted blood-brain barrier, increased cerebral metabolism, or cerebral cell swelling. CSD does not explain the appearance of premonitory symptoms or allodynia, long before the actual onset of aura, and there is a variation in the pain mechanisms of migraine and CSD.[35] Lastly, some researchers believe that the mechanism of aura in migraine is initiated by waves of CSD, leading to delayed activation of the trigeminovascular pathway underlying the headache phase.[36]

2.3 The trigeminovascular system and migraine

There is growing evidence that migraine pathophysiology may, in part, include dysfunction of subcortical structures. These include diencephalic and brainstem nuclei that can modulate the perception of activation of the trigeminovascular system, which carries sensory information from the cranial vasculature to the brain.[37] Dysfunction of these nuclei, and their connections to other key brain centers, may contribute to the cascade of events that results in other symptoms of migraine, such as light and sound sensitivity, thus providing a comprehensive explanation of the neurobiology of the disorder.[37] The premonitory phase begins as early as 3 days before the headache phase, and involves a complex interplay between various cortical and subcortical brain regions, including the hypothalamus and brainstem nuclei that modulate nociceptive signaling.[27] The headache phase involves activation of the trigeminovascular system,[7] which conveys nociceptive information from the meninges to the brain. Once the headache develops, it typically throbs, intensifies with an increase in intracranial pressure, and presents itself in association with nausea, vomiting, and abnormal sensitivity to light, noise, and smell. In 30% of patients, an aura phase may occur during some attacks and likely correlates with a CSD-like event.[27] Migraine can also be accompanied by abnormal skin sensitivity and muscle tenderness.

Moskowitz et al. published a novel hypothesis proposing an integral role for the neuropeptide-containing trigeminal nerve.[38] The activation of trigeminal sensory fibers leading to a painful neurogenic inflammation, including vasodilatation, leakage of plasma protein, and mast cell degranulation, within the meningeal (dural) vasculature is mediated by inflammatory neuropeptide (CGRP; substance P (SP); and neurokinin-A (NK-A)) release from trigeminal sensory fibers (Fig. 2).[6, 39–42] Antimigraine drugs such as sumatriptan and dihydroergotamine block the development of plasma extravasation and ultrastructural changes, as well as plasma CGRP increase in the superior sagittal sinus following electrical trigeminal ganglion stimulation.[39] Both sumatriptan and dihydroergotamine bind with high affinity to the 5-HT1D/1B receptors, implicating that their neurogenic antiinflammatory activity is mediated by activation of 5-HT autoreceptors present on sensory fibers innervating blood vessels in dura matter.[39] Selective serotonin receptor agonists (triptans) offer three main mechanisms of action: cranial vasoconstriction, peripheral trigeminal inhibition, and inhibition of the transmission through second-order neurons of the trigeminocervical complex. As such, triptans are routinely used for acute migraine.[42]

At present, the most widely accepted model of migraine is the neurovascular theory centering on the activation of the trigeminovascular system, and proposing that migraine is mediated by prolonged activation of meningeal nocicepters, which are located in the dura matter and vessels.[43] The trigeminovascular system is thought to comprise the pseudounipolar neurons located in the trigeminal ganglion that innervate the pial and dural meningeal vessels, and efferent projections synapsing with second-order neurons in the spinal cord trigeminal nucleus caudalis, providing projections to several high brain centers, including the posterior thalamus, hypothalamus, and cortex.[43] Activation of perivascular trigeminal nerves within meninges causes the release of CGRP, phosphorylation of extracellular signal-regulated kinase (ERK), SP, and NKA (Fig. 2), leading to a series of peripheral and central events, including inflammation and peripheral/central sensitization.[44, 45] In addition to CGRP, other neuropeptides and enzymes (e.g., cyclooxygenase 2 (COX2)) are important peripheral and central mediators of inflammation and pain.[45, 46] Prostaglandin E 2 (PGE2) is a COX product and the best known lipid mediator with a pronociceptive role, contributing to inflammatory pain. PGE2 has been reported to provoke immediate migraine-like attacks in patients with migraine without aura, suggesting that it may play an important role in the pathogenesis of migraine.[47] However, a highly specific and potent EP4 receptor antagonist (BGC20-1531) did not attenuate PGE2-induced headache and vasodilatation of both intra- and extra-cerebral arteries in healthy volunteers, suggesting that a single receptor blockade may not be sufficient to block PGE2-induced responses.[47] Therefore other PGE2 receptor subtype antagonists should be investigated as potential drug targets for the treatment of migraine.

This hypothesis transforms the migraine field and the understanding of its key concepts (such as the role of neuropeptides and their release from meningeal trigeminal nerve endings) into the mechanism of migraine, blocking neuropeptide release by administering antimigraine drugs, and activating and sensitizing the trigeminal afferents by meningeal inflammatory stimuli and upstream role of intense brain activity.[48] The theory underlying the role of CGRP within the context of trigeminovascular system is widely accepted in the pathophysiology of migraine. CGRP is abundant in trigeminal ganglion neurons, and is released from the peripheral nerve and central nerve terminals as well as being secreted within the trigeminal ganglion. Blocking sensitization of the trigeminal nerve by attenuating CGRP activity in the periphery may be sufficient to block a migraine attack, but the site of action of CGRP to initiate and maintain migraine prevention still remains open to debate.

2.4 Microbiota gut-brain axis and migraine

GBA and migraine is the most recent theory of migraine that may be successful in providing novel targets for prevention and treatment of migraine headache. However, further research studies are required to substantiate the clinical use of probiotics, prebiotics, and fecal microbiota transplantation (FMT) to validate this theory.[17] The next section discusses the underlying molecular mechanism of the microbiota-gut-brain axis and migraine theory after reviewing the importance of gut microbiota dysbiosis, gastrointestinal (GI) diseases, and neurological disorders including migraine headache.

3 Gut microbiota dysbiosis and migraine

Migraine headache is often associated with some GI symptoms, including nausea, vomiting, diarrhea, constipation, and dyspepsia.[49] Several studies have suggested that inflammation and neuroimmune modulation in the GI tract could play an important role in the pathogenesis of migraine headache.[50, 51] However, it still remains unclear how gut microbiota contribute.

The GI tract is colonized by trillions of microorganisms, collectively known as gut microbiota. A growing number of studies have reported the importance of intestinal microorganisms for regulation of host health and diseases. Alterations in gut microbiota composition results in dysbiosis, causing modulation of GI function, reduction in immune system, change in host metabolism, and disruption of gut-brain axis, leading to the pathogenesis of metabolic and neurological diseases. in host metabolism, and disruption of gut-brain axis, leading to the pathogenesis of metabolic and neurological diseases.[52]

However, exactly how the gut microbiome is involved in migraine headache is still unclear. Possible underlying mechanisms of migraine and gut intestinal diseases could increase gut permeability, allowing bacterial translocation as well as systemic diffusion of microbial products or metabolites inducing inflammation, leading to gut microbiota dysbiosis.

4 Gastrointestinal disorders, gut microbiota dysbiosis, and migraine

An increased frequency of GI disorders (such as *Helicobacter pylori* infection, irritable bowel syndrome (IBS), gastroparesis, hepatobiliary disorders, and celiac disease (CD)) has been demonstrated in patients with migraine compared with the general population (Table 1).[53–71]

Table 1 Gastrointestinal disorders associated with migraine and gut microbiota dysbiosis.

GI disorders	Prevalence of migraine	Gut microbiota dysbiosis	References
Helicobacter pylori infection (HPI)	Higher prevalence of *H. pylori* infection in PWM than in controls (Infection rate: 45% in PWM vs 33% in controls)	Gut microbial shift in HPI patients (high level of *Prevotella)*	53, 54
IBD	The prevalence of migraine in patients with IBD is significantly greater than in normal people	Increased intestinal permeability and inflammation	55, 56
IBS	Greater prevalence of migraine in IBS patients than control. Two to three times more prevalence in women than men	Increased intestinal permeability and inflammation	57, 58
CD	Greater prevalence of CD in migraine patients than controls (evident in both children and adults, but the greater figures are in the latter); migraine in PWM may be relieved by gluten-free diet	Increased intestinal permeability and inflammation	59–62
Gastroparesis	Present during migraine attacks, also possible in interictal periods	Increased gut permeability and inflammation, abnormal ANS	63–66
Infantile colic	Mothers with migraine are 2.6 times more likely to have infants with colic than mothers without migraine	↑ Twofold in Proteobacteria in infants with colic Significant ↓ in Bifidobacteria, *Lactobacilli*, and Bacteroidetes in infants with colic	67–71

ANS, autonomic nervous system; *GI*, gastrointestinal; *PWM*, patient with migraine; Both during and after migraine attack (ictal and interictal period in migraine).

The following section focuses on the relationship among GI disorders, gut microbiota dysbiosis, and migraine headache.

4.1 *H. pylori* infection, gut microbiota dysbiosis, and migraine

The relationship between infectious diseases such as *H. pylori* infection and migraine headache has been the focus of many research studies. *H. pylori* is a Gram-negative organism that causes chronic active gastric inflammation. It is strongly related to the outbreak and severity of migraine headaches.[53] The relationship between migraine and *H. pylori* infection can be based on inflammation, oxidative stress, and nitric oxide imbalance, as well as via bacterium-released toxins that may alter the host immune response. Thus association between migraine and *H. pylori* infection may include induced immune, inflammatory, and vascular responses and subsequent release of immune cells and inflammatory and vasoactive agents into the gastric mucosa, which may lead to hypersensitivity of brain pain-sensitive structures.[53] *H. pylori* treatment has been reported to significantly reduce migraine headaches. An *H. pylori*-positive group was reported to have a greater proportion of enterotype 1 (dominated by high level of *Prevotella*), whereas a healthy *H. pylori*-negative group had a greater proportion of enterotype 2 (dominated by *Bacteroides*). These results indicate alterations in gut microbial species due to *H. pylori* infection.[54] This shift in gut microbiota levels in the *H. pylori* group confirms the occurrence of gut microbiota dysbiosis due to *H. pylori* infection.

4.2 Inflammatory bowel disease, gut microbiota dysbiosis, and migraine

The two main forms of Inflammatory bowel disease (IBD) are ulcerative colitis and Crohn's disease, which are characterized by defects in the barrier function of the intestinal epithelial layer and the mucosal immune system. Factors such as antibiotics, nonsteroidal antiinflammatory drugs, stress, and infection are thought to trigger IBD. They decrease the mucosal barrier integrity, modulate the immune response, and change the luminal microenvironment, resulting in susceptibility to inflammation. The prevalence of migraine in IBD patients is 30%,[55,56] which is higher than the population basal rate of 15%–18% for women and 6%–8% men.[57]

4.3 IBS, gut microbiota dysbiosis, and migraine

IBS is a common GI disorder whose underlying pathophysiology is poorly understood. Similar to migraine, alteration in gut microbiota composition (dysbiosis) has been reported in IBS, implicating the potential disturbance in the GBA due to enhanced inflammation, influencing nociceptive response in the trigeminal pathway. IBS refers to several long-term conditions that involve inflammation of the digestive tract or gut. Migraine without aura is frequently reported in female patients with IBS.[57] Young female patients with IBS and migraine without aura have been reported to display more severe associated issues, including anxiety, gut microbiota dysbiosis, and a range of other combordities.[57] Moreover, diarrhea-predominant IBS patients display increased intestinal permeability associated with an increased Functional Bowel Disorder Severity Index score, and increased hypersensitivity to visceral and thermal nociceptive pain stimuli.[58] Both migraine and IBS are more prevalent among females, implicating that sex hormones may also play some role in these disorders. 5-HT is a major signaling molecule in the central nervous system (CNS) as well as in the gut. Therefore therapeutic agents modulating serotoninergic receptors (agonists and antagonists) are effective for IBS and migraine therapy.

4.4 CD, gut microbiota dysbiosis, and migraine

CD is another condition with similar symptoms to IBD. It is also characterized by inflammation in the intestine. CD is an inflammatory response to gluten (a mixture of seed storage proteins found in grains such as wheat, rye, barley, and oat).[59] The symptoms of CD go away after starting a gluten-free diet, suggesting that a gluten-free diet is the primary treatment for CD.[60] Headaches associated with CD are predominantly migraines.[61] The inflammatory reaction is associated with intestinal damage, including dysfunction of the tight junctions resulting in an increased intestinal permeability. Moreover, CD is an autoimmune disease that occurs in genetically predisposed people where the ingestion of gluten leads to damage in the small intestine. Enhanced pro-inflammatory immune responses in intestinal disorders like CD and IBS result in increased intestinal permeability, implicating that gut microbiota dysbiosis may be a potent mediator in migraine headache.[62]

4.5 Gastroparesis, gut microbiota dysbiosis, and migraine

Gastroparesis is a chronic disorder manifested by delayed emptying of solids and liquids from the stomach in the absence of mechanical obstruction. Its clinical manifestations include nausea, vomiting, bloating, and weight loss, among other symptoms.[63] Migraine attacks are often associated with delayed gastric emptying, as well as during interictal periods. Gastric emptying time is delayed in both ictal and interictal periods, suggesting an alteration in enteric autonomic function (ANS).[64] Gastric stasis during migraine may be attributed to increased sympathetic nervous system activity, decreased parasympathetic nervous system activity, or both.[64] Pharmokinetic studies and studies measuring gastric motor function suggest that gastric emptying may delay absorption of an orally administered drug by delaying its effectiveness. Migraine-associated gastroparesis is a problem when treating migraine with oral medicines, like oral triptans.[65] The Wistar-Kyoto (WKY) rat strain is commonly used in stress-associated functional GI disorders. A group of researchers analyzed gastroparesis and lipid metabolism-associated dysbiosis in delayed gastric emptying of WKY rats. They found less abundant cecal *Ruminococcus*, *Roseburia*, and unclassified *Lachnospiraceae*, and more abundant minor taxa *Dorea*, *Turicibacter*, and *Lactobacillus*, compared to the Sprague-Dawley strain (control) of rats. These results implicate gastroparesis and lipid metabolism-associated dysbiosis in WKY rats.[66] Thus, similar to other GI disorders, gastroparesis may also be linked with migraine via gut microbiota dysbiosis.

4.6 Infantile colic, gut microbiota dysbiosis, and migraine

Infantile colic is a common cause of inconsolable crying during the first months of life. Colic has been attributed to a number of causes such as stress during pregnancy, negative experiences during childbirth, allergy to milk proteins, and an infant's need for additional months of womb-like nurturing. It has often been referred to as a pain syndrome like migraine. Infantile colic has common pathogenic roots with migraine. The treatment of infantile colic includes drug therapy, nutritional modifications, and behavioral strategies. Gelfand et al. conducted a study in a group of infant-mother pairs. This group reported that infants with a maternal history of migraine were 2.6 times as likely to have colic as infants without a maternal history of migraine.[67] Dysbiosis generally refers to a difference in microbial community composition when comparing a diseased group with healthy controls. Infants with abdominal colic have reported less intestinal microbiota diversity and stability compared to control infants in the first weeks of life. De Weerth et al. provided a comprehensive analysis of the fecal microbiota in infants with colic and a control group.[68] According to this study, Proteobacteria were significantly increased in colic infants compared with control infants, whereas, Bifidobacteria and *Lactobacilli* were significantly reduced in colic infants.[68] Moreover, the colic phenotype was positively correlated with specific groups of Proteobacteria, including bacteria related to *Escherichia*, *Klebsiella*, *Serratia*, *Vibrio*, *Yersinia*, and *Pseudomonas*, but negatively with bacteria belonging to the Bacteroidetes and Firmicutes phyla; bacteria in the latter phyla produce butyrate and lactate.[68] Another group tested the hypothesis that colic-related dysbiosis is associated with visceral hypersensitivity. This study found that fecal samples taken from infants suffering from colic (with the longest crying time) led to visceral hyperalgesia (linked to *Bacteroides vulgatus* and *Bilophila wadsworthia*) in recipient mice, possibly as a result of gut microbiota dysbiosis.[69] Partty et al. demonstrated the presence of low-grade, systemic inflammation in colic babies, based on circulating levels of interleukin-8, macrophage inhibitory factor-1-β, and monocyte chemotactic peptide-1.[70] Rhoads et al. confirmed that infant colic is associated with gut inflammation (as determined by fecal calprotectin) linked to intestinal dysbiosis (fewer *Bifidobacilli*).[71]

All GI disorders discussed in this section are shown to be associated with migraine headache (Table 1).[53–71] The next section discusses the molecular mechanism underlying migraine headache via microbiome-gut-brain axis.

5 The gut-brain axis and migraine

The GBA is defined as a two-way communication system that allows intestinal microbes to communicate with the brain and vice versa. This bi-directional communication between the CNS and the enteric nervous system (ENS) links the emotional and cognitive centers of the brain with the peripheral intestinal functions.[72] Moreover, the GBA plays a role in monitoring and integrating gut functions as well as in liking the emotional and cognitive centers of the brain with the peripheral intestinal functions and mechanisms, including immune activation, intestinal permeability, enteric reflex, and entero-endocrine signaling. In clinical practice, evidence of microbiota-GBA interactions are the result of the association of dysbiosis with CNS disorders and GI disorders. Actually, such communication network includes the CNS (both brain and spinal cord), the autonomic nervous system (ANS), ENS, and the hypothalamic pituitary adrenal (HPA) axis (Fig. 3).[17, 72–74]

As a key component of the GBA, gut microbiota influence the development and maturation of the hypothalamic-pituitary-adrena (HPA) axis, affecting the development and function of the immune system, regulating the blood-brain barrier, modulating the synthesis and recognition of neurotransmitters, regulating neurogenesis, forming myelination, and supporting the development and function of the brain. Dysregulation of the GBA has been implicated not only in the pathogenesis and pathophysiology of

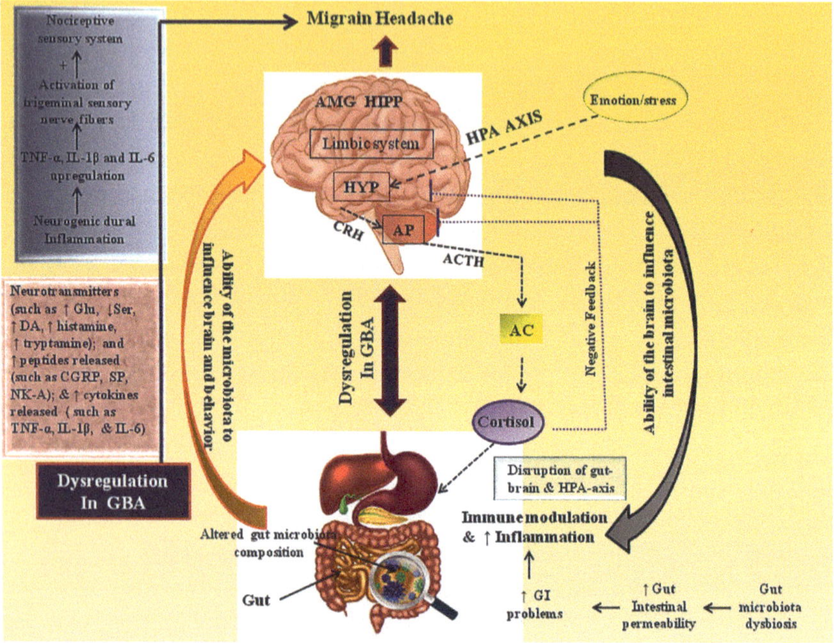

Fig. 3 The dysregulation of the GBA may be the molecular mechanism underlying the migraine headache. *5-HT*, serotonin; *AC*, adrenal cortex; *ACTH*, adrenocorticotropic hormone; *AMG*, amygdala; *CGRP*, calcitonin gene-related peptide; *CRH*, corticotrophin-releasing hormone; DA, dopamine; *GBA*, gut-brain axis; *Glu*, glutamate; *HIPP*, hippocampus; *HPA*, hypothalamic-pituitary-adrenal; *HYP*, hypothalamus; *IL-1β*, interleukin-1 beta; *IL-6*, interleukin-6; *NK-A*, neurokinin-A; *SP*, substance P; *TNF-α*, tumor necrosis factor alpha.

classic brain-gut disorders (IBS, IBD, CD) and other functional GI disorders,but also a growing list of psychiatric and neurologic pathologies including mood and anxiety disorders, depression, autistic spectrum disorders, multiple sclerosis, Alzheimer disease, Parkinson disease, and migraine headache.[17, 70, 72, 74]

Microbiota may interact with GBA through different mechanisms. The principal one likely being modulation of the intestinal barrier (leaky gut), whose perturbation can influence all the underlying compartments. Intestine and brain communicate through the GBA, which is of bidirectional nature. This signaling pathway involves neuronal, endocrine, and immunological factors. The GBA plays an essential role in the regulation of GI motility, secretion, digestion, absorption, and blood flow; in the regulation of appetite, energy balance, and metabolism; and in the modulation of mucosal immunity. In turn, the brain receives interoceptive input from the

GI tract, which integrates this information with other systems input, and transmits integrated feedback back to the end-organs of the GI tract. Thus regulation of the GBA is essential for maintaining homeostasis of the CNS and GI tract. However, the GBA dysfunction underlies the clinical signs of the functional GI and CNS disorders. Several neurotransmitters may play a role in this process, including serotonin, dopamine, gamma-aminobutyric acid, and CGRP.[72–75] Therefore, due to the bidirectional association between permeability of the gut and inflammatory state, augmented gut permeability can stimulate inflammatory and immune responses through lipopolysaccharide leakage and subsequently proinflammatory cytokines can reinforce gut permeability.[76, 77] As mentioned earlier, proinflammatory cytokines (TNF-α, IL-1β, and IL-6) could also affect nociceptive responses in the trigeminal pathway and play a role in triggering migraine headache.[78–83] Fig. 3 depicts several mechanisms underlying the bidirectional relationships between the gut and the brain (from gut microbiota to brain and from brain to gut microbiota) by means of neural, endocrine, immune, and humoral links. In a healthy person, gut bacteria keep normal balance of the GBA via indirect signaling in which these inflammatory cytokines play an important role. However, dysregulation of GBA allows unhealthy gut microbiota to modulate the intestinal barrier, which can influence all the underlying compartments.

Germ-free (GF) mice are bred in isolators that fully block exposure to microorganisms, with the intent of keeping them free of detectable bacteria. Gut microbiota deprivation in GF mice has been reported to prolong nitroglycerine (NTG)-induced migraine-like pain. This pain prolongation is completely reversed by gut colonization with fecal microbiota from WT mice.[16] Moreover, the chronic use of antibiotics has been reported to produce chronicity of NTG-induced migraine-like pain in a mouse model due to disturbance in gut microbiota and recovering microbiomes with probiotics preventing the chronicity of such pain,[16] implicating that gut microbiota dysbiosis is critical for migraine-like pain chronicity. TNFα (a proinflammatory cytokine) plays an important role in the development of chronic pain.[84–86] It increases α-amino-3-hydroxy-5-methyl-4-isoxazolepropionic acid receptor (AMPAR) GluA1 phosphorylation and trafficking in the spinal dorsalhorn neurons, contributing to inflammatory pain. AMPAR GluA1 Ser831 phosphorylation is critical for NTG-induced migraine-like pain.[87] Thus activation of the TNFα-AMPAR pathway in the trigeminal nociceptive system is likely one of the underlying mechanisms for gut microbiota dysbiosis-enhanced migraine-like pain.

Gut microbiota can regulate microglia maturation and function by producing short chain fatty acids (SCFAs).[88] Gut microbiota dysbiosis may activate microglia and secrete more TNFα to enhance migraine-like headache. The exact pathogenesis of migraine is still undefined. However, several factors, including gut microbiota, seem to play an important and proven role via the GBA.[27] Intestinal microbiota composition and function are related to nervous system activity and, in turn, communicating with the GBA through endocrine, immunological, and neural messages. The brain normally regulates movements and functions of the GI tract. Hormonal factors through the HPA axis mediate stress responses on gut functions. The GI system is believed to be able to affect the CNS. A number of brain functions such as cognition, behavior, and even nociception are under the influence of the gut system. Inflammation and neuroimmune modulation in the GI tract could play an important role in the pathogenesis of migraine headache, but how gut microbiomes contribute to migraine headache via dysregulation of GBA is still unclear.[17, 49, 73, 89, 90] However, the tentative mechanism (Fig. 3) may be that increased intestinal epithelium permeability results in gut microbiota dysbiosis, which induces the release of proinflammatory cytokines, activating the trigeminal pathway and stimulating the nociceptive sensory system, thus leading to migraine headache.

6 Conclusion

Migraine is a chronic neurological disorder that presents with significant challenges due to its unpredictable onset, episodic pattern, intensity, and duration. The literature findings included in this chapter confirm the association between migraine and functional GI disorders. Most patients with a migraine experience GI disturbances including nausea, vomiting, abdominal pain, or diarrhea, in addition to various neurological and vascular symptoms. The involvement of the gut microbiota via the GBA suggests that gut microbiota dysbiosis can be associated with migraine, which may be influenced by multiple factors such as gut microbiota profile, neuropeptides and serotonin pathway, inflammatory mediators (IL-1β, IL-6, IL-8, and TNF-α), stress hormones, and nutritional substances. Several other factors including environmental factors (such as noise, odors, bright or flickering light), lifestyle (unhealthy diet, stress, sleep deficit), and hormones (medications including menopausal hormone therapy, oral contraceptives) are known to be triggers of migraine headache. Some research studies have shown that migraine sufferers have greater levels of bacteria, which are involved in processing nitrates and typically found in processed meats and some wines.

Use of a probiotic mouthwash, which may be an option in the near future, appropriate consumption of a controlled diet that contains high-fiber, low-glycemic index food, and supplementation with vitamins and minerals (vitamin D, omega-3, and probiotics) may help improve migraine-associated features. Probiotics may help by restoring tight-junction integrity and protecting the intestinal barrier thus preventing changes in hippocampal neurogenesis and expression of hypothalamic genes involved in synaptic plasticity. However, the precise mechanisms and pathways related to the dysregulation of GBA resulting in migraine headache still need to be fully elucidated.

References

1. Global Burden of Disease Study. Global, regional, and national incidence, prevalence, and years lived with disability for 301 acute and chronic diseases and injuries in 188 countries, 1990–2013: a systematic analysis for the Global Burden of Disease Study 2013. *Lancet*. 2015;386:743–800.
2. Chen J, Wang Q, Wang A, Lin Z. Structural and functional characterization of the gut microbiota in elderly women with migraine. *Front Cell Infect Microbiol*. 2020;9:470.
3. Katsarava Z, Buse DC, Manack AN, Lipton RB. Defining the differences between episodic migraine and chronic migraine. *Curr Pain Headache Rep*. 2012;16:86–92.
4. Headache Classification Committee of the International Headache Society (IHS). The international classification of headache disorders, 3rd edition (beta version). *Cephalalgia*. 2013;33:629–808.
5. Lipton RB, Silberstein SD. Episodic and chronic migraine headache: breaking down barriers to optimal treatment and prevention. *Headache*. 2015;55(Suppl 2):103–122.
6. Farooqui T. Trace amines and their potential role in primary headaches: an overview. In: Farooqui T, Farooqui AA, eds. *Trace Amines and Neurological Disorders*. Oxford: Academic Press; 2016:349–366.
7. Victor TW, Hu X, Campbell JC, Buse DC, Lipton RB. Migraine prevalence by age and sex in the United States: a life-span study. *Cephalalgia*. 2010;30:1065–1072.
8. Nazari F, Safavi M, Mahmudi M. Migraine and its relation with lifestyle in women. *Pain Pract*. 2010;10(3):228–234.
9. Gonzalez A, Hyde E, Sangwan N, Gilbert JA, Viirre E, Knight R. Migraines are correlated with higher levels of nitrate-, nitrite-, and nitric oxide-reducing oral microbes in the American Gut Project Cohort. *mSystems*. 2016;1(5).
10. Samsum M, Coveñas R, Ahangari R, Yajeya J, Narváez JA. Role of neuropeptides in migraine: where do they stand in the latest expert recommendations in migraine treatment? *Drug Dev Res*. 2007;68:294–314.
11. Aggarwal M, Puri V, Puri S. Serotonin and CGRP in migraine. *Ann Neurosci*. 2012;19(2):88–94.
12. Deen M, Hansen HD, Hougaard A, et al. High brain serotonin levels in migraine between attacks: A 5-HT_4 receptor binding PET study. *Neuroimage Clin*. 2018;18:97–102.
13. Neeb N, Reuter U. Nitric oxide in migraine. *CNS Neurol Disord Drug Targets*. 2007;6(4):258–264.
14. Sparaco M, Feleppa M, Lipton R, Rapoport A, Bigal M. Mitochondrial dysfunction and migraine: evidence and hypotheses. *Cephalalgia*. 2006;26(4):361–372.
15. Nattagh-Eshtivania E, Mahmood Sanib MA, Dahria M, et al. The role of nutrients in the pathogenesis and treatment of migraine headaches: review. *Biomed Pharmacother*. 2018;102:317–325.

16. Tang Y, Liu S, Shu H, Lora Yanagisawa L, Tao F. Gut microbiota dysbiosis enhances migraine-like pain via TNFα upregulation. *Mol Neurobiol.* 2020;57(1):461–468.
17. Arzani M, Jahromi SR, Ghorbani Z, et al. Gut-brain axis and migraine headache: a comprehensive review. *J Headache Pain.* 2020;21(1):15.
18. Willis T. *The Anatomy of the Brain and Nerves.* Montreal: McGill University Press; 1964.
19. Wolff HG. *Headache and Other Head Pain.* New York: Oxford University Press; 1948.
20. Negro A, Koverech A, Martelletti P. Serotonin receptor agonists in the acute treatment of migraine: a review on their therapeutic potential. *Pain Res.* 2018;8(11):515–526.
21. Linde M, Mellberg A, Dahlöf C. Subcutaneous sumatriptan provides symptomatic relief at any pain intensity or time during the migraine attack. *Cephalalgia.* 2006;26:113–121.
22. Lassen LH, Haderslev PA, Jacobsen VB, Iversen HK, Sperling B, Olesen J. CGRP may play a causative role in migraine. *Cephalalgia.* 2002;22:54–61.
23. Rahmann A, Wienecke T, Hansen JM, Fahrenkrug J, Olesen J, Ashina M. Vasoactive intestinal peptide causes marked cephalic vasodilatation but does not induce migraine. *Cephalalgia.* 2007;28:226–236.
24. Schytz HW, Birk S, Wienecke T, Kruuse C, Olesen J, Messoud A. PACAP38 induces migraine-like attacks in patients with migraine without aura. *Brain.* 2009;132(1):16–25.
25. Goadsby P. The vascular theory of migraine—a great story wrecked by the facts. *Brain.* 2009;132:6–7.
26. Leậo AAP. Spreading depression of activity in the cerebral cortex. *J Neurophysiol.* 1944;7:359–390.
27. Dodick DW. A phase-by-phase review of migraine pathophysiology. *Headache.* 2018;58(Suppl 1):4–16.
28. Charles AC, Baca SM. Cortical spreading depression and migraine. *Nat Rev Neurol.* 2013;9:637–644.
29. Russell MB, Olesen J. A nosographic analysis of the migraine aura in a general population. *Brain.* 1996;119:355–361.
30. Lashley KS. Patterns of cerebral integration indicated by the scotomas of migraine. *Arch Neurol Psychiatr (Chicago).* 1941;46:331–339.
31. Lauritzen M. Pathophysiology of migraine aura: the spreading depression theory. *Brain.* 1994;117:199–210.
32. Milner PM. Note on a possible correspondence between the scotomas of migraine and spreading depression of Leậo. *Electroencephalogr Clin Neurophysiol Suppl.* 1958;10:705.
33. Costa C, Tozzi A, Rainero I, et al. Cortical spreading depression as a target for anti-migraine agents cortical spreading depression as a target for anti-migraine agents. *J Headache Pain.* 2013;14:62.
34. Durham PL, Garrett FG. Neurological mechanisms of migraine: potential of the gap-junction modulator tonabersat in prevention of migraine. *Cephalalgia.* 2009;29(Suppl 2):1–6.
35. Borgdorff P. Arguments against the role of cortical spreading depression in migraine. *Neurol Res.* 2018;40(3):173–181.
36. Pietrobon D, Moskowitz MA. Pathophysiology of migraine. *Annu Rev Physiol.* 2013;75:365–391.
37. Akerman S, Holland PR, Peter J, Goadsby PJ. Diencephalic and brainstem mechanisms in migraine. *Nat Rev Neurosci.* 2011;12(10):570–584.
38. Moskowitz MA, Reinhard JF, Romero J, Melamed E, Pettibone DJ. Neurotransmitters and the fifth cranial nerve: is there a relation to the headache phase of migraine? *Lancet.* 1979;2:883–885.
39. Buzzi MG, Moskowitz MA. The trigemino-vascular system and migraine. *Pathol Biol.* 1992;40(4):313–317.
40. Williamson DJ, Hargreaves RJ. Neurogenic inflammation in the context of migraine. *Microsc Res Tech.* 2001;53(3):167–178.

41. Peroutka SJ. Neurogenic inflammation and migraine: implications for the therapeutics. *Mol Interv.* 2005;5(5):304–311.
42. Araki N. Migraine. *Jpn Med Assoc J.* 2004;47(3):124–129.
43. Kaise EA, Russo EF. CGRP and migraine: could PACA play a role too? *Neuropeptides.* 2013;47:451–461.
44. Ramachandran R, Bhatt DK, Ploug KB, et al. Nitric oxide synthase, calcitonin gene-related peptide and NK-1 receptor mechanisms are involved in GTN-induced neuronal activation. *Cephalgia.* 2014;34(2):136–147.
45. Dong X, Hu Y, Jing L, Chen J. Role of phosphorylated extracellular signal-regulated kinase, calcitonin gene-related peptide and cyclooxygenase-2 in experimental rat models of migraine. *Mol Med Rep.* 2015;12(2):1803–1809.
46. Kawabata A. Prostaglandin E2 and pain—an update. *Biol Pharm Bull.* 2011;34(8):1170–1173.
47. Antonova M. Prostaglandins and prostaglandin receptor antagonism in migraine. *J Headache Pain.* 2013;14(Suppl 1):P114.
48. Ashina M, Hansen JM, Do TP, Melo-Carrillo A, Burstein R, Moskowitz MA. Migraine and the trigeminovascular system—40 years and counting. *Lancet.* 2019;18(8):P795–P804.
49. Hindiyeh N, Aurora SK. What the gut can teach us about migraine. *Curr Pain Headache Rep.* 2015;19(7):33.
50. Camara-Lemarroy CR, Rodriguez-Gutierrez R, Monreal-Robles R, Marfil-Rivera A. Gastrointestinal disorders associated with migraine: a comprehensive review. *World J Gastroenterol.* 2016;22(36):8149–8160.
51. van Hemert S, Breedveld AC, Rovers JM, et al. Migraine associated with gastrointestinal disorders: review of the literature and clinical implications. *Front Neurol.* 2014;5:241.
52. Quigley EMM. Gut bacteria in health and disease. *Gasteroenterol Hepatol.* 2013;9(9):560–569.
53. Su J, Zhou XY, Zhang GX. Association between Helicobacter pylori infection and migraine: a meta-analysis. *World J Gastroenterol.* 2014;20(40):14965–14972.
54. Wang D, Li Y, Zhong H, Qiuxia Ding Q, Lin Y, Tang S. Alterations in the human gut microbiome associated with *Helicobacter pylori* infection. *FEBS Open Bio.* 2019;9(9):1552–1560.
55. Ford S, Finkel AG, Isaacs KL. Migraine in patients with inflammatory bowel disorders. *J Clin Gastroenterol.* 2009;43(5):499.
56. Cheraghi SC, Daryani NE, Ghabaee M. A survey on migraine prevalence in patients with inflammatory bowel disease—a single centre experience. *Middle East J Dig Dis.* 2016;8(4):282–288.
57. Zhou Q, Zhang B, Verne GN. Intestinal membrane permeability and hypersensitivity in the irritable bowel syndrome. *Pain.* 2009;146:41–46.
58. Li C, Yu S, Li H, et al. Clinical features and risk factors for irritable bowel syndrome in migraine patients. *Pak J Med Sci.* 2017;33(3):720–725.
59. Dimitrova AK, Ungaro RC, Lebwohl B, et al. Prevalence of migraine in patients with celiac disease and inflammatory bowel disease. *Headache.* 2013;53(2):344–355.
60. Rashtak S, Murray JA. Review article: coeliac disease, new approaches to therapy. *Aliment Pharmacol Ther.* 2012;35:768–781.
61. Zis P, Julian T, Hadjivassiliou M. Headache associated with coeliac disease: a systematic review and meta-analysis. *Nutrients.* 2018;10:1445.
62. Rahmoune H, Boutrid N. Migraine, celiac disease and intestinal microbiota. *Pediatr Neurol Briefs.* 2017;31(2):6.
63. Parkman HP. Migraine and gastroparesis from a gastroenterologist's perspective. *Headache.* 2013;53(Suppl 1):4–10.
64. Aurora SK, Papapetropoulos S, Kori SH, Kedar A, Abell TA. Gastric stasis in migraineurs: etiology, characteristics, and clinical and therapeutic implications. *Cephalalgia.* 2013;33(6):408–415.
65. Newman LC. Why triptan treatment can fail: focus on gastrointestinal manifestations of migraine. *Headache.* 2013;53(Suppl 1):11–16.

66. Dalziel JE, Fraser K, Young W, et al. Gastroparesis and lipid metabolism-associated dysbiosis in Wistar-Kyoto rats. *Am J Physiol Gastrointest Liver Physiol*. 2017;313(1):G62–G72.
67. Gelfand AA, Thomas KC, Goadsby PJ. Before the headache: infant colic as an early life expression of migraine. *Neurology*. 2012;79:1392–1396.
68. de Weerth C, Fuentes S, Puylaert P, de Vos WM. Intestinal microbiota of infants with colic: development and specific signatures. *Pediatrics*. 2013;131(2):e550–e558.
69. Eutamène H, Garcia-Rodenas CL, Yvon S, et al. Luminal contents from the gut of colicky infants induce visceral hypersensitivity in mice. *Neurogastroenterol Motil*. 2017;29(4).
70. Partty A, Kalliomaki M, Salminen S, Isolauri E. Infantile colic is associated with low-grade systemic inflammation. *J Pediatr Gastroenterol Nutr*. 2016;64:691–695.
71. Rhoads JM, Collins J, Fatheree NY, et al. Infant colic represents gut inflammation and dysbiosis. *J Pediatr*. 2018;203:55–61.
72. Carabotti M, Scirocco A, Maselli MA, Severi C. The gut-brain axis: interactions between enteric microbiota, central and enteric nervous systems. *Ann Gastroenterol*. 2015;28:203–209.
73. Mayer EA, Tillisch K, Gupta A. Gut/brain axis and the microbiota. *J Clin Invest*. 2015;125(3):926–938.
74. Skonieczna-Żydecka K, Marlicz W, Misera A, Koulaouzidis A, Łoniewski I. Microbiome-the missing link in the gut-brain axis: focus on its role in gastrointestinal and mental health. *J Clin Med*. 2018;7(12):521.
75. Martin CR, Osadchiy V, Kalani A, Mayer EA. The brain-gut-microbiome axis. *Cell Mol Gastroenterol Hepatol*. 2018;6(2):133–148.
76. Marchiando AM, Graham WV, Turner JR. Epithelial barriers in homeostasis and disease. *Annu Rev Pathol*. 2010;5:119–144.
77. Mennigen R, Bruewer M. Effect of probiotics on intestinal barrier function. *Ann NY Acad Sci*. 2009;1165:183–189.
78. Yücel M, Kotan D, Gurol Çiftçi G, Çiftçi IH, Cikriklar HI. Serum levels of endocan, claudin-5 and cytokines in migraine. *Eur Rev Med Pharmacol Sci*. 2016;20(5):930–936.
79. Leung L, Cahill CM. TNF-alpha and neuropathic pain—a review. *J Neuroinflammation*. 2010;7(1):27.
80. Bruno PP, Carpino F, Carpino G, Zicari A. An overview on immune system and migraine. *Eur Rev Med Pharmacol Sci*. 2007;11(4):245–248.
81. Perini F, D'Andrea G, Galloni E, et al. Plasma cytokine levels in migraineurs and controls. *Headache*. 2005;45(7):926–931.
82. Dominguez C, Vieites-Prado A, Perez-Mato M, et al. Role of adipocytokines in the pathophysiology of migraine: a cross-sectional study. *Cephalalgia*. 2018;38(5):904–911.
83. Martami F, Jahromi SR, Togha M, Ghorbani Z, Seifishahpar M, Saidpour A. The serum level of inflammatory markers in chronic and episodic migraine: a case-control. *Neurol Sci*. 2018;39(10):1741–1749.
84. Calvo M, Dawes JM, Bennett DL. The role of the immunesystem in the generation of neuropathic pain. *Lancet Neurol*. 2012;11(7):629–642.
85. de Miguel M, Kraychete DC, Meyer Nascimento RJ. Chronic pain: cytokines, lymphocytes and chemokines. *Inflamm Allergy Drug Targets*. 2014;13(5):339–34954.
86. Ren K, Dubner R. Interactions between the immune and nervous systems in pain. *Nat Med*. 2010;16(11):1267–1276.
87. Tang Y, Liu S, Shu H, Xing Y, Tao F. AMPA receptor GluA1Ser831 phosphorylation is critical for nitroglycerin-induced migraine-like pain. *Neuropharmacology*. 2018;133:462–469.
88. Erny D, Hrabe de Angelis AL, Jaitin D, et al. Host microbiotaconstantly control maturation and function of microglia in the CNS. *Nat Neurosci*. 2015;18(7):965–977.
89. Asano Y, Hiramoto T, Nishino R, et al. Critical role of gut microbiota in the production of biologically active, free catecholamines in the gut lumen of mice. *Am J Physiol Gastrointest Liver Physiol*. 2012;303(11):G1288–G1295.
90. Barrett E, Ross RP, O'Toole PW, Fitzgerald GF, Stanton C. γ-Aminobutyric acid production by culturable bacteria from the human intestine. *J Appl Microbiol*. 2012;113(2):411–417.

CHAPTER 16

Insights on the modulatory role of Ayurveda-based herbal preparations on gut microbiome and neuroprotection

P.S. Rajini[a], M.M. Srinivas Bharath[b], and M. Muralidhara[c]
[a]Food Protectants and Infestation Control, CSIR-CFTRI, Mysore, India
[b]Department of Clinical Psychopharmacology and Neurotoxicology, National Institute of Mental Health and Neuro Sciences (NIMHANS), Bangalore, India
[c]Department of Biochemistry and Nutrition, CSIR—Central Food Technological Research Institute (CFTRI), Mysuru, Karnataka, India

1 Introduction

Ayurveda came into existence around 900 BCE and is accepted as one of the oldest treatises on medicine.[1] The term "Ayurveda" derived from Sanskrit comprises two words: "Ayus" meaning life and "Veda" meaning learning or science. Hence, "Ayurveda" signifies "Knowledge of life." The origin of Ayurveda can be traced from ancient knowledge in *Rig Veda* and *Atharva Veda*[2,3] and comprises a complete self-sustainable system of medicine involving a scientific tradition of harmonious living. Ayurvedic medicine is a personalized system of traditional medicine native to India (4500–1500 BCE) and is becoming increasingly popular globally, largely due to its profound conceptual basis, holistic therapeutic practice, and survival of its medicines since prehistoric times.[4]

Ayurvedic herbal remedies are historically considered among the most powerful means of maintaining human health and homeostasis. The basic tenet of Ayurvedic treatment is to "prevent the cause of the disease" and to "bring awareness in the patient about the cause".[5,6] In Ayurveda, the three "Doshas" (*Vata*, *Pitta*, and *Kapha*) are well-organized elements, which are described as highly susceptible to imbalance and derangement.[7] Various qualities and functions have been attributed to each dosha. The kinetic mechanism of a system has been attributed to *Vata*, the metabolic mechanism to *Pitta*, and the structural and stability mechanism to *Kapha*.[8]

Gut Microbiota in Neurologic and Visceral Diseases
https://doi.org/10.1016/B978-0-12-821039-0.00020-4

The Tridoshas are considered regulatory control factors for the basic physiologic processes in living systems that maintain their identity throughout biologic history. *Vata* and its sub doshas regulate input/output processes and motion; *Pitta* and its sub doshas regulate throughput, turnover, and energy; and *Kapha* and its sub doshas regulate storage, structure, and lubrication.[9] All these constitute the *Prakriti* of an individual, which determines the physical as well as mental characteristics of the human. A balance between these three doshas is fundamental to health, whereas imbalance causes disease/s. Ayurveda pursues to normalize body functions with several techniques encompassing advice on food and activity, internal herbal preparations, purification treatments (*panchakarma*), and surgical methods (*shalyachikitsa*).

In the past two decades, there has been a spectacular growth in the market for herbal medicinal products throughout the world. In developing countries, about 80% of people are known to rely on traditional herbal medicine for their primary health care. The Indian subcontinent is one of the major biodiversity centers with about 45,000 plant species, of which about 15,000 medicinal plants have been recorded in India.[10] Nearly 7500 plants are being employed for curing different diseases and only 1200 to 1500 have been incorporated into the official Ayurvedic pharmacopeia.[10] Current demands for herbal medicines have resulted in an annual market of $1.5 billion.[11] Initially, the rationale for the utilization of Ayurvedic medicinal plants had depended mostly on traditional usage, presumed safety, less cost, relatively easy access, and efficacy. However, in recent years, not only the pharmacological/toxicological and mechanistic studies have begun to be examined, but also both preclinical and clinical studies are being conducted to verify their claimed pharmacological/therapeutic effects.[12]

To visualize the extent (nature and degree) of neuroprotection, which may be possible to achieve by Ayurveda-herbal formulations (AHF) in humans, it is imperative to understand the physiological role as well as the underlying mechanism/s related to the implications of gut microbiota (GM) in brain health/function. Further, it is helpful to examine the various evidence related to the involvement of GM in neurodegenerative disorders (NDD) and their modulation by natural compounds and specific polyphenols. Most importantly, comprehensive data on the direct modulatory propensity of such AHF against neurological disorders in humans are limited. Hence it is plausible that such an approach would enable us in the future to exploit various AHF and achieve maximum neuroprotection against NDD among the vulnerable population at a relatively lower cost.

2 Gut microbes and brain health

2.1 Gut microbiome (GM)

The human gastrointestinal tract (GIT) is known to harbor a plethora of microorganisms, most of which belong to the domain Bacteria. Due to their manifold effects on host physiology and host health, there is tremendous interest and focused research to elucidate the role and function of gut microbial communities. The gut contains 500–1000 different microbial species, with the greatest concentrations in the colon. The GM in the first years of life is unstable and largely dominated by *Clostridium leptum* and *C. coccoides* species.[13] Among healthy adults, phyla such as Firmicutes, Bacteroidetes, and Actinobacteria are dominant. The population of Actinobacteria (especially Bifidobacteriaceae), although found in small concentrations, are very important as modulators of intestinal barrier function.[14–16] Considerable variations between the number of facultative anaerobes in adults and the elderly have been reported.[17,18] While *Bifidobacterium* spp. are known to enhance the barrier function of the gut epithelium and gut health, a reduction in their population is shown to result in chronic low-grade inflammation. In the elderly, a significant reduction of Bacteroidetes and *Bifidobacterium* is accompanied by a decrease in *Lactobacilli*.[19] Firmicutes outnumber Bacteroidetes in the gut in adulthood, whereas Bacteroidetes outnumber Firmicutes in the elderly. There exists a significant difference in the GM between elderly and younger adults; it is less diverse and resilient and can be modulated by environmental factors/interventions in the elderly.[20] The GIT is populated by more than 10^{14} bacteria cells, the majority of which are located in the ileum and colon.[21] Microbes belonging to the genera *Lactobacillus* and *Bifidobacterium* are known to be beneficial. The presence of *Lactobacillus rhamnosus* is associated with increased intestinal barrier function, expression of epithelial tight junction proteins, and decreased serum levels of inflammatory cytokines.[22]

Brain function is reported to be significantly affected by one of many different pathways, including the neuroanatomical pathway of the gut-brain axis, the neuroendocrine–hypothalamic-pituitary-adrenal axis, the immune system, and the microbiota metabolism in the gut, the intestinal mucosal barrier, and the blood-brain barrier (BBB).[23] Intestinal health can potentially affect the CNS via the physiological effects of intestinal barrier function and peripheral neuron activity. Being bidirectional, the communication pathways also extend downstream from the brain to the gut, with changes in neurophysiological behaviors affecting the GM.[24,25]

2.2 Gut Microbiota and NDD

The GM possesses the propensity to influence various aspects of CNS biology through multiple mechanisms. While it is reported that GM influences numerous metabolic and immunologic aspects in health and disease, our understanding of precisely how the GM modulates brain function before and during the progression of the disease is limited. Studies implicating the microbiome-gut-brain axis have mainly described associations between GM composition and disease symptomology. Studies have focused on the impact of gut microbes, and the metabolites they generate from dietary components, on BBB integrity (BBI).[26–28] Given this, there has been focused research to obtain basic insights into how these metabolites travel throughout the body along bidirectional neural and circulatory routes.

Multiple pieces of evidence have demonstrated the therapeutic potential of manipulating the GM on brain health through the delivery of prebiotics, probiotics, and fecal microbiota transplantation. Although these approaches are effective in treating GIT disorders, evidence on their suitability and efficacy to treat NDD are limited.[29] Abundant data suggest that GM is likely to play a key role in a range of neurological conditions, such as autism spectrum disorder (ASD), anxiety, depression, chronic pain, stress, Alzheimer's disease (AD), and Parkinson's disease (PD).[30–32] The GM has been shown to contribute significantly to the onset and severity of these neurological disorders. Further, the GM can also play an important role in the effectiveness or side effects of medications used to treat these NDD.[33] Due to ethical considerations, along with the tremendous genetic and lifestyle heterogeneity among populations, mechanistic studies of the microbiome are difficult to perform in humans. Evidence on the influence of GM on the brain is largely derived from animal studies, in particular germ-free rodents. Studies have revealed that the absence of gut microbes is associated with structural alterations in the BBB, characterized by decreased expression of tight junction proteins, leading to increased permeability compared to conventional specific pathogen-free (SPF) mice.[34] A causal relationship between gut microbes and BBBI has been demonstrated in studies i involved colonizing germ-free mice using fecal material from SPF mice.[35]

2.3 GM-derived metabolites

Impaired motor function and/or dementia are the major causes of disability and dependency among older people worldwide. Chronic neuroinflammation, a characteristic feature of all dementias and NDD, involves over-activation and dysregulation of microglia, the resident macrophage-like immune cells of the brain.[36] The morphology of activated microglia changes in association with elevated secretion of pro-inflammatory cytokines (e.g., IFN-γ, IL-6, and TNF-α) and the release of reactive oxygen and nitrogen species, which can potentially lead to neuronal cell death, loss of BBBI, and brain damage.[37,38]

The GM is known to metabolize dietary components (macro and micronutrients, fibers, and polyphenols) into a range of compounds such as short-chain fatty acids (SCFAs), trimethylamines (TMAs), amino acid derivatives, and vitamins.[39] These metabolites can modulate the disease severity of NDD via two mechanisms: (1) immune-mediated neurodegeneration and (2) direct effects of microbiome-derived metabolites on cells of the CNS. The microbially derived metabolites and dietary components are known to possess essential metabolic and signaling functions that can modulate host homeostasis, including BBBI and brain function.[40] Brief accounts of some of these metabolites are presented in the sections that follow.

Short-chain fatty acids (SCFAs): SCFAs in concentrations of approximately 100 mM in the colonic lumen stimulate colonic blood flow and upper-gut motility, and influence water and salt uptake.[41,42] In animal models that exhibit synaptic and neuronal loss and learning, administration of butyrate or enhanced intake of dietary soluble fibers is shown to improve neuroinflammation, while intracerebroventricular injections of butyrate are reported to modify histone acetylation in the hippocampus resulting in improved memory and learning.[43] High brain levels of propionate exacerbated symptoms in a mouse model of autism, which can be mitigated with butyrate supplementation.[44] Providing mixtures of SCFAs or simply increasing dietary fibers proved to rescue the hypometabolism associated with neuronal dysfunction in neurodegenerative conditions.[45]

Trimethylamine (TMA): GM is known to produce TMA as a metabolite from metabolism of dietary choline, lecithin, carnitine, and trimethylamine-N-oxide (TMAO), all of which are present in a variety of foods. The genera *Anaerococcus*, *Clostridium*, *Escherichia*, *Proteus*, *Providencia*, and *Edwardsiella* are involved in the production of TMA. TMAO has been shown to restore the ability of mutant tau protein and promote microtubule assembly and neuron death, which are the hallmark pathological features of AD.[46–48]

Amino acid metabolites: Products of amino acid catabolism by GM influence the balance of production of excitatory and inhibitory neurotransmitters, which is critical for proper brain function. *Lactobacillus* and *Bifidobacterium* species are known to metabolize glutamate, the excitatory neurotransmitter in the brain, to produce γ- aminobutyric acid (GABA), a major inhibitory neurotransmitter.[49] Decarboxylases secreted by *C. sporogenes* is known to contribute towards converting tryptophan to the neurotransmitter tryptamine, which is involved in the release of serotonin by cells of the enteroendocrine and enteric nervous systems (ENS).[50–52] *B. infantis* can increase plasma levels of tryptophan.[19] Bacterial conversion of tryptophan to indoles, ligands for aryl hydrocarbon receptors can decrease brain inflammation. Tryptophan is a precursor of many other diverse microbial and host metabolites including kynurenic acid, which has anti-inflammatory properties in the GI tract and is considered to be neuroprotective.[53]

Vitamins: GM is known to produce several vitamins that play a significant role in NDD. Major vitamins produced by GM are vitamin K by *Escherichia coli*, *Klebsiella pneumoniae*, *Propioni bacterium*, and *Eubacteriu*; vitamin B2 (riboflavin) by *Bacillus subtilis* and *E. coli*; B9 (folic acid) by *Bifidobacterium*, *Lactococcus lactis*, and *Streptococcus thermophilus*; and B12 (cobalamin) by *L. reuteri* and *Propionibacterium freudenreichii*. Low levels of vitamin K have been correlated with apolipoprotein Eε4 allele, a risk factor linked to AD.[54] Vitamin K has a beneficial role in the modulation of α-synuclein fibrillization associated with PD.[55] Increased dietary vitamin K intake has been associated with less severe subjective memory complaints in the elderly.[56] While a deficiency in vitamin B synthesis is associated with several neurological disorders such as beriberi, polyneuropathy, and cerebellar ataxia, vitamin B9 has a preventive role in the development of neural tube defects, coronary heart disease, cancer, and neuropsychiatric disorders.[57–59] Folate deficiency has been associated with depression, with methyl folate being one of the only medical foods licensed for treating depression.[60] High-dose B-vitamin treatments, including B6, B9, and B12, can decrease levels of homocysteine, which slows the atrophy of specific regions of the brain associated with cognitive decline in AD patients.[61]

2.4 GM modulation: Existing approaches

It is hypothesized that a complex interplay of signaling between epithelial, immune, and neural cell signaling networks is involved in translating microbial metabolite changes in the gut and the brain. Hence altering or shaping the GM is considered a viable strategy to prevent or treat various extra-intestinal diseases.[62–64] Current therapeutic approaches involve modification of the existing microbial composition in the gut by altering nutrient availability to promote the growth of particular classes or species of bacteria (prebiotics) or by introducing or expanding "beneficial" species (probiotics) or by fecal microbial transplantation. These approaches and their principle mechanism/s of action are briefly discussed in the sections that follow.

Prebiotics: Several plant polysaccharide complexes present in foods cannot be digested by human enzymes due to their insolubility or to a deficiency of human-encoded hydrolytic enzymes. These food substrates are called prebiotics, which are nondigestible short-chain carbohydrates that, when ingested selectively, have the potential to stimulate the development of some health-promoting colonic bacteria.[65] Common prebiotics include disaccharides (lactulose), oligosaccharides (fructooligosaccharides [FOS], galactooligosaccharides [GOS]), and polysaccharides (inulin).[66] Other compounds that are considered prebiotics are resistant starches, pectin, whole grains, and polyphenols. Prebiotic compounds are metabolized by the GM and resulting metabolites, such as SCFAs, act on the brain through different pathways contributing to an effect on BBB permeability and neurological health. Several animal studies have examined the effect of prebiotics on brain function such as mood modulation and stress responses.[67] Supplementation with long-chain FOS or short-chain GOS, which promotes the growth of *Bifidobacteria*, significantly reduced anxiety-related behaviors and increased social behavior in mice. However, several studies investigating the benefit of prebiotics on human brain function are so far inconclusive.[68]

Probiotics: Probiotics are nonpathogenic live microorganisms added to, or naturally occurring in food products, which confer a health benefit to the host when present in adequate amounts.[69] The major extensively commercialized probiotics are composed of species of *Lactobacillus*, *Bifidobacterium*, *Lactococcus*, *Streptococcus*, and *Enterococcus*.[70] Further, few recent studies have also explored the beneficial features of next-generation probiotics or live therapeutic products, namely, *Akkermansia muciniphila*, *Faecalibacterium prausnitzii*, species of *Bacteroides*, and genetically modified *Lactococcus* strains.[71–74]

In general, probiotic treatments have been associated with modulating mood and anxiety in animal models and humans. *B. longum* reduced 24-h urinary-free cortisol, a biomarker for stress response, among healthy human volunteers.[75] Consumption of fermented milk containing *L. casei* among healthy humans significantly lowered plasma cortisol levels,[76] while a probiotic strain of *L. rhamnosus* exhibited a protective effect on symptoms of postpartum depression.[77] Some of the speculated mechanism/s includes enhanced production of neurotransmitters (GABA and glutamate) and brain-derived neurotrophic factor (BDNF) implicated in learning processes and control of the fear altered balance of circulating pro and anti-inflammatory cytokines and restoration of inflammation-induced BBB permeability.[78] Interestingly, few studies have shown that probiotic modulation significantly alleviates GI symptoms of PD patients.[79] In animal models of AD, administration of *Bifidobacteria/Lactobacilli* mixtures induced a significant improvement in memory and learning, attenuation of oxidative stress, reduction in histopathological markers in the brain, and restoration of neuronal cell proteolytic functions.[80,81] Results of two randomized, double-blinded, placebo-controlled clinical trials have shown improved Mini-Mental State Examination (MMSE) scores, reduced serum C-reactive protein, and improved insulin metabolism in patients receiving probiotic mixture containing *Bifidobacteria* and *Lactobacilli* species.[82,83]

Fecal microbiota transplantation (FMT): This approach involves the administration of fecal material containing GM from a healthy donor to a recipient with a dysbiosis-related condition, to restore normal diversity and function in the microbial community. Few studies reported that the physiological and behavioral features of depression in GM-depleted rats after transplantation with fecal microbiota from depressed patients, as well as a dysregulation in tryptophan metabolism, similar to that reported in autism, schizophrenia, and NDD.[84–86]

Another dietary approach, "personalized nutrition" (PN), utilizes a large specific population's metadata and bioinformatics tools, allowing for personalized dietary interventions that can modify the microbiome by affecting metabolic homeostasis.[87,88] While personalized strategies to propose optimized nutrition are being investigated, our knowledge related to nutrition derived from microbiome studies is limited.[88] Similar attempts are also being made in the field of Ayurveda.

2.5 Modulation of GM by phytochemicals

Phytochemicals are known to exert neuroprotective effects in vivo by a variety of targets. They protect neuronal cells by modification of etiopathogenic factors in NDD such as oxidative stress, mitochondrial dysfunction, and activated mitochondrial apoptosis cascades, deficits of neurotrophic factors, and protein aggregation and accumulation. Further, they can modify cellular signal pathways to induce the expression of neuroprotective genes.[89] Many researchers have reviewed the gamut of mechanisms based on evidence gathered both in in vitro and preclinical models.[90]

It is well documented that GM plays an important role in maintaining intestinal integrity and function and a loss in microbial balance causes severe damage at the local and systemic levels.[91] Since dietary interventions have the propensity to alter the GM composition and activity, numerous researchers have explored this as a therapeutic tool to modulate the intestinal microbiota to alleviate NDD.[92–94] Several phytochemical compounds have been documented to favor the growth of specific beneficial bacteria in the intestine (Table 1). These can potentially reduce the proliferation of pathogenic bacteria in the gut, mediate microbial metabolic processes, and regulate signal translation or genetic expression with phospholipid cell membranes, thus increasing the permeability and loss of cellular constituents and destruction or inactivation of genetic material.[95] Phytochemicals comprehensively studied in various models belong to three large groups, based on their biosynthetic origins: phenolic compounds, terpenoids, and nitrogen compounds or alkaloids. Further, these compounds are also classified based on their biosynthetic pathway and chemical structure: terpenoids, alkaloids, betalains, glucosinolates, cyanogenic glycosides, polyacetylenes, anthocyanins, and other flavonoids.[96–98]

Polyphenols (PP) are a large family of phytochemicals with high chemical diversity and constitute the active principles of plant-derived nutraceuticals and herbal medicinal products.[99–102] PP, the naturally

Table 1 List of phytochemicals that have shown potential to modulate GM in human or mice models.

In humans Phenolic acids; stilbenoids; flavonols; dihydroflavonols; and anthocyanins	↑ *Enterococcus* spp., *Prevotella* spp., *Bifidobacterium* spp., *B. uniformis*, *E. lento*, *B. coccoides*, *E. rectale*
In humans Catechins; flavan-3-ols and monomeric flavan- 3-ol rich	↓ *S. mutans*, *Shigella*, *V. cholerae*, *C.histolyticum*, *C. coccoides-E. rectale* ↑ *Lactobacillus* spp.
In mice models Anthocyanidins and Flavonoids	↑ *Lactobacillus* spp., *Bifidobacterium* ↓ *E. coli*, *Salmonella* spp.
In mice models Chlorogenic acid-polyphenols	↑ *Bacteroides*, *Bifidobacterium*, *Lactobacillus*, *Ruminococcaceae*
In mice models Syringe; *p*-coumaric; 4-hydroxybenzoic; and vanillic	↑ *Bifidobacterium*, *Enterococcus*, *Eggerthellalenta*
In mice models Procyanidins; catechin and epicatechin	↑ *Bifidobacterium* spp.
In mice models Gallic acids	↓ *E. coli*, *P. aeruginosa*

(Adapted from referenec Karole S, Shrivastava S, Thoams S, Soni B, et al. Polyherbal formulation concept for synergistic action: a review. J Drug Delivery Therap. 2019;9:453–466.)

occurring micronutrients present in many food sources, are known to possess strong antioxidant as well as anti-inflammatory properties. However, the precise molecular and cellular mechanisms underlying their biological effects are currently unclear despite significant advances in our understanding of the biology of PP. Several recent studies suggest that PP exert action by modulation of signaling pathways and mediators in cells and these processes act together to maintain brain homeostasis.[103,104]

Dietary flavonoids represent a diverse range of polyphenolic compounds that occur naturally in plant foods. They are present in considerable quantities in commonly consumed fruit, vegetables, grains, herbs, and beverages. Based on their structural complexity, they are subclassified as flavonols, flavones, flavanones, flavan-3-ols, isoflavones, and anthocyanins. The antineuroinflammatory properties and their ability to suppress neuronal apoptosis and stimulate pro-survival signaling cascade mechanisms are well documented.[105,106] Dietary PP possess the propensity to modulate GM composition and function, interfering with bacterial quorum sensing and membrane permeability, as well as sensitizing bacteria to xenobiotics. Flavonoids can significantly affect gut metabolism and immunity and exert antiinflammatory properties. Because of this, various research groups are focusing on obtaining basic insights on polyphenol-mediated attenuation of neuro-inflammatory processes and their role as prophylactic agents, impact on cognitive functions, learning, memory, neuronal plasticity, and neuroprotection. Currently, studies to decipher their role as modulators of the gut-brain axis are also receiving wide attention.[107,108]

Colonic microflora may transform the PP into bioactive compounds, which can influence intestinal ecology and affect human health. Evidence from preclinical and clinical studies suggests that PP can express prebiotic properties and exert antimicrobial activities against pathogenic GM. Dietary PP has shown significant benefits in specific neurological disorders, accompanied by a major impact on GM towards symbiosis.[109] However, their selective permeability across the BBB, poor absorption, rapid metabolism, and systemic elimination limit their bioavailability and therapeutic efficacy.[110] Hence the beneficial effects of orally administered PP in the CNS remain a subject of debate. The proposed benefits of PP either as protective/prophylactic agents or as therapeutic molecules may be achieved by the consumption of a natural PP-enriched diet, by their use as food supplements, or with formulations as pharmaceutical drugs/nutraceuticals. While the health effects of PP depend on the consumed amount and their bioavailability, their overconsumption has raised safety concerns due to the accumulation of high levels of these molecules.[111]

3 Evidence of GM involvement in specific NDD

Evidence gathered from both preclinical and clinical studies in the past decade indicates that GM does play a vital role in aging as well as the development of neurological disorders such as stroke, PD, and AD.[112] While several studies have shown specific alterations in GM in experimental models of NDD, data in humans is rather limited. Some of these studies that have suggested a good correlation between alterations in GM and NDD are presented in the sections that follow.

3.1 Evidence of involvement GM-PD and HD

Since gastrointestinal dysfunctions are consistently associated with PD, they are considered as early biomarkers and may precede the classical motor manifestations.[113] Abnormal GM is often documented among PD patients and an increase in Gram-negative bacteria is consistently reported.[114] However, a wide variation in the proportion of affected phyla, families, genera, and species of gut bacteria is common. PD patients have shown increased gut permeability, which allows an enhanced translocation of bacteria and bacterial products such as lipopolysaccharides (LPS) to the systemic circulation. Other major associated abnormalities reported are altered intestinal permeability, decreased serum levels of LPS binding protein, and altered blood microbiome suggesting a "leaky gut" syndrome condition. LPS can potentially translocate and reach the systemic circulation, and the CNS and is shown to activate brain microglia by binding to toll-like receptor type-4, leading to the expression of a pro-inflammatory microglial profile with the release of reactive oxygen species (ROS) and inflammatory mediators, in the substantia nigra. Dysfunction of the BBB facilitates entry of LPS into the CNS and interestingly this abnormality is corroborated in PD patients by functional neuroimaging techniques.[115] Additionally, LPS seems to increase nitration and oligomerization of α-synuclein by local induction of the enzyme nitric oxide synthase, which supports the notion that LPS besides enhancing the inflammatory response of the CNS in PD patients, it can accelerate the neurodegenerative process.[116,117] Studies also show that abnormalities in GM may lead to deficient production of butyrate (a histone deacetylase inhibitor) implicated in the proper function of intercellular tight junctions. Patients with PD have shown reduced fecal excretion of SCFAs and less abundant butyrate-producing bacteria compared to healthy controls. Among PD patients, there is a decrease of butyrate, which protects dopaminergic neurons from degeneration by upregulating the neurotrophic factors, such as BDNF and GDNF.[118]

Among PD patients, researchers have examined if there occurs a progressive change in GM over the years. Lower counts of *Bifidobacterium*, *B. fragilis*, and *C. leptium* at baseline along with their progressive increase over time were associated with progressive worsening of motor and nonmotor symptoms among PD patients. While one study reported an abundance of Enterobacteriaceae correlated with postural instability and gait difficulties,[38] another study reported a greater severity of motor symptoms correlated with the abundance of *Bacteroides*.[119] However, most studies have not reported a correlation between motor function and abnormalities of GM.

Further, among PD patients, the impairment of the epithelial barrier has been associated with a decrease of Prevotellaceae, which are the main producers of mucin, a protein that protects the epithelium from pathogens.[120] *Prevotella* is a commensal microbe in the colon that is reported to degrade several plant polysaccharides and mucin glycoproteins in the mucosal layer of the gut as well as interact with the immune system.[121,122] Reduced abundance of Prevotellaceae in PD patients is associated with an abundance of Enterobacteriaceae. Prevotellaceae is found to be associated with greater levels of health-promoting neuroactive SCFAs and a large capacity for biosynthesis of thiamine and folate.[123] In addition, reduced prevalence of this enterotype is consistent with decreased levels of the vitamins in PD patients.[124] Decreased abundance of Prevotellaceae also fits well with observations of increased gut permeability in PD since low *Prevotella* levels may indicate decreased mucin synthesis.[40,125,126] Decreased abundance of Prevotellaceae and increased abundance of Lactobacillaceae were also associated with decreased levels of the gut hormone ghrelin, which can regulate nigrostriatal dopamine function and may reduce neurodegeneration in PD.[127] Lactobacilli have also been found to modulate the activity of ENS neurons and vagal afferents, which may affect cellular alpha-synuclein secretion.[119]

It is speculated that GM could also play a role in Huntington's disease (HD) since presymptomatic HD subjects display serum metabolomic shifts that suggest changes in gut microbial-derived metabolites.[128] Studies in HD animal models and humans also reveal gastrointestinal dysfunctions, which may contribute to the worsening of the disease. The consistent loss of body weight, the most important nonneurological complication of HD and a direct consequence of gastrointestinal defects, has been associated with impaired gut mobility and malabsorption. Few studies have reported decreased levels of Prevotella and increased abundance of Enterobacteriaceae in feces of autistic children, suggesting the role of these bacteria in CNS disorders.[129,130]

3.2 Evidence of involvement of GM: Alzheimer's disease (AD)

Several reports suggest that compromised gut microbiota dysbiosis may represent an important co-factor in AD.[131,132] Studies in animal models of AD have revealed significant alterations in intestinal microflora, establishing evidence for the involvement of GM in the pathogenesis of AD.Various bacterial species have been shown to produce or aggravate the production of Ab plaques. The neuroinflammatory response that causes misfolding of neuronal amyloid-b and α-synuclein is determined by gut dysbiosis, priming the innate immune system by microbiota.[133] Studies have demonstrated different intestinal microbial populations both in human patients and in animal models of AD.[134] Microbiota profiles of AD patients have shown decreased numbers of Firmicutes and Actinobacteria, and an increase in Bacteroidetes. Within the Firmicutes, the families of Ruminococcaceae, Turicibacteraceae, Clostridiaceae, and *C. sensustricto* were less abundant among AD patients.[135] Firmicutes, but more Bacteroidetes, in transgenic mice Ad models are associated with chronic neuroinflammation.[136]

Bacteroidetes, a family of Gram-negative bacteria, have LPS as the major outer membrane component, which can trigger systemic inflammation and promote the release of pro-inflammatory cytokines. LPS is associated with AD pathology since it can potentiate amyloid fibrillogenesis[137,138] in humans. Both LPS and Gram-negative *E. coli* fragments found to be greater in AD showed co-localization with amyloid plaques in the brains of AD patients.[139] Animal studies showed that *Akkermansia* and seven other bacterial genera were correlated with levels of cerebral soluble Aβ42. *Akkermansia* and *Butyricicoccus*, which can increase gut barrier integrity, were negatively correlated with the number of pathogenic Aβ42 in the brain.[140,141]

Few studies have reported elevated fecal calprotectin levels in patients with AD, indicating a leaky intestinal barrier.[139] Thus, increased abundance of Bacteroides may result in increased translocation of LPS from the intestines into the systemic circulation and the brain, and thus contribute to or exacerbate AD pathology. In human studies, postmortem examinations of the brains of AD patients showed increased histone deacetylase levels, which are associated with butyrate.[140] Decreased amounts of enteric *Bifidobacterium* in pulmonary alveolar proteinosis (PAP) transgenic mice were found to be correlated with decreased levels of GABA, resulting in CNS dysfunction. Moreover, GABA levels were decreased in the cortices of AD patients in a postmortem study.[141] Interestingly, dysbiosis was also found to decrease the levels of BDNF in the hippocampus and cortex.[142] Further, dysbiosis of the intestinal microbiome has also been implicated in stroke and is correlated with increased vascular and epithelial permeability.[143,144]

Few researchers have suggested that changes in the GM could have a direct effect on the CNS via the gut-brain axis with chronic mild systemic inflammation contributing to pathogenesis. Cyanobacteria present in small numbers in the GI tract are thought to produce β-*N*-methylamino-L-alanine (BMAA), which is elevated in the brains of AD, PD, and amyotrophic lateral sclerosis (ALS) patients. BMAA is an excitotoxin that activates metabotropic glutamate receptors, thereby inducing depletion of the major antioxidant glutathione. As a result, neurons and glial cells are unable to effectively regulate ROS RNS production in the brain.[145] Support for this possibility is provided by clinical studies that have shown that supplementation with *Lactobacilli*- and *Bifidobacteria*-based probiotics significantly improve cognitive, sensory, and emotional functions among AD subjects. More importantly, certain herbal products, in particular dietary PP, have shown the propensity to restore dysbiosis.[146]

3.3 Flavonoids and GM

Flavonoids, a class of polyphenolic compounds, have been proposed as novel agents capable of attenuating neurodegenerative pathology.[147] A body of evidence suggests that anthocyanins (often consumed in greater amounts than other flavonoids) interact with neurons and microglia, facilitating synaptic connectivity under both normal and pathologic conditions, and speculates that they exert their effects through modulation of the GM.[95,148] While the bioavailability of anthocyanins is considered to be low, high amounts of these compounds reach the colon after consumption of anthocyanin-rich foods. Anthocyanins are extensively metabolized by the gut bacteria and these compounds can predominate for several days in the organism given the enterohepatic recirculation.[149] Anthocyanins are utilized by these microorganisms and are shown to selectively stimulate the growth of some bacterial groups. Growth of *Lactobacillus/Enterococcus* spp. and *Bifidobacterium* spp.[150] are significantly enhanced with the consumption of anthocyanins. In animal models, anthocyanins were shown to attenuate the negative impact of high-fat (HF) diets on neuroinflammation.[96] Supplementation with blackberry anthocyanins (BE) was shown to prevent some of the features of HF diet-induced dysbiosis. Further, anthocyanins have been shown to alter host tryptophan metabolism, generating metabolites responsible for the control of CNS inflammation, which may be a possible mechanism underlying their anti-neuroinflammatory properties. Anthocyanins may act, therefore, as mediators of the microbiota-gut-brain axis, allowing the control of neuroinflammation by GM modulation.[151–153]

3.4 Curcumin, GM, and NDD

Curcumin represents one of the most studied herbal remedies, and its neuroprotective potential in various models has been extensively researched and reviewed.[154–156] It is generally considered the main component of turmeric responsible for different pharmacological activities. However, curcumin is characterized by low systemic bioavailability and rapid metabolism.[156] Curcumin in the gut is shown to favor the growth of beneficial bacteria strains such as *Bifidobacteria* and *Lactobacilli* with a reduction of pathogenic strains.[157,158] Besides its various direct effects, curcumin treatment decreases microbial richness and diversity and hence it is reasonably speculated that curcumin modulates the homeostasis of the gut-brain axis, which may explain its neuroprotective properties.[156,159]

3.5 Spice herbs and neuroprotection

Several pieces of evidence in preclinical studies have demonstrated the neuroprotective potential of a variety of spice plants that are extensively used globally.[160] Some of the major spice plants are *Crocus sativus* (*C. sativus*), *Nigella sativa* (*N. sativa*), *Coriandrum sativum* (*C. sativum*), *Ferula assafoetida* (*F. assafoetida*), *Thymus vulgaris* (*T. vulgaris*), *Zataria multiflora* (*Z. multiflora*), and *Curcuma longa* (*C. longa*). These are traditionally used for dietary, additive, spice, and medicinal purposes. Many of these are also employed in various AHF. The major components of these herbs are carotenoids, monoterpenes, and polyphenol compounds, which possess significant neuroprotective propensity.[103–106] These medicinal plants enhance antioxidant defense mechanisms, reduce oxidant levels, and inhibit acetylcholinesterase activity in the CNS. Further, their efficacy is also mediated via reduced proinflammatory cytokines such as IL-6, IL-1b, TNF-α, and total nitrite generation.

4 Ayurveda and NDD

From time immemorial, humans have sought a cure for NDD from numerous natural herbs.[161] Ayurveda first seeks to diagnose patients' conditions before treating them with internal preparations, diet, and habit restrictions. Since plant-based preparations play a major role in the Ayurvedic healing process, numerous studies have attempted to characterize phytochemicals from medicinal as well as dietary plants with protective effects on the nervous system.[162] However, the efficacy of herbal bioactive components is known to be restricted by various factors such as low bioavailability, fast metabolism, permeability, and lack of stability within the CNS.[163]

Globally, the prevalence of NDD is predicted to increase largely due to population growth and enhanced longevity. Owing to their impact on the quality of life, there is a growing demand for the development of novel therapies and neuroprotective agents to prevent as well as to retard their progression.[164] The secondary metabolites of medicinal plants that are of greater pharmaceutical importance represent either byproducts of anabolic or catabolic processes or direct products of metabolism, having some significant biological functions.[165] Accordingly, the neuroprotective potential of various secondary metabolites has been the subject of intense research for several decades.[166,167] These metabolites demonstrate the ability to act simultaneously on various targets and major mechanism/s underlying their antioxidant, protein misfolding, neuro-inflammation, anti-apoptotic, neurotropic, inhibition of monoamine oxidase/acetylcholinesterase enzyme, and antithrombotic effects.

Although Ayurvedic herbal preparations have been used for many years, only recently has there been a focus on mechanistic studies to examine their role in CNS disorders and dementias. Phytochemical analysis of herbal preparations has revealed the presence of many potent compounds, such as lignans, flavonoids, tannins, polyphenols, triterpenes, sterols, and alkaloids,[168,169] which possess a wide spectrum of pharmacological activities such as anti-inflammatory, anti-amyloidogenic, anti-cholinesterase, hypolipidemic, and antioxidant effects (Table 1).

4.1 Ayurvedic preparations: Concept of Triphala

Nervous system disorders, called "Vata Vyadhi" in Sanskrit, are thought to be caused by imbalances of *Vata*, the biological air humor, the energy that moves through the brain, and the nerves controlling both voluntary and involuntary functions. Hence, *Vata* derangements always involve some weakness, disturbance, or hypersensitivity of the nervous system.[170] Since both Ayurveda and Western medicine essentially agree that health and disease begin in the gut, the concept of *Tridosha rasayana* has been extensively employed in Ayurvedic medicine, as it promotes longevity and rejuvenation in patients of all constitutions and ages. Triphala represents an essential foundational formula, as it promotes efficient digestion, absorption, elimination, and rejuvenation.[171] Triphala is revered as a multiuse therapeutic and perhaps even panacea historically as in well-respected Ayurvedic medical texts. Accordingly, a substantial number of studies have validated the wide spectrum uses of Triphala and demonstrated its various properties.[94,172,173]

Various factors such as variability in herbal source, processing, bioavailability, digestion, and absorption of herbal components determine the true efficacy of herbs on human health to be highly variable. Interestingly, this variability is known to be at least partially due to inherent variation in the GM that acts on the ingested components of herbal remedies and transforms them into compounds with increased bioabsorption and bioactivity. These features have confounded the true efficacy of herbal remedies as it pertains to the maintenance of human health and/or the ability to reverse chronic disease states. While numerous studies have shown that altered GM is a feature among several NDD, very few clinical studies have evaluated the efficacy of Ayurvedic herbal preparations in modulating or restoring the GM.[174] The majority of evidence that documents altered GM with herbal preparations were obtained from animal models of NDD either with pure phytochemical/s or with polyherbal formulations (PHFs). However, there is still considerable debate as to whether or not the GM changes are core to the pathophysiology of such conditions or are mere epiphenomena. The efficacy of AHF will be subjected to extensive validation clinically to intervene in the progression of a few neurological disorders and it is plausible that the neuroprotective efficacy may be mediated by via their effect on GM (Scheme 1).

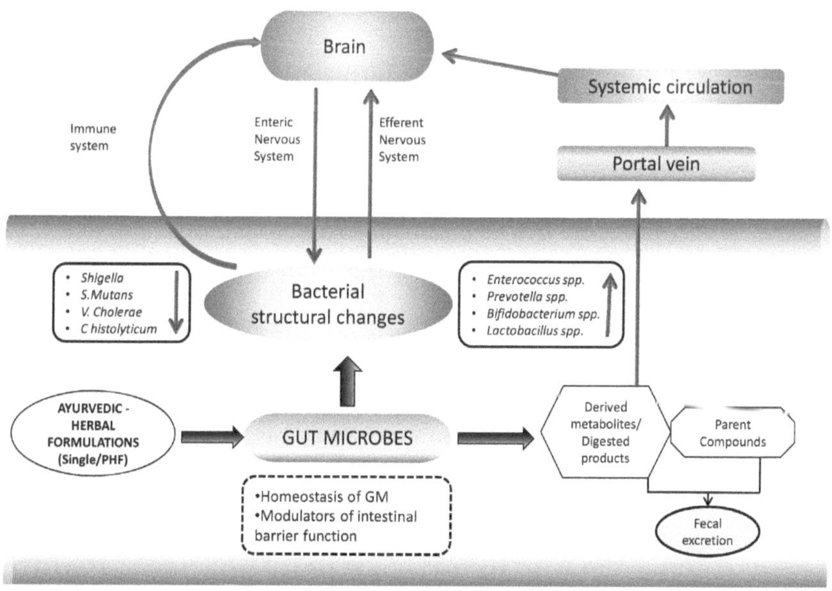

Scheme 1 Scheme depicting the pattern of bidirectional AHF—gut microbiota interactions. The potential mechanisms involve: (1) the direct effect of AHF on the structure of gut microbiota, (2) effect thru GM-derived metabolites, and (3) a direct effect on immune cells to produce cytokines (this aspect has not been discussed here). *(Adapted and modified from reference Chen F, Wen Q, Jiang J, Li HL, et al. Could the gut microbiota reconcile the oral bioavailability conundrum of traditional herbs. J Ehtnopharamcol 2016;179:253–264.)*

4.2 Polyherbalism—Polyherbal formulations (PHFs)

In the Ayurvedic system, naturally occurring herbs/herbal ingredients incorporated into a certain formula, designated as a PHF, has been demonstrated to possess potential interactive effects. PHFs can be used for the treatment of various diseases, as their ingredients work together in a dynamic way to produce maximum therapeutic efficacy with minimum side effects.[175] A major advantage of PHFs over conventional single-component drugs is the presence of multiple active compounds that together can provide a potentiating effect that may not be achievable by any single compound. Based on the nature of the interaction, two mechanisms of synergism have been speculated: (1) pharmacokinetic synergism, which describes the ability of the herb to facilitate the absorption, distribution, metabolism, and elimination of the other herbs and (2) pharmaco-dynamic synergism, which is the synergistic effect when active constituents with similar therapeutic activity are targeted to a similar receptor or physiological system.

PHFs provide improved convenience for patients by eliminating the need of taking more than one different single herbal formulation at a time, which indirectly leads to better compliance and therapeutic effect. These benefits have resulted in the popularity of PHFs in the market when compared to single herbal formulations. PHFs are relatively cheaper, eco-friendly, and more readily available than allopathic drugs. Their better affordability and greater accessibility account for increasing demand globally, especially in rural areas and some developing countries where expensive modern treatments are not available. While PHFs are commonly used globally, several herbal therapies are still under in vivo evaluation and have not been evaluated by clinical trials for their efficacy and safety. Several of them have been tested in preclinical settings in various animal models, but only a few of them have been evaluated in humans.[176]

Extensively used medicinal plants and formulations in Ayurveda to slow down brain aging and enhance memory include Brahmi (*Bacopa monnieri*), Ashwagandha (*Withania somnifera*), Turmeric (*Curcuma longa*), Shankhpushpi (*Convolvulus pluricaulis*, *Evolvulu salsinoides*, and other species), gotu kola (*Centella asiatica*), and guggulu (*Commiphora mukul* and related species). A formulation known as *Brahmi Ghrita*, containing Brahmi, Vaca (*Acorus calamus*), Kushta (*Saussurea lappa*), Shankhpushpi, and Purana Ghrita (old clarified butter/old ghee), is very popular. These formulations not only reduce brain aging and induce anti-stress and memory-enhancing effects that help in the regeneration of neural tissues, but they also induce antioxidant, antiinflammatory, antiamyloidogenic, nutritional, and immune-supportive effects in the human body. Some of these have been tested both in preclinical and clinical models for their neuroprotective efficacy in NDD[177–181] (Table 2).

Table 2 Some polyherbal formulations employed in preclinical settings in NDD.

Name	Medicinal plant	Model
BR-16A	***Bacopa monnieri*, *Centella asiatica*, *Acorus calamus*, *Withania somnifera*, *Tinospora cordifolia*, *Embelica officinalis***, *Evolvulus aisinoides*, *Saussurea lappa*, *Terminalia belerica*, ***Terminalia chebula***, *T.arjuna*	Neuroleptics-induced catalepsy in mice
Wanderer plus	*Paeonia lactiflora*, *Poria cocos*, *Atractylodes macroephala*, *Paeonia suffruticosa*, *Gardenia jasminoides*, *Zingiber officinale*, *Glycyrrhiza uralensis*, *Bupleurum chinense*, *Anglica sinensis*, *Mentha haplocalyx*	Metaanalysis of randomized controlled trials
Smrithi	***Bacopa monniera***, *Hydrocotyle asiatica*, ***Acorus calamus*, *Asparagus racemosus*, *Emblica officinalis***	Forced swimming, tail suspension test)
Perment	*Clitoria ternatea*, ***Withania somnifera*, *Asparagus racemosus*, *Bacopa monniera***	Stress induced depressive model
Siotone	***Withania somnifera***, *Ocimum sanctum*, ***Asparagus racemosus***	Foot shock stress induced model
Catuama	*Paullinia cupana*, *Trichilia catigua*, *Ptychopetalum olacoides*, *Zingiber officinale*	Rabbit corpus cavernosum
Epic-Q	***Withania somnifera*, *Acorus calamus***, *Evolvulusal sinoides*, *Nardostachys jatamansi*, ***Centella asciatica*, *Glycyrrhiza glabra***	Forced swim, tail suspension test, MES, INH induced seizure
Medhagulika	***Acorus calamus***, *Clitoriaternatea*, *Glycyrrhiza glabra*, *Elaeocarpussphaericus*, ***Bacopa monnieri*, *Centella asiatica***, *Sarcostemma acidum*	Forced swim, tail suspension test etc.
Mamsyadi Kwatha	*Nardostachys jatamansi*, ***Withania somnifera***, *Hyocymus niger*	Behavioral despair test, anti-reserpine, CFS test

Continued

Table 2 Some polyherbal formulations employed in preclinical settings in NDD—cont'd

Name	Medicinal plant	Model
RO13	*Eclipta alba*, *Glycyrrhiza glabra*, *Hemidesmus indicus*, *Hibiscus rosa-sinensis*, *Nelumbo nucifera*, *Quercus infectoria*, *Rosa damascena*, ***T. chebula***, *Z. officinalis*	Tail suspension, Force swim test
Bramhi Ghrita	***Bacopa monneri***, *Evolvulus alsinoids*, ***Acorus calamus***, *Saussurea lappa*, cow's ghee	Tail flick method

(Adapted from reference MartínezY, Más D. Role of herbs and medicinal spices as modulators of gut microbiota 2020 chapter in book: herbs and spices publisher: Intech Open (n.d.). https://doi.org/10.5772/intechopen.91208.)

4.3 Triphala and modulation of GM

Triphala is a highly efficacious polyherbal Ayurvedic medicine consisting of fruits of the three plant species *Emblica officinalis* (Euphorbiaceae family), *Terminalia bellerica* (Combretaceae family), and *Terminalia chebula* (*Combretaceae* family).[182] Triphala possesses significant antibacterial action against a variety of Gram-positive and Gram-negative bacteria (e.g., *S. mutans*). It is speculated that PP present in Triphala can promote the growth of beneficial *Bifidobacterium* and *Lactobacillus* species and simultaneously inhibit the growth of undesirable intestinal residents, such as *E. coli*, which may induce inflammatory reaction. Further, the lactic acid bacteria possess enzymatic activity to degrade plant tannins such as gallic acid contained in Triphala. More importantly, Triphala-derived PP, such as chebulinic or ellagic acids, can be transformed by the human intestinal microbiota into various active metabolites including urolithins, which modulate the inflammatory process by generating anti-inflammatory compounds and prevent oxidative injuries of enterocytes.[183]

Thus several researchers opine that the possible manipulation of GM by Triphala extracts may represent a new strategy in the development of functional foods or nutraceuticals and become an effective way of restoring GM in patients with impaired lower GI function. Although speculative, similar mechanisms may explain at least in part the neuroprotective properties of several other PHF. Nevertheless, there is considerable scope to validate the use of Triphala in other neurological conditions and establish its neuroprotective efficacy.[184] Interestingly, Triphala-induced benefit is well documented in persons of all ages and studies have shown that the benefits may be enhanced by co-administration of specific probiotic species. Thus, probiotic formulations consisting of bacterial species capable of mediating the increased digestion, bioabsorption, and bioactivity of Triphala may increase and make more uniform the response and impact of Triphala treatment on human populations.[185]

4.4 Nanoformulations of herbal extracts and NDD

Owing to the fact that the presence of the BBB is the main obstacle for NDD treatment, strategies such as engineered polymeric nanoparticles (PNPs) have been used as drug delivery systems. This approach effectively enhances the passage of drugs or herbal extracts through the BBB and increases the absorption of the compound/s to the brain.[186,187] PNP-loaded herbal extracts have been shown to have enhanced effects on NDD by improving drug biodegradability and biocompatibility, ameliorating therapeutic efficacy, removing pharmacokinetics restrictions, reducing side effects, controlling release, and site targeting.[188] Nanoformulations of several herbal medicines and natural compounds have also been attempted in many animal models of NDD. Nanoformulations of several natural products such as curcumin, quercetin, resveratrol, piperine, gallic acid, Epigallocatechin gallate, and ferulic acid have been developed and tested in preclinical studies.[189–191] Despite the improved performance of NP formulations in preclinical models of neurological disorders, further comprehensive studies are needed to address the safety issues related to these systems. Further, the clinical efficacy of NPs in the area of neurological medicine needs long-term assessments. Currently, there have been no studies that have examined if nanoformulations of herbal compounds have any specific effect on the microbiome in preclinical settings. It is also possible that the improved therapeutic efficacy of such formulations may be at least in part mediated by the gut microbiome.

5 Recent approaches in microbiome research

5.1 Gut microbiome in human populations

Mechanistic studies of the microbiome are typically difficult to perform in humans because of tremendous genetic and lifestyle heterogeneity as well as ethical issues associated with colonizing human subjects with microbes that are hypothesized to cause disease.[192] In this regard, recent approaches have attempted to complement observations in humans with interventions in animal models. These studies have produced new insight into the microbial origins of disease that cannot be acquired from human studies alone. Currently, Microbiome Wide Association Studies (MWAS) are revolutionizing clinical investigations by providing greater patient stratification and new biomarkers of disease.[193] Employing omics tools, there is an ongoing concerted effort to better standardize and integrate methods and data resources to improve our ability to understand microbial dynamics in human systems. Population-based microbiome studies complemented with mechanistic experimental work in mice can use microbial associations with phenotype in humans to identify bacteria or compounds that then can be tested in intervention studies to reveal causal pathways.

5.2 Ayurgenomics: Ayurvedic approach

The therapeutic principles of Ayurveda focus on the principles of *Prakriti* (each individual's unique constitution) and *tridoshas*. Thus *Prakriti* determines the characteristic response of each individual to medications, environmental conditions, and dietary factors. Several studies have led to the association of *Prakriti* with molecular, biochemical, and genomic entities.[194] Ayurgenomics, a recently introduced research field, attempts to bridge this gap between genomics and Ayurveda and serves as an aid in the understanding of inter-individual differences in responses to therapies in various diseases.[195] It focuses on studying inter-individual variances in patients from identical ethnic backgrounds. Studies involving subjects of various *Prakriti* types (*Vata*, *Pitta*, and *Kapha*) have been carried out to identify molecular differences that affect susceptibility and responses of individuals to various environmental or disease conditions.[195] A classification method for the human population, with respect to DNA methylation signatures, is reported based upon the traditional Ayurveda concept of *Prakriti*.[196]

Attempts are also being made to understand the molecular determinants of the Ayurvedic way of *endo*-phenotyping individuals into distinct constitution types termed "*Prakriti*," which forms the basis of personalized treatment.[197] The predicted functional signatures corresponded well with the phenotypes associated with the *Prakriti* groups. For example, *Pitta* individuals are known to be more prone to inflammation, whereas *Kapha* individuals are found to possess functional signatures associated with obesity and soft tissue cancer. *Vata* individuals harbor functional signatures related to neurological and stress-related disorders. Interestingly, this study found the presence of Bacteroidetes and Firmicutes as major GM components in varying composition across *Prakriti* groups. Multiple species of the core microbiome showed differential abundance within *Prakriti* types, with gender-specific signature taxons. This study highlights the importance of *Prakriti*-based endo-phenotypes to explain the variability among healthy individuals in GM flora that have important consequences for an individual's health, disease, and treatment.

6 Conclusions and future perspective

Extensive data suggest that GM can profoundly affect the function and structure of the brain via the intestine-brain axis, although the specific mechanisms remain unclear. Studies have established that dysbiosis of the intestinal microbiota can induce brain pathology. Dysbiosis of the intestinal microbiome facilitates the entry of LPS and pathogens into the circulatory system and CNS, which in turn may lead to a vicious cycle of neuronal destruction. Current data on the modulatory role of Ayurvedic herbal preparations on the GM is limited, although neuroprotective effects documented with several Ayurvedic herbal formulations may be partly related to their potential to modulate GM. Accordingly, with AHF there is a considerable scope to comprehensively assess the extent of their potential to modulate GM either directly or through derived metabolites as well as examine their efficacy to serve as prophylactic agents against common NDD. Further, there is an urgent need to address the possibilities of employing them as neuro-therapeutics at the early stages of any specific NDD and their potential to slow the progression of disease.

With a better understanding of the bidirectional communication between the brain and microbiota, it is plausible to focus our future Ayurvedic research on designing and conducting studies on how precisely dysbiosis is implicated in the pathogenesis of CNS disorders. This would enable us to develop and implement strategies to re-establish a normal microbiota, which would be desirable even in the early phases of disease. Doing so will help us understand the responses of different diets and Ayurvedic herbal formulations administered to the *Prakriti* groups and explain the personalized medicinal systems of Ayurveda.

Acknowledgment

The authors thank Dr. Greeshma Mohandas for drawing the scheme.

References

1. Kapoor LD. Handbook of Ayurvedic Medicinal Plants. vol. 2. *Herbal Reference Library.* Boca Raton, FL, USA: CRC Press; 2000.
2. Mukherjee PK, Ranjit N, Harwansh K, et al. Development of Ayurveda—tradition to trend. *J Ethnopharmacol.* 2017;97:10–24.
3. Balkrishna A, Misra LN. Ayurvedic plants in brain disorders: the herbal hope. *J Tradit Med Clin Nature.* 2017;6:2–12.
4. Chopra A, Saluja M, Tillu G. Ayurveda-modern medicine interface: a critical appraisal of studies of Ayurvedic medicines to treat osteoarthritis and rheumatoid arthritis. *J Ayurveda Integr Med.* 2010;17:190–198.

5. Mukherjee PK, Harwansh RK, Bahadur S, et al. Evidence based validation of Indian traditional medicine-way forward. *World J Tradit Chin Med.* 2016;2:48–61.
6. Sen S, Chakraborty R. Revival, modernization and integration of Indian traditional herbal medicine in clinical practice: importance, challenges and future. *J Tradit Complement Med.* 2017;7:234–244.
7. Karole S, Shrivastava S, Thomas S, et al. Polyherbal formulation concept for synergic action: a review. *J Drug Delivery Therapeut.* 2019;9:453–466.
8. Rammanohar P. *Clinical Evidence in the Tradition of Ayurveda: Evidence-Based Practice in Complementary and Alternative Medicine.* Berlin, Germany: Springer-Verlag; 2012:70.
9. Lakhotia SC. Translating Ayurveda's Dosha-Prakriti into objective parameters. *J Ayurveda Integr Med.* 2014;3:176–178.
10. Kamboj VP. Herbal medicine. *Curr Sci (India).* 2000;78:35–39.
11. Chauhan NS, Pandey R, Mondal AK, et al. Western Indian rural gut microbial diversity in extreme Prakriti endo-phenotypes reveals signature microbes. *Front Microbiol.* 2018;9:118–128.
12. Harvey AL, Edrada-Ebel R, Quinn RJ. The re-emergence of natural products for drug discovery in the genomics era. *Nat Rev Drug Discov.* 2015;14:111–129.
13. Sharon G, Sampson TR, Geschwind DH, Mazmanian SK. The central nervous system and the gut microbiome. *Cell.* 2016;167:915–935.
14. Claesson MJ, Cusack S, O'Sullivan O, et al. Composition, variability, and temporal stability of the intestinal microbiota of the elderly. *Proc Natl Acad Sci U S A.* 2011;108:4586–4591.
15. Cani PD, Neyrinck AM, Fava F, et al. Selective increases of bifidobacteria in gut microflora improve high fat-diet-induced diabetes in mice through a mechanism associated-with endotoxaemia. *Diabetologia.* 2007;50:2374–2383.
16. Harmsen HJ, de Goffau MC. The human gut microbiota. *Adv Exp Med Biol.* 2016;902:95–108.
17. Backhed F, Ley R, Sonnenburg J, et al. Host-bacterial mutualism in the human intestine. *Science.* 2005;307:1915–1920.
18. Arboleya S, Watkins C, Stanton C, Ross RP. Gut bifidobacteria populations in human health and aging. *Front Microbiol.* 2016;7:1204.
19. Wang Y, Wang Z, Wang Y, et al. The gut-microglia connection: implications for central nervous system diseases. *Front Immunol.* 2018;9:e2325.
20. Richard ML, Sokol H. The gut mycobiota: insights into analysis, environmental interactions and role in gastrointestinal diseases. *Nat Rev Gastroenterol Hepatol.* 2019;16:331–345.
21. Harach T, Marungruang N, Dutilleul N, et al. Reduction of Alzheimer's disease beta-amyloid pathology in the absence of gut microbiota. *arXiv:150902273v2.* 2015.
22. Parker A, Fonseca S, Carding SR. Gut microbes and metabolites as modulators of blood-brain barrier integrity and brain health. *Gut Microbes.* 2020;11:135–157.
23. Spielman LJ, Klegeris A. The role of insulin and incretins in neuroinflammation and neuro-degeneration. *Immunoendocrinology.* 2014;1, e391.
24. Huttenhower C, Gevers D, Knight R, et al. Structure, function and diversity of the healthy human microbiome. *Nature.* 2012;86:207–214.
25. Bose S, Cho J. Targeting chaperones, heat shock factor-1, and unfolded protein response: promising therapeutic approaches for neurodegenerative disorders. *Ageing Res Rev.* 2017;35:155–175.
26. Cryan JF, Dinan TG. Mind-altering microorganisms: the impact of the gut microbiota on brain and behavior. *Nat Rev Neurosci.* 2012;3:701–712.
27. Collins SM, Kassam Z, Bercik P. The adoptive transfer of behavioral phenotype via the intestinal microbiota: experimental evidence and clinical implications. *Curr Opin Microbiol.* 2013;16:240–245.
28. Cryan JF, O'Riordan KJ, Cowan CSM, et al. The microbiota-gut-brain axis. *Physiol Rev.* 2019;99:1877–2013.

29. Gilbert JA, Blaser MJ, Caporaso JG, Jansson JK, Lynch SV, Knight R. Current understanding of the human microbiome. *Nat Med.* 2018;24:392–400.
30. Benakis C, Martin-Gallausiaux, Trezzi J, et al. The microbiome-gut-brain axis in acute and chronic brain diseases. *Curr Opin Neurobiol.* 2020;61:1–9.
31. Chandra S, Alam MT, Dey J, et al. Healthy gut, healthy brain: the gut microbiome in neurodegenerative disorders. *Curr Top Med Chem.* 2020;20:1142–1153.
32. Dinan TG, Cryan JF. Gut instincts: microbiota as a key regulator of brain development, ageing and neurodegeneration. *J Physiol.* 2017;595:489–503.
33. Walsh J, Griffin BT, Clarke G, Hyland NP. Drug–gut microbiota interactions: implications for neuropharmacology. *Br J Pharmacol.* 2018;175:4415–4429.
34. Braniste V, Al-Asmakh M, Kowal C, et al. The gut microbiota influences blood-brain barrier permeability in mice. *Sci Transl Med.* 2014;6:263ra158.
35. Fröhlich EE, Farzi A, Mayerhofer R, et al. Cognitive impairment by antibiotic-induced gut dysbiosis: analysis of gut microbiota-brain communication. *Brain Behav Immun.* 2016;56:140–155.
36. Block ML, Hong JS. Microglia and inflammation-mediated neurodegeneration: multiple triggers with a common mechanism. *Prog Neurobiol.* 2005;76:77–98.
37. Block ML, Zecca L, Hong JS. Microglia-mediated neurotoxicity: uncovering the molecular mechanisms. *Nat Rev Neurosci.* 2007;8:57–69.
38. Scheperjans F, Aho V, Pereira PA, et al. Gut microbiota are related to Parkinson's disease and clinical phenotype. *Mov Disord.* 2015;30:350–358.
39. Sampson TR, Debelius JW, Thron T, et al. Gut microbiota regulate motor deficits and neuroinflammation in a model of Parkinson's disease. *Cell.* 2016;167:1469–1480. e12.
40. Wu GD, Chen J, HoMC, et al. Linking long-term dietary patterns with gut microbial enterotypes. *Science.* 2011;334:105–108.
41. Hoyles L, Snelling T, Umlai UK, et al. Microbiome-host systems interactions: protective effects of propionate upon the blood-brain barrier. *Microbiome.* 2018;6:55.
42. Ganapathy V, Thangaraju M, Prasad PD, et al. Transporters and receptors for short-chain fatty acids as the molecular link between colonic bacteria and the host. *Curr Opin Pharmacol.* 2013;13:869–874.
43. Matt SM, Allen JM, Lawson MA, et al. Butyrate and dietary soluble fiber improve neuroinflammation associated with aging in mice. *Front Immunol.* 2018;9:1832.
44. MacFabe DF, Cain DP, Rodriguez-Capote K, et al. Neurobiological effects of intraventricular propionic acid in rats: possible role of short chain fatty acids on the pathogenesis and characteristics of autism spectrum disorders. *Behav Brain Res.* 2007;176:149–169.
45. Zilberter Y, Zilberter M. The vicious circle of hypometabolism in neurodegenerative diseases: ways and mechanisms of metabolic correction. *J Neurosci Res.* 2017;95:2217–2235.
46. Matsumoto T, Nagase Y, Hirose J, et al. Regulation of bone resorption and sealing zone formation in osteoclasts occurs through protein kinase B-mediated microtubule stabilization. *J Bone Miner Res.* 2013;28.1191–1202.
47. Tseng HC, Graves DJ. Natural methylamine osmolytes, trimethylamine N oxide and betaine, increase tau-induced polymerization of microtubules. *Biochem Biophys Res Commun.* 1998;250:726–730.
48. Yang DS, Yip CM, Huang TH, Chakrabartty A, Fraser PE. Manipulating the amyloid-beta aggregation pathway with chemical chaperones. *J Biol Chem.* 1999;274:32970–32974.
49. Barrett E, Ross RP, O'Toole PW, Fitzgerald GF. Etal. Gamma-aminobutyric acid production by culturable bacteria from the human intestine. *J Appl Microbiol.* 2012;113:411–417.
50. Agus A, Planchais J, Sokol H. Gut microbiota regulation of tryptophan metabolism in health and disease. *Cell Host Microbe.* 2018;23:716–724.
51. Williams BB, Van Benschoten AH, Cimermancic P, et al. Discovery and characterization of gut microbiota decarboxylases that can produce the neurotransmitter tryptamine. *Cell Host Microbe.* 2014;16:495–503.

52. Yano JM, Yu K, Donaldson GP, et al. Indigenous bacteria from the gut microbiota regulate host serotonin biosynthesis. *Cell.* 2015;161:264–276.
53. Stone TW, Darlington LG. The kynurenine pathway asa therapeutic target in cognitive and neurodegenerative disorders. *Br J Pharmacol.* 2013;169:1211–1227.
54. Allison AC. The possible role of vitamin K deficiency in the pathogenesis of Alzheimer's disease and in augmenting brain damage associated with cardiovascular disease. *Med Hypotheses.* 2001;57:151–155.
55. Da Silva FL, Coelho Cerqueira E, de Freitas MS, et al. Vitamins K interact with N-terminus alpha-synuclein and modulate the protein fibrillization in vitro. Exploring the interaction between quinones and alpha-synuclein. *Neurochem Int.* 2013;62:103–112.
56. Soutif-Veillon A, Ferland G, Rolland Y, et al. Increased dietary vitamin K intake is associated with less severe subjective memory complaint among older adults. *Maturitas.* 2016;93:131–136.
57. Dietiker C, Kim S, Zhang Y, Christine CW. Characterization of vitamin B12 supplementation and correlation with clinical outcomes in a large longitudinal study of early parkinson's disease. *J Mov Disord.* 2019;12:91–96.
58. Douaud G, Refsum H, de Jager CA, et al. Preventing Alzheimer's disease-related gray matter atrophy by B-vitamin treatment. *Proc Natl Acad Sci U S A.* 2013;110:9523–9528.
59. Langohr HD, Petruch F, Schroth G. Vitamin B1, B2 and B6 deficiency in neurological disorders. *J Neurol.* 1981;225:95–108.
60. Finglas PM, Wright AJ, Wolfe CA, et al. Is there more to folates than neural-tube defects? *Proc Nutr Soc.* 2003;62:591–598.
61. Owen RT. Folate augmentation of antidepressant response. *Drugs Today (Barc).* 2013;49:791–798.
62. Vieira AT, Fukumori C, Ferreira CM. New insights into therapeutic strategies for gut microbiotamodulation in inflammatory diseases. *Clin Transl Immunol.* 2016;5:e87.
63. Martín R, Langella P. Emerging health concepts in the probiotics field: streamlining the definitions. *Front Microbiol.* 2019;10:1047.
64. Bindels LB, Delzenne NM, Cani PD, Walter J. Towards a more comprehensive concept for prebiotics. *Nat Rev Gastroenterol Hepatol.* 2015;12:303–310.
65. Gibson GR, Roberfroid MB. Dietary modulation of the human colonic microbiota: introducing the concept of prebiotics. *J Nutr.* 1995;125:1401–1412.
66. Iannitti T, Palmieri B. Therapeutical use of probiotic formulations in clinical practice. *Clin Nutr.* 2010;29:701–725.
67. Savignac HM, Corona G, Mills H, et al. Prebiotic feeding elevates central brain derived neurotrophic factor, N-methyl-D-aspartate receptor subunits and D-serine. *Neurochem Int.* 2013;63:756–764.
68. de Cossío LF, Fourrier C, Sauvant J, et al. Impact of prebiotics on metabolic and behavioural alterations in a mouse model of metabolic syndrome. *Brain Behav Immun.* 2017;64:33–49.
69. FAO/WHO. Expert consultation on evaluation of health and nutritional properties of probiotics in food including powder milk with live lactic acid bacteria. *Córdoba (Argentina).* 2001;1–4.
70. Kunes J, Prazienkova V, Popelova A, et al. Prolactin-releasing peptide: a new tool for obesity treatment. *J Endocrinol.* 2016;230:R51–R58.
71. Cani PD, de Vos WM. Next-generation beneficial microbes: the case of *Akkermansia muciniphila*. *Front Microbiol.* 2017;8:1765.
72. Breyner NM, Michon C, de Sousa CS, et al. Microbial anti-inflammatory molecule (MAM) from *Faecalibacterium prausnitzii* shows a protective effect on DNBS and DSS-induced colitis model in mice through inhibition of NF-kappa B pathway. *Front Microbiol.* 2017;8:114.

73. Hamady ZZ, Scott N, Farrar MD, et al. Xylan regulated delivery of human keratinocyte growth factor-2 to the inflamed colon by the human anaerobic commensal bacterium *Bacteroides ovatus*. *Gut*. 2010;59:461–469.
74. Saarela MH. Safety aspects of next generation probiotics. *Curr Opin Food Sci*. 2019;30:8–13.
75. Messaoudi M, Lalonde R, Violle N, et al. Assessment of psychotropic-like properties of a probiotic formulation (*Lactobacillus helveticus* R0052 and *Bifidobacterium longum* R0175) in rats and human subjects. *Br J Nutr*. 2011;105:755–764.
76. Kato-Kataoka A, Nishida K, Takada M, et al. Fermented milk containing *Lactobacillus casei* strain Shirota prevents the onset of physical symptoms in medical students under academic examination stress. *Benefic Microbes*. 2016;7:153–156.
77. Slykerman RF, Hood F, Wickens K, et al. Effect of *Lactobacillus rhamnosus* HN001 in pregnancy on postpartum symptoms of depression and anxiety: a randomised double-blind placebo-controlled trial. *EBioMedicine*. 2017;24:159–165.
78. Lu Z, Zhang W, Zhang N, et al. The expression of glutamate transporters in chest compression-induced audiogenic epilepsy: a comparative study. *Neurol Res*. 2008;30:915–919.
79. Tamtaji OR, Taghizadeh M, Daneshvar Kakhaki R, et al. Clinical and metabolic response to probiotic administration in people with Parkinson's disease: a randomized, double-blind, placebocontrolled trial. *Clin Nutr*. 2019;38:1031–1035.
80. Nimgampalle M, Kuna Y. Anti-Alzheimer properties of probiotic, *Lactobacillus plantarum* MTCC 1325 in Alzheimer's disease induced Albino rats. *J Clin Diagn Res*. 2017;11:KC01–KC05.
81. Athari Nik Azm S, Djazayeri A, Safa M, et al. *Lactobacilli* and bifidobacteria ameliorate-memory and learning deficits and oxidative stress in beta-amyloid (1–42) injected rats. *Appl Physiol Nutr Metab*. 2018;43:718–726.
82. Akbari E, Asemi Z, Daneshvar KR, et al. Effect of probiotic supplementation on cognitive function and metabolic status in Alzheimer's disease: a randomized, double-blind and controlled trial. *Front Aging Neurosci*. 2016;8:256.
83. Tamtaji OR, Heidari-Soureshjani R, Mirhosseini N, et al. Probiotic and selenium co-supplementation, and the effects on clinical, metabolic and genetic status in Alzheimer's disease: a randomized, double-blind, controlled trial. *Clin Nutr*. 2018.
84. Kelly JR, Borre Y, Brien CO, et al. Transferring the blues: depression-associated gutmicrobiota induces neurobehavioural changes in the rat. *J Psychiatr Res*. 2016;82:109–118.
85. McFarlane HG, Kusek GK, Yang M, et al. Autism-like behavioral phenotypes in BTBR T+tf/J mice. *Genes Brain Behav*. 2008;7:152–163.
86. Schwarcz R, Bruno JP, Muchowski PJ, Wu HQ. Kynurenines in the mammalian brain: when physiology meets pathology. *Nat Rev Neurosci*. 2012;13:465–477.
87. David LA, Maurice CF, Carmody RN, et al. Diet rapidly and reproducibly alters the human gut microbiome. *Nature*. 2014;505:559–563.
88. Kolodziejczyk AA, Zheng D, Elinav E. Diet–microbiota interactions and personalized nutrition. *Nat Rev Microbiol*. 2019;17:742–753.
89. Dugger BN, Dickson DW. Pathology of neurodegenerative diseases. *Cold Spring Harb Perspect Biol*. 2017;9:a028035.
90. Martínez Y, Más D, Betancur C, et al. Role of the phytochemical compounds like modulators in gut microbiota and oxidative stress. *Curr Pharm Des*. 2020;26:1–15.
91. De Filippis F, Vitaglione P, Cuomo R, et al. Dietary interventions to modulate the gut microbiome-how far away are we from precision medicine. *Inflamm Bowel Dis*. 2018;24:2142–2154.
92. Farooqui AA. Effect of lifestyle, aging, and phytochemicals on the onset of neurological disorders. In: *Phytochemicals, Signal Transduction, and Neurological Disorders*. New York, NY, USA: Springer; 2012:1–29.

93. Newman DJ, Cragg GM. Natural products as sources of new drugs from 1981 to 2014. *J Nat Prod*. 2016;79:629–661.
94. Belapurkar P, Goyal P, Tiwari-barua P, et al. Immunomodulatory effects of triphala and its individual constituents: a review. *Indian J Pharm Sci*. 2014;76:467–475.
95. Morais CA, de Rosso VV, Estadella D, Pisani LP. Anthocyanins as inflammatory modulators and the role of the gut microbiota. *J Nutr Biochem*. 2016;33:1–7.
96. Faria A, Fernandes I, Norberto S, et al. Interplay between anthocyanins and gut microbiota. *J Agric Food Chem*. 2014;62:6898–6902.
97. Velmurugan BK, Rathinasamy B, Lohanathan BP, et al. Neuroprotective role of phytochemicals. *Molecules*. 2018;23:2485.
98. Sharifi-Rad M, Lankatillake C, Dias DA, et al. Impact of natural compounds on neurodegenerative disorders: from preclinical to pharmacotherapeutics. *J Clin Med*. 2020;9:1061.
99. Spencer JP. Flavonoids and brain health: multiple effects underpinned by common mechanisms. *Genes Nutr*. 2009;4:243–250.
100. Landete JM. Updated knowledge about polyphenols: functions, bioavailability, metabolism, and health. *Crit Rev Food Sci Nutr*. 2012;52:936–948.
101. Di Meo F, Valentino A, Petillo O, et al. Bioactive polyphenols and neuromodulation: molecular mechanisms in neurodegeneration. *Int J Mol Sci*. 2020;21:2564.
102. Godos J, Castellano S, Ray S, et al. Dietary polyphenol intake and depression: results from the mediterranean healthy eating, lifestyle and aging (MEAL) study. *Molecules*. 2018;23:999.
103. Kawabata K, Yoshioka Y, Terao J. Role of intestinal microbiota in the bioavailability and physiological functions of dietary polyphenols. *Molecules*. 2019;24:370.
104. Silva RFM, Pogačnik L. Polyphenols from food and natural products: neuroprotection and safety. *Antioxidants (Basel)*. 2020;9:61.
105. Poti F, Santi D, Spaggiari G, et al. Polyphenol health effects on cardiovascular and neurodegenerative disorders: a review and meta-analysis. *Int J Mol Sci*. 2019;20:351.
106. Costa SL, Silva VD, Dos Santos SC, et al. Impact of plant-derived flavonoids on neurodegenerative diseases. *Neurotox Res*. 2016;30:41–52.
107. Szwajgier D, Borowiec K, Pustelniak K. The neuroprotective effects of phenolic acids: molecular mechanism of action. *Nutrients*. 2017;9:477.
108. Tresserra-Rimbau A, Lamuela-Raventos RM, Moreno JJ. Polyphenols, food and pharma. Current knowledge and directions for future research. *Biochem Pharmacol*. 2018;156:186–195.
109. Filosa S, Di Meo F, Crispi S. Polyphenols-gut microbiota interplay and brain neuromodulation. *Neural Regen Res*. 2018;3:2055–2059.
110. Di Meo F, Donato S, Di Pardo A, et al. New therapeutic drugs from bioactive natural molecules: the role of gut microbiota metabolism in neurodegenerative diseases. *Curr Drug Metab*. 2018;19:478–489.
111. Pandareesh MD, Mythri RB, Srinivas Bharath MM. Bioavailability of dietary polyphenols: factors contributing to their clinical application in CNS diseases. *Neurochem Int*. 2015;89:198–208.
112. Chu C, Murdock MH, Jing D, et al. The microbiota regulate neuronal function and fear extinction learning. *Nature*. 2019;574:543–548.
113. Ilieva H, Maragakis NJ. Motor neuron disease: basic science. *Adv Neurobiol*. 2017;15:163–190.
114. Baizabal-Carvallo JF, Alonso-Juarez M. The link between gut dysbiosis and neuroinflammation in Parkinson's disease. *Neuroscience*. 2020;15(432):160–173.
115. Poirier AA, Aubé B, Côté M, Morin N, Di Paolo T, Soulet D. Gastrointestinal dysfunctions in Parkinson's disease: symptoms and treatments. *Parkinsons Dis*. 2016;676:2528.
116. Cho I, Blaser MJ. The human microbiome: at the interface of health and disease. *Nat Rev Genet*. 2012;13:260–270.

117. Keshavarzian A, Green SJ, Engen PA, et al. Colonic bacterial composition in Parkinson's disease. *Mov Disord.* 2015;30:1351–1360.
118. Wu X, Chen PS, Dallas S, et al. Histone deacetylase inhibitors up-regulate astrocyte GDNF and BDNF gene transcription and protectdopaminergic neurons. *Int J Neuropsychopharmacol.* 2008;11:1123–1134.
119. Lin CH, Chen CC, Chiang HL, et al. Altered gut microbiota and inflammatory cytokine responses in patients with Parkinson's disease. *J Neuroinflammation.* 2019;16:129.
120. Pathak-Gandhi N, Vaidya ADB. Management of parkinson's disease in ayurveda: medicinal plants and adjuvant measures. *J Ethnopharmacol.* 2017;197:46–51.
121. Arumugam M, Raes J, Pelletier E, et al. Enterotypes of the human gut microbiome. *Nature.* 2011;473:174–180.
122. Scher JU, Sczesnak A, Longman RS, et al. Expansion of intestinal *Prevotella copri* correlates with enhanced susceptibility to arthritis. *Elife.* 2013;2:e01202.
123. Sheflin A, Melby CL, Carbonero F, Weir TL. Linking dietary patterns with gut microbial composition and function. *Gut Microbes.* 2017;8:2113–2129.
124. dos Santos EF, Busanello EN, Miglioranza A, et al. Evidence that folic acid deficiency is a major determinant of hyperhomocysteinemia in Parkinson's disease. *Metab Brain Dis.* 2009;24:257–269.
125. Luong KV, Nguyen LT. The beneficial role of thiamine in Parkinson disease. *CNS Neurosci Ther.* 2013;19:461–468.
126. van Ijzendoorn SCD, Derkinderen P. The intestinal barrier in Parkinson's disease: current state of knowledge. *J Parkinsons Dis.* 2019;9:S323–S329.
127. Queipo-Ortuno MI, Seoane LM, Murri M, et al. Gut microbiotacomposition in male rat models under different nutritional statusand physical activity and its association with serum leptin and ghrelin levels. *PLoS One.* 2013;8:e65465.
128. Van Den Berge N, Ferreira N, Gram H, et al. Evidence for bidirectional and trans-synaptic parasympathetic and sympathetic propagation of alpha-synuclein in rats. *Acta Neuropathol.* 2019;138:535–550.
129. Kang DW, Park JG, Ilhan ZE, et al. Reduced incidence of prevotella and other fermenters in intestinal microflora of autistic children. *PLoS ONE.* 2013;8:e68322.
130. De Angelis M, Piccolo M, Vannini L, et al. Fecal microbiota and metabolome of children with autism and pervasive developmental disorder not otherwise specified. *PLoS ONE.* 2013;8, e76993.
131. He Y, Li B, Sun D, Chen S. Gut microbiota: implications in Alzheimer's disease. *J Clin Med.* 2020;9:2042.
132. Li Z, Zhu H, Zhang L, Qin C. The intestinal microbiome and Alzheimer's disease: a review. *Animal Model Exp Med.* 2018;1:180–188.
133. Vogt NM, Kerby RL, Dill-McFarland KA, et al. Gut microbiome alterations in Alzheimer's disease. *Sci Rep.* 2017;7(1):13537.
134. Harach T, Marungruang N, Duthilleul N, et al. Reduction of Abeta amyloid pathology in APPPS1 transgenic mice in the absence of gut microbiota. *Sci Rep.* 2017;7:41802.
135. Asti A, Gioglio L. Can a bacterial endotoxin be a key factor in the kinetics of amyloid fibril formation? *J Alzheimers Dis.* 2014;39:169–179.
136. Sheng JG, Bora SH, Xu G, et al. Lipopolysaccharide-induced neuroinflammation increases intracellular accumulation of amyloid precursor protein and amyloid betapeptide in APPswe transgenic mice. *Neurobiol Dis.* 2003;14:133–145.
137. Zhan X, Stamova B, Jin LW, et al. Gramnegative bacterial molecules associate with Alzheimer disease pathology. *Neurology.* 2016;87:2324–2332.
138. Zhao Y, Lukiw WJ. Microbiome-generated amyloid and potential impact on amyloidogenesis in Alzheimer's disease (AD). *J Nat Sci.* 2015;1:e138.
139. Leblhuber F, Geisler S, Steiner K, et al. Elevated fecal calprotectin in patients with Alzheimer's dementia indicates leaky gut. *J Neural Transm (Vienna).* 2015;122:1319–1322.

140. Graff J, Rei D, Guan JS, et al. An epigenetic blockade of cognitive functions in the neurodegenerating brain. *Nature.* 2012;483:222–226.
141. Solas M, Puerta E, Ramirez MJ. Treatment options in Alzheimer's disease: the GABA story. *Curr Pharm Des.* 2015;21:4960–4971.
142. Bercik P, Denou E, Collins J, et al. The intestinal microbiota affect central levels of brain-derived neurotropic factor and behavior in mice. *Gastroenterology.* 2011;141:599–609.
143. Zhu L, Liu W, Alkhouri R, et al. Structural changes in the gut microbiome of constipated patients. *Physiol Genomics.* 2014;679–686.
144. Mancuso C, Santangelo R. Alzheimer's disease and gut microbiota modifications: the long way between preclinical studies and clinical evidence. *Pharmacol Res.* 2018;129:329–336.
145. Jiang C, Li G, Huang P, Liu Z, Zhao BJ. The gut microbiota and Alzheimer's disease. *Alzheimers Dis.* 2017;58:1–15.
146. Angelucci A, Cechova K, Amlerova J, Hort J. Antibiotics, gut microbiota, and Alzheimer's disease. *J Neuroinflammation.* 2019;16:108.
147. Friedland RP. Mechanisms of molecular mimicry involving the microbiota in neurodegeneration. *J Alzheimers Dis.* 2015;45:349–362.
148. Solanki I, Parihar P, Mansuri ML, Parihar MS. Flavonoid-based therapies in the early management of neurodegenerative diseases. *Adv Nutr.* 2015;6:64–72.
149. Vernocchi P, Chierico FD, Putignani L. Gut microbiota metabolism and interaction with food components. *Int J Mol Sci.* 2020;21:3688.
150. Meireles M, Marques C, Norberto S, et al. Anthocyanin effects on microglia M1/M2 phenotype: consequence on neuronal fractalkine expression. *Behav Brain Res.* 2016;305:223–228.
151. Kay CD, Pereira-Caro G, Ludwig IA, et al. Anthocyanins and flavanones are more bioavailable than previously perceived: a review of recent evidence. *Annu Rev Food Sci Technol.* 2017;8:155–180.
152. Hidalgo M, Oruna-Concha MJ, Kolida S, et al. Metabolism of anthocyanins by human gut microflora and their influence on gut bacterial growth. *J Agric Food Chem.* 2012;60:3882–3890.
153. Marques C, Meireles M, Faria A, Calhau C. High-fat diet-induced dysbiosis as a cause of neuroinflammation. *Biol Psychiatry.* 2016;80:e3–e4.
154. Mythri R, Srinivas Bharath MM. Curcumin: a potential neuroprotective agent in Parkinson's disease. *Curr Pharm De.* 2012;18:91–99.
155. Muralidhara M. Neuroprotective propensity of curcumin: evidence in animal models. In: Pradeep H, Yenisetti S, Rajini PS, Farroqui T, Farooqui AA, eds. *Mechanisms, and Its Potential Therapeutic Value: IN Curcumin for Neurological and Psychiatric Diseases—Neurochemical and Pharmacological Properties.* Academic Press; 2019:514.
156. Shen L, Liu L, Ji HF. Regulative effects of curcumin spice administration on gut microbiota and its pharmacological implications. *Food Nutr Res.* 2017;61:1361780.
157. McFadden RM, Larmonier CB, Shehab KW, et al. The role of curcumin in modulating colonic microbiota during colitis and colon cancer prevention. *Inflamm Bowel Dis.* 2015;21:2483–2494.
158. Meo D, FMargarucci S, Galderisi U, et al. Curcumin, gut microbiota, and neuroprotection. *Nutrients.* 2019;1:2426.
159. Zam W. Gut microbiota as a prospective therapeutic target for curcumin: a review of mutual influence. *J Nutr Metab.* 2018;136:79–84.
160. Khazdair MR, Anaeigoudari A, Hashemzehi M, Mohebbati R. Neuroprotective potency of some spice herbs, a literature review. *J Tradit Complement Med.* 2019;98:e105.
161. Jaiswal YS, Williams LL. A glimpse of Ayurveda—the forgotten history and principles of Indian traditional medicine. *J Tradit Complement Med.* 2017;7:50–53.

162. Naoi M, Shamoto-Nagai M, Maruyama W. Neuroprotection of multifunctional phytochemicals as novel therapeutic strategy for neurodegenerative disorders: antiapoptotic and antiamyloidogenic activities by modulation of cellular signal pathways. *Future Med.* 2019;14:FNL9.
163. Davinelli S, Maes M, Corbi G, Zarrelli A, et al. Dietary phytochemicals and neuroinflammaging: from mechanistic insights to translational challenges. *Immun Ageing.* 2016;13:16.
164. Mannangatti P, Naidu KN. Indian herbs for the treatment of neurodegenerative disease. *Adv Neurobiol.* 2016;12:323–336.
165. Parasuraman S, Thing GS, Dhanaraj SA. Polyherbal formulation: concept of Ayurveda. *Pharmacogn Rev.* 2014;8:73–80.
166. Kumar GP, Khanum F. Neuroprotective potential of phytochemicals. *Pharmacogn Rev.* 2012;6:81–90.
167. Chen F, Wen Q, Jiang J, et al. Could the gut microbiota reconcile the oral bioavailability conundrum of traditional herbs. *J Ehtnopharamcol.* 2016;179:253–264.
168. Muralidhara M, Yenisetti SV, Yenisetti SC. Current understanding on the beneficial role of nutrition in Parkinson's disease—an overview. *J Aging Sci.* 2017;5:1–10.
169. Kumar S, Dobos GJ, Rampp T. The significance of Ayurvedic medicinal plants. *J Evidence-Based Complementary Altern Med.* 2017;22:494–501.
170. Howes MJ, Simmonds MS. The role of phytochemicals as micronutrients in health and disease. *Curr Opin Clin Nutr Metab Care.* 2014;17:558–566.
171. Pulok K, Mukherjee SR, Bhattacharyya S, et al. Clinical study of 'Triphala'—a well known phytomedicine from India. *Iran J Pharmacol Ther.* 2005;5:51–54.
172. Gowda DV, Muguli G, Rangesh PR, Deshpande RD. Phytochemical and pharmacological actions of triphala: ayurvedic formulation—a review. *Int J Pharm Sci Rev Res.* 2012;15:61–65.
173. Tarasiuk A, Mosińska P, Fichna J. Triphala: current applications and new perspectives on the treatment of functional gastrointestinal disorders. *Chin Med.* 2018;13:39.
174. Kumar NS, Nair AS, Nair AM, Murali M. Pharmacological and therapeutic effects of triphala—a literature review. *J Pharmacogn Phytochem.* 2016;23:23–27.
175. Mukherjee PK, Bahadur S, Harwansh RK, et al. Development of traditional medicines: globalizing local knowledge or localizing global technologies. *Pharm Times.* 2013;45:39–42.
176. Mukherjee PK, Rai S, Bhattacharya S, et al. Marker analysis of polyherbal formulation, Triphala—a well-known Indian traditional medicine. *Indian J Tradit Knowl.* 2008;7:379–383.
177. Martínez Y, Más D. Role of herbs and medicinal spices as modulators of gut microbiota 2020 chapter in book: herbs and spices publisher. *Herbs and Spices.* Intech Open; 2020. https://doi.org/10.5772/intechopen.91208.
178. Shinomol GK, Muralidhara M, Srinivas Bharath MM. Exploring the role of "Brahmi" (*Bacopa monnieri* and *Centella asiatica*) in brain function and therapy. *Recent Pat Endocr Metab Immune Drug Discovery.* 2011;5:33–49.
179. Yenisetti SC, Manjunath MJ, Muralidhara M. Neuropharmacological properties of *Withania somnifera*—Indian ginseng: an overview on experimental evidence with emphasis on clinical trials and patents. *Recent Pat CNS Drug Discov.* 2016;14:1–15.
180. Gray NE, Magana AA, Lak P, et al. *Centella asiatica*—phytochemistry and mechanisms of neuroprotection and cognitive enhancement. *Phytochem Rev.* 2018;17:161–194.
181. Gokul K. Muralidhara. Dietary supplements as cognitive enhancers: role of flavonoid-rich foods and their relevance in age-related neurodegeneration. In: Watson RR, Preedy V, eds. *Bioactive Nutraceuticals and Dietary Supplements in Neurological and Brain Disease, Prevention and Therapy.* 1st ed. Academic Press; 2014:512.
182. Peterson CT, Denniston K, Chopra D. Therapeutic uses of Triphala in Ayurvedic medicine. *J Altern Complement Med.* 2017;23:607–614.

183. Selma MV, Beltrán D, Luna MC, et al. Isolation of human intestinal bacteria capable of producing the bioactive metabolite isourolithin from ellagic acid. *Front Microbiol.* 2017;8:1521.
184. Baliga MS, Meera S, Mathai B, et al. Scientific validation of the ethnomedicinal properties of theayurvedic drug triphala: a review. *Chin J Integr Med.* 2012;18:946–954.
185. Save SN, Choudhary S. Effects of triphala and guggul aqueous extracts on inhibition of protein fibrillation and dissolution of preformed fibrils. *RSC Adv.* 2017;7:20460–20468.
186. Moradi SZ, Momtaz S, Bayrami Z, Farzaei MH. Nanoformulations of herbal extracts in treatment of neurodegenerative disorders. *Front Bioeng Biotech.* 2020;8.Article 238.
187. Caruso G, Caffo M, Alafaci C, et al. Could nanoparticle systems have a role in the treatment of cerebral gliomas? *Nanomed Nanotechnol Biol Med.* 2011;7:744–752.
188. Ratheesh G,Tian L,Venugopal JR, et al. Role of medicinal plants in neurodegenerative diseases. *Biomanuf Rev.* 2017;2:2.
189. Rehman MU,Wali AF, Ahmad A, et al. Neuroprotective strategies for neurological disorders by natural products: an update. *Curr Neuropharmacol.* 2019;17:247–267.
190. Sandhir R,Yadav A, Mehrotra A, et al. Curcumin nanoparticles attenuate neurochemical and neurobehavioral deficits in experimental model of Huntington's disease. *Neuromol Med.* 2014;16:106–118.
191. Niu X, Chen J, Gao J. Nanocarriers as a powerful vehicle to overcome blood-brain barrier in treating neurodegenerative diseases: focus on recent advances. *Asian J Pharm Sci.* 2019;14:480–496.
192. Gilbert JA, Blaser MJ, Caporaso JG, et al. Current understanding of the human microbiome towards mechanistic studies of the microbiome. *Nat Med.* 2018;24:392–400.
193. Sinha R, Abu-Ali G,Vogtmann E, et al. Assessment of variation in microbial community amplicon sequencing by the microbiome quality control (MBQC) project consortium. *Nat Biotechnol.* 2017;35:1077–1086.
194. Prasher B,Varma B, Kumar A, et al. TRISUTRA Ayurgenomics consortium, Whole genome expression and biochemical correlates of extreme constitutional types defined in Ayurveda. *J Ethnopharmacol.* 2017;197:274–293.
195. Mukerjee PK, Nema NK,Venkatesh P. Changing scenario for promotion and development of Ayurveda-way forward. *J Ethnopharmacol.* 2012;143:424–434.
196. Prasher B, Gibson G, Mukerji M. Genomic insights into Ayurvedic and Western approaches to personalized medicine. *J Genet.* 2016;95:209–228.
197. Chaudhari D, Dhotre D, Agarwal D, et al. Understanding the association between the human gut, oral and skin microbiome and the Ayurvedic concept of prakriti. *J Biosci.* 2019;44:112.

CHAPTER 17

The emerging roles of gut microbiome on neurotoxic outcomes: Implications for neurological disorders

Gokul Krishna[a], Meghashri Sridhar[b], and M. Muralidhara[c]
[a]Department of Child Health, University of Arizona College of Medicine—Phoenix, Phoenix, AZ, United States
[b]Wiley APAC Services LLP, New Delhi, India
[c]Department of Biochemistry and Nutrition, CSIR—Central Food Technological Research Institute (CFTRI), Mysuru, Karnataka, India

1 Introduction

The rapid rise of industrialization in modern society has a major impact on human health. A variety of chemicals in the atmosphere, drinking water, and occupational settings have contributed to an increased public health problem as a result of widespread human exposures associated with detrimental insults to the brain and nervous system.[1,2] The neurotoxic outcomes include a panoply of biochemical, morphological, behavioral, and physiological dysfunction, with a long-lasting impact on the developing and mature brain.[3] Based on the magnitude of exposure, the toxic agents may promote developmental disabilities with consequences that might emerge as age advances. The major classes of environmental pollutants such as pesticides, metals, inorganic compounds, and organic substance pollutants directly act on the targets of the nervous system or even affect other biological functions. In particular, environmental neurotoxicant exposures throughout life play a causal role in the development of neurodegenerative diseases as diverse as Parkinson's disease (PD), Alzheimer's disease (AD), and Huntington's disease (HD).[4] Further, human exposure through dietary contaminants can be a major contributor to certain neurological disorders with diverse clinical manifestations.[5] The neuropathological and functional alterations as a result of chemical toxicant-induced pathogenic processes underlie fluctuations in neurological function and associated behavioral impairments.[6]

Neurotoxic chemicals usually initiate their effect at the molecular level by a variety of different mechanisms, either by deleterious effects on

Gut Microbiota in Neurologic and Visceral Diseases
https://doi.org/10.1016/B978-0-12-821039-0.00022-8

important processes or by promoting toxic cascades. The common mechanisms of neurotoxicant-induced pathologies identified include blood–brain barrier disruption, oxidative stress, mitochondrial dysfunction, synaptic toxicity, neuroinflammation, and protein aggregation.[4] Such alterations proceed with cellular damage and ultimately failure of neurological function that manifests as behavioral deficits.[7] In particular, the developing brain is much more vulnerable to damage by toxic agents compared to the adult brain because several chemicals can to pass through the human placenta and induce developmental insult.[8] Despite the presence of endogenous defense and repair processes provided by supporting cells and structures, the chemicals can disrupt the antioxidant machinery that can lead to cellular demise. There is an immense effort underway to understand the action of neurotoxic hazards of environmental contaminants and their dose and regimen of exposure, many of which share common features and mechanisms to induce neurological impairments.

Although neurological disorders have traditionally been studied within the central nervous system (CNS), peripheral biological system contributions have been observed to contribute to disease progression.[9] It is emerging that chemical agents exert neurotoxicity by compromising the systemic physiology by altering the homeostasis of indigenous gut microbiota harboring the host digestive tract, thereby modulating the multidirectional pathways of the intricate dynamic interactions between the gut and brain. In particular, the gut serves as an important primary exposure route for several chemicals that are widely disseminated in the environment. The microbiota is well recognized as an essential mediator for its ability to influence the brain and behavior, increasing the risk of neurodevelopmental, psychiatric, and neurodegenerative diseases.[10] There is increasing in understanding the action of microbiota on neurotoxic outcomes, with evidence supporting environmental origins of devastating neurodegenerative diseases.[11, 12] In particular, there is intriguing evidence from animal studies to implicate the gut microbiome in neurotoxic outcomes during development and later stages in life.

We are beginning to learn that microbiota influence neurological processes including neuroinflammation, synaptic plasticity, and neurotransmitter signaling.[13, 14] The mechanisms underlying interactions between microbial dysbiosis (alterations in microbial composition) and the nervous system are under continuous investigation. Microbiota interactions with neurophysiology and behavior have been supported by human studies showing that perturbations in the microbiota are linked to several neurological disorders (Table 1). It is becoming evident that members belonging to several taxa

Table 1 Reported microbiome alterations in neurological disorders.

Sample type	Characterization method	Microbiota changes at various granular taxonomic levels	References
Human, Parkinson's disease (PD)			
Feces	16S rRNA amplicon sequencing	Genus: *Lactobacillus*, *Bifidobacterium Christensenella*, *Catabacter*, *Oscillospira* ↑; *Bacteroides* and *Prevotella* ↓ Species: *Lactobacillus mucosae*, *Ruminococcus bromii*, *Palpillibacter cinnamivorans*, *Christensenella minuta*, *Catabacter hongkongensis* ↑; *Bacteroides massiliensis*, *Bacteroides coprocola*, *Bacteroides dorei*, *Bacteroides plebeus*, *Prevotella copri*, *Ruminococcus Callidus* ↓	15
Feces	16S rRNA amplicon sequencing	*Bifidobacteriaceae*, *Eubacteriaceae* and *Desulfovibrionaceae* ↑ *Lachnospiraceae*, *Streptococcaceae*, *Pasturellaceae*, *Comamonadaceae*, *Methylobacteriaceae*, *Actinomycetaceae*, *Halmonadaceae*, *Xanthomonadaceae*, *Aphingomonadaceae*, *Micrococcaceae* ↓	16
Feces	16S rRNA amplicon sequencing	Family: *Bifidobacteriaceae*, *Tissierellaceae*, *Lactobacillaceae*, *Christensenellaceae*, and *Verrucomicrobiaceae* ↑; *Pasteurellaceae and Lachnospiraceae* ↓ Genus: *Bifidobacterium*, *Lactobacillus*, *Akkermansia*, *Christensenellaceae* ↑; *Lachnospiraceae*, *Faecalibacterium*, *Blautia*, *Roseburia* ↓	17
Feces	16S rRNA amplicon sequencing	*Verrucomicrobiaceae*, *Bifidobacteriaceae*, *Streptococcaceae*, *Desulfohalobiaceae* ↑ *Bacteroidaceae*, *Lachnospiraceae*, *Brevibacteriaceae*, *Sphingobacteriaceae* ↓	18
Human, Alzheimer's disease (AD)			
Feces	16S rRNA gene sequencing	Phylum: *Bacteriodetes* ↑; *Firmicutes and Actinobacteria* ↓ Family: *Bacteroidaceae*, *Gemellaceae*, and *Rikenellaceae* ↑; *Bifidobacteriaceae*, *Ruminococcaceae*, *Clostridiaceae* and *Turicibacteraceae* ↓ Genus: *Bacteroides*, *Alistipes*, *Bilophila*, *Phascolarctobacterium*, *Blautia*, and *Gemella* ↑; *Bifidobacterium*, *Clostridium*, *Turicibacter* ↓	19

Continued

Table 1 Reported microbiome alterations in neurological disorders—cont'd

Sample type	Characterization method	Microbiota changes at various granular taxonomic levels	References
Feces	16S rRNA gene sequencing	Phylum: *Bacteroidetes* ↓ Class: *Actinobacteria* and *Bacilli* ↑; *Negativicutes* and *Bacteroidia* ↓ Order: *Lactobacillales* ↑; *Bacteroidales* and *Selenomonadales* ↓ Family: *Ruminococcaceae*, *Enterococcaceae* and *Lactobacillaceae* ↑; *Lanchnospiraceae*, *Bacteroidaceae*, and *Veillonellaceae* ↓	20
Human, Autism spectrum disorder (ASD)			
Feces	16S rRNA gene sequencing	Genus/species: *Faecalibacterium prausnitzii* and *Haemophilus parainfluenzae* ↓	21
Rectal biopsy	16S rRNA gene sequencing	Species: *Clostridium lituseburense; Lachnoclostridium bolteae; Flavonifractor plautii* ↑; *Dorea formicigenerans*, *Blautia luti* and *Sutterella* sp. ↓	22

of microbiota can engage in the metabolism of neurotoxicants and regulate mucosal and systemic immune responses to influence inflammation.[13, 23, 24] Gut bacteria additionally have the capacity to influence numerous aspects by directly producing active metabolites that subsequently interact with the nervous system.[25] Diet has been a major driving force contributing to the composition and diversity of trillions of microorganisms harbored in the gut.[26] The new field of research is now focused on dietary manipulation with nutritional tools such as prebiotics and probiotics to modulate microbial ecosystems and potentially mitigate neurotoxicity. Animal models have been invaluable to neurotoxicity research essentially using neurotoxin models providing proof-of-principle data replicating key pathological features of neurological diseases and several studies have aided in understanding the gut microbiome as a modifier of neurotoxicity.

2 The link between gut and brain

The human body is colonized by a large variety of microbes, the majority of which reside within the gastrointestinal (GI) tract. The bidirectional gut–brain axis involving microbes consists of a complex ecosystem, predominantly bacteria with archaea, fungi, protozoa, and viruses forms the center of the interface between the brain and gut. The gut microbiome has been implicated in several neurodegenerative and neurodevelopmental disorders as diverse as PD, AD, and autism spectrum disorder (ASD).[27–29] Altered bacterial composition in these disorders implicates that equilibrium of gut homeostasis is essential for normal brain function and behavior. The precise mechanisms of how microbiota impact brain function remain largely varied and complex. Such interactions involve the capacity of intestinal microbes to influence the immune system that can then alter how the immune components subsequently interact with the nervous system.[30] Further, the direct connection of the enteric system with the CNS occurs through the vagus nerve-dependent signaling pathway.[31] These interactions form the components of the gut–brain axis and microbiota as a critical node that establishes communication along this axis and regulates health and disease. Environmental toxin exposure appears to greatly alter microbiota composition and function, providing the crucial role of microbiota influence on the disastrous neurotoxic consequences of a large array of compounds. Microbiota dysbiosis after chemical toxicity can adversely affect gut physiology with loss of barrier function to promote inappropriate brain–gut axis communication and associated consequences for brain function.

Exposure to toxic agents during pregnancy increases the risk of neurodevelopmental disorders.[32] This heightened vulnerability is compounded with greater absorption rates and a reduced detoxification process compared to adults.[33] Further, maternal microbiota has been increasingly recognized as an important regulator of newborn neurodevelopment. Prenatal and postnatal brain developmental stages involving rapid changes in neuronal organization are influenced by the microbiota of the mother for brain maturation,[10] and exposure to environmental agents during pregnancy has been reported to increase the risk for developmental neurotoxicity.[2] Studies are beginning to show the impact of microbiota on developmental programming and the beneficial effects of modulating maternal gut microflora to shape fetal brain development. Interest in the role of the microbiome on neurotoxic outcomes has increased over the past few years, harnessing advanced high-throughput technologies to define the identities of microbial communities. Many studies have demonstrated the application of genomic technologies to reveal several molecules derived or modulated by the microbiome. Such studies have been extended to understand the products of microbial community metabolism generated by one or more microbes to understand the functional roles for many such microbial molecules in the host. In this chapter, we discuss the potential mechanisms by which environmental toxin-induced gut dysbiosis contributes to neurotoxic outcomes, current evidence for neurotoxicant-induced microbiome alterations, and dietary strategies manipulating the gut microbiota to mitigate neurotoxicity (Fig. 1).

3 Neurotoxicity–microbiota interactions

Epidemiological, animal, and in vitro reports strongly support association of chemical exposure and neurotoxicity. The functional significance of the relationship between neurotoxicity and microbiota has received increased attention with a growing body of reports on mechanisms that underlie the role of the microbiome and the neurotoxicity of chemicals. Microbiota-mediated neurotoxicant biotransformation, production of neuroactive metabolites short-chain fatty acids (SCFAs), intestinal barrier dysfunction, and immune system function are all critical for the microbial influence of neurotoxic outcomes.

3.1 Microbial transformation of neurotoxicants

The microbiota encodes several enzymes, many of which are exclusively microbial dependent, that are capable of metabolizing environmental

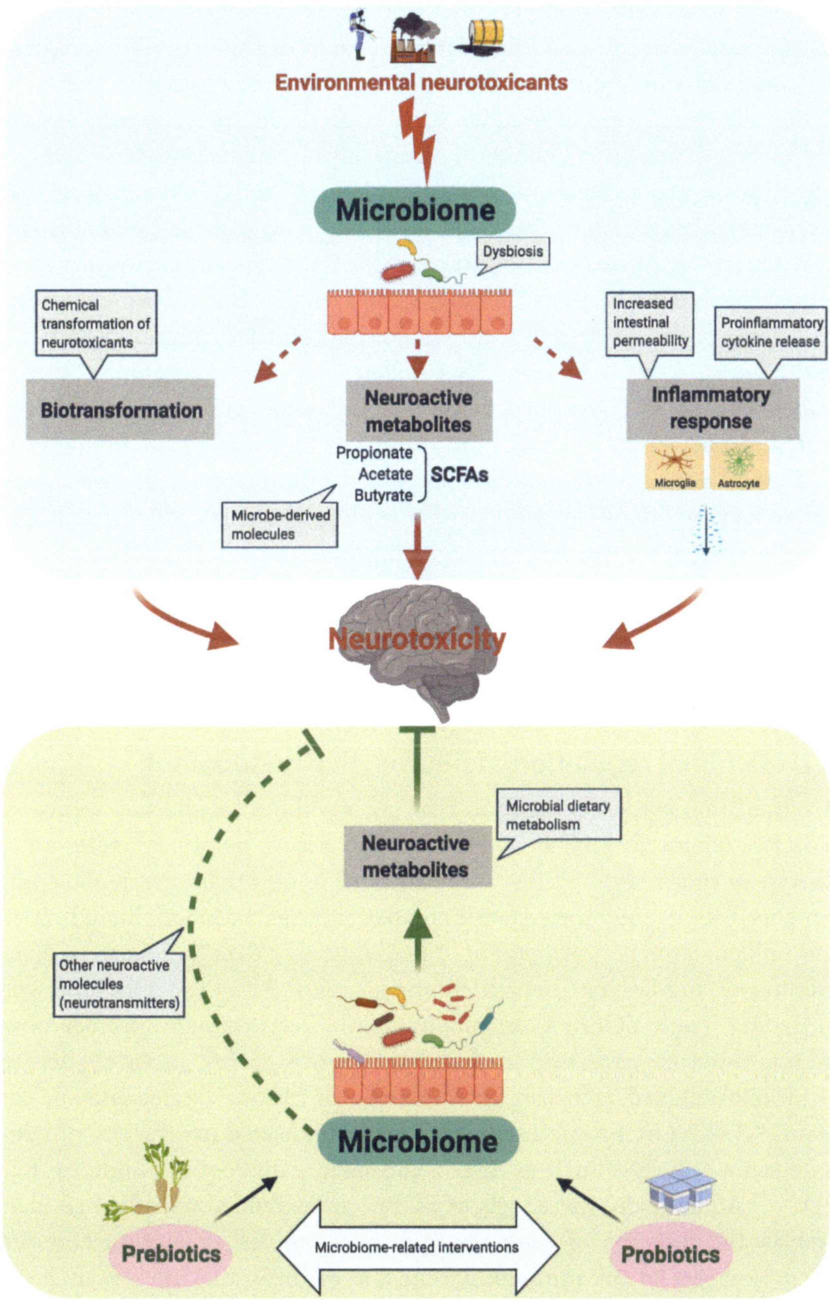

Fig. 1 Gut microbial regulation on mechanisms and interventions for neurotoxic outcomes. Created with BioRender.com

chemicals to modify or amplify, thus potentially influencing their bioavailability. Early investigations have identified bacterial β-glucuronidase, azoreductases, and nitroreductases from gut microbes that have the ability to chemically transform compounds such as pesticides, polychlorobiphenyls, and metals.[34] Since then, chemical metabolism by microbiota has been extended to several compounds with neurological impact. Chemical metabolism by the microbiome has been identified with organic pollutants and heavy metals with neurological impact.[34–36] Such chemical transformation largely involving structural alterations can change pharmacokinetic properties crucial for prodrug activation, altering their toxicities and lifetimes in the human body.[37] Additionally, the herbicide glyphosate inhibit microbial enzyme EPSP (5-enolpyruvylshikimate-3-phosphate) synthase.[38] Some representatives of gut microbes can transform foodborne contaminants including aflatoxins to alter their toxicities.[39] Future studies on examining microbially mediated chemical transformations in the gut are warranted to provide comprehensive insights regarding how microbiota influences systemic concentrations of either parent compounds or their toxic metabolites. It is imperative to gain more information on gut microbial chemical metabolism with enzymes and knowledge of enzyme mechanism to facilitate toxicological risk assessment.

3.2 Microbial regulation of neuroactive metabolites

Accumulating evidence indicates that gut microbes are the key source of bioactive chemicals such as SCFAs that may act on neuronal health either directly or indirectly.[40] SCFAs are recognized as important metabolic components that can enter the systemic circulation, cross the blood–brain barrier, and influence the nervous system.[41] The most abundant SCFAs are acetate, propionate, and butyrate, which comprise almost 95% of SCFAs produced in the gut. These SCFAs exert physiological effects through monocarboxylate transporters expressed in the blood–brain barrier[30] and activation of G protein-coupled receptors or inhibition of histone deacetylases in the brain.[42] Altered SCFA production has been implicated in a variety of neurobehavioral diseases such as ASD[43] and neuropathological condition like PD.[13, 44] Additionally, the effects of SCFAs are extended to influence maturation and function of microglia,[45] implicating the interaction between microbiota, peripheral immune system, and brain. Studies are revealing the influence of SCFAs on a range of psychological functioning including anxiety and depressive processes.[46] The beneficial effects of soluble dietary fibers are often considered to be mediated by these metabolic products that can modulate brain and behavior.[47] Due to the realization of the importance of the pleiotropic nature of SCFAs, studies are currently underway to understand microbiome–brain signaling by microbiota-generated metabolites.

3.3 Microbial influence on neuroimmune function

The immune system acts as a critical node for brain function including inflammation. The differentiation and function of immune cells are influenced by gut microbes and have significant implications on human health. Moreover, the immune cell pathway within the brain and periphery has been proposed as a mechanism regulating microbial influence on the brain,[48] making neuroimmune interactions crucial elements to understand neuroinflammatory mechanisms of environmental neurotoxicants. Gut microbes influence peripheral immune cell activation with important consequences for neuroinflammation.[49] In fact, it is recognized that microbiota affects the maturation and function of brain immune cells (microglia and astrocytes) and the effect of microbial-derived substances on distinct signaling pathways.[49] Dysbiosis of the gut microbiota after chemical exposure induces intestinal barrier dysfunction to promote local inflammation, increasing proinflammatory cytokine production in the intestine and brain.[48, 50] Since food contaminant exposure occurs mainly through diet, the GI tract provides a unique environment susceptible to adverse effects of chemicals. In vitro findings demonstrate that foodborne chemicals induce shifts in microbial volatiles and metabolic pathways to promote proinflammatory responses.[51] Additionally, it has been reported that gut microbes influence the activation of microglia, contributing to neuroinflammation and α-synuclein-mediated motor deficits,[13] with persistent proinflammatory signaling linked to the severity of neurodegenerative diseases.[52]

4 Microbiome and neurotoxicants

4.1 Pesticides

Several classes of pesticides including insecticides, herbicides, and fungicides are used in agriculture, gardening, and pest control. These substances bioaccumulate and persist in the environment. Humans are commonly exposed to these substances through the environment (e.g., rural residence), well water, farming, and consumption of contaminated foods. Evidence for environmental etiologies suggests that cumulative exposure to pesticides generates lasting neurotoxic effects and contributes to the development of neurodegenerative diseases such as PD, AD, and amyotrophic lateral sclerosis

(ALS).[53, 54] Decades of research have revealed several important cellular and molecular correlates of these diseases as a result of chronic pesticide exposure.[55] Experimental animal studies across pesticide models of neurodegenerative diseases show disruptions in the gut microbiota impacting manifestation of neurological and behavioral abnormalities.

4.1.1 MPTP (1-methyl-4-phenyl-1,2,3,6-tetrahydropyridine)

MPTP is a lipophilic compound that can cross the blood–brain barrier to be converted to intermediate 1-methyl-4-phenyl-2,3-dihydropyridium ($MPDP^+$) and then to active pyridinium species, 1-methyl-4-phenylpyridinium (MPP^+) by glial enzyme monoamine oxidase-B (MAO-B).[56] MPTP induces Parkinson-like neuropathology by selectively targeting dopamine neurons in the substantia nigra in several species.[47,57,58] Mice intraperitoneally administered with MPTP (18 mg/kg) twice a week for 5 weeks produced increased inducible nitric oxide synthase (iNOS), increased tyrosine hydroxylase (TH) and α-synuclein expression in the ileum, decreased abundance of *Lachnospiraceae*, and increased abundance of *Erysipelotrichaceae* and *Prevotellaceae*.[59] Subacute MPTP exposure caused GI alterations with reduced fecal water content, reduced TH content and expression in the ileum, damage to the intestinal epithelial barrier, and changes in bacterial taxa, most notably increased *Clostridiales*, *Erysipelotrichales*, and *Desulfovibrionales* and decreased *Prevotellaceae* and *Porphyromonadaceae*.[60] MPTP exposure reduced *Clostridium* and *Bacteroides* spp. and *Parabacteroides* with the administration of antibiotics (ampicillin, neomycin sulfate, and metronidazole).[61] Greater abundances of Firmicutes and Tenericutes, and lesser abundance of *Proteobacteria* with altered concentration levels of SCFAs, including propionic acid and isobutyric acid with altered fecal metabolomic profile, have been observed with MPTP.[61] Transplantation of fecal microbiota to MPTP-administered mice reduced microbial dysbiosis, alleviated motor impairment, and enhanced striatal dopaminergic and serotonergic alterations.[62]

4.1.2 Rotenone

Rotenone is a naturally occurring botanical compound derived from the roots of a plant belonging to *Lonchocarpus* species and is widely regarded as an environmental contributor to PD. Rotenone is a classical, high-affinity mitochondrial complex I inhibitor known to cause oxidative stress, α-synuclein phosphorylation, DJ-1 modifications, and inhibition of ubiquitin-proteasome system.[63,64] Rotenone induces functional alterations in GI function to promote gastroparesis[65] and neuropathology.[66] Interestingly, rotenone promotes PD-like progression in mice through release of aggregated protein by enteric neurons.[67] Chronic administration of rotenone has been shown to reduce colon motility, inflammation, and α-synuclein overexpression with a decrease in fecal microbiota diversity and increased Firmicutes:Bacteroidetes ratio.[68] GI dysfunction has been observed with rotenone with delayed gastric emptying, mucosal thickening and hyperplasia of goblet cells, and changes in microbial diversity.[69] Mice treated with rotenone at 10 mg/kg body weight daily by oral gavage for 28 days displayed changes in gut commensals associated with species richness and diversity. Reductions were observed with beneficial *Bifidobacterium* while *Rikenellaceae*, *Ruminococcaceae*, and *Allobaculum* were increased in rotenone-treated mice.[70] Our studies in pregnant rats indicate the harmful impact of rotenone exposure during GD 6–19 on beneficial bacteria such as *Bifidobacteria* and *Lactobacilli* in the gut[71] as well as increased markers related to brain oxidative stress and mitochondrial dysfunction.

4.1.3 Paraquat

Paraquat (PQ, 1,1′-dimethyl-4,4′-bipyridinium dichloride) is a bipyridal oxidant-producing herbicide widely used and has been associated with an increased risk of PD in several epidemiological studies. PQ occurs as a divalent cation molecule that undergoes redox cycling with cellular diaphorases to form the monovalent cation (PQ^+) generating large amounts of reactive species. The cytotoxic effect of PQ involves superoxide generation accompanied by oxidative stress-mediated c-Jun N-terminal kinase activation and Bak-dependent cell death mechanisms.[72, 73] The significance of gut microbial colonization has been shown in germ-free *Drosophila* parkin mutants showing resistance to PQ with longer survival rates.[74] In the mutant A53T transgenic mouse model, which expresses human α-synuclein, oral PQ administration induces earlier expression of Ser129 phosphorylated α-synuclein in the enteric nervous system.[75]

4.1.4 Organophosphates

Organophosphate (OP) insecticides include chlorpyrifos (CPF), malathion, diazinon, methyl parathion, and others. Although some of these substances are banned, organophosphates are widely used for pest control in agricultural and residential communities. Organophosphates have shown potential neurotoxic effects on fetal and childhood neurodevelopmental stages[76, 77] and even later during brain development, including glial cell functioning.[78] Although several restrictions have been placed on the use of organophosphates, CPF continues to be one of the widely used OP compounds owing to its persistence. CPF damages neurotransmitter systems, particularly inhibiting cholinesterase activity, and behavioral symptoms occur as a result of acetylcholine accumulation at the synapse.[79] The developmental neurotoxicity of CPF exposure has been demonstrated in rats with gestational exposure showing changes in neural protein markers in the brains of offspring.[80] Studies linking microbiota changes to CPF exposure in mice have shown a significant decrease in the relative abundance of *Firmicutes* and *Lactobacillaceae* in feces.[25] Further, it is observed that CPF causes dysbiosis of gut bacteria in rats fed with a high-fat diet, showing a reduction in alpha diversity and changes in microbes that have the ability to generate fermentation products (SCFAs) and those that influence inflammation and immune response.[81] Daily CPF administration for 9 weeks in rats on a normal fat diet reduced GI-related peptides and increased fecal *Allobaculu*, *Candidatus*, *Saccharimonas*, *Coprococcus*, *Anaeroplasma*, *Roseburia*, and Sutterella while decreasing the abundance of *Pseudoflavonifractor*, *Anaerosporobacter*, *Aerococcus*, *Brevundimonas*, and *Trichococcus*.[82]

4.2 Metals and inorganic compounds

Metals are a major source of global neurotoxicity that affect brain growth leading to life-long deficits. Many epidemiologic studies have documented association of heavy metal exposure to development of neurodegenerative and neurodevelopmental disorders later in life.

4.2.1 Manganese

Despite being an essential trace element with important physiological functions, inorganic manganese (Mn) is identified as a neurotoxicant originating from inorganic oxides and salts. Mn neurotoxic effects have been linked to motor abnormalities and higher-order cognitive deficits associated with disruption of dopaminergic function.[83] The effect on neurotransmitters has been noted in studies showing microbiota can influence their production.[14] Studies of mouse models evaluated the sex-dependent effects of Mn exposure on gut microbiota via drinking water for 13 weeks. Results showed that Mn exposure altered bacterial abundance along with sex-specific effect enrichment of genes related to tryptophan and γ-aminobutyric acid (GABA) synthesis.[84] Gut microbiota has been recognized as an integral mediator of neuroimmune interactions regulating responses to neuroinflammation.[85] Fecal microbiota transplantation studies in rats reduced Mn neurotoxicity by reducing expression of amyloid-β (Aβ), proinflammatory cytokines NOD-, LRR- and pyrin domain-containing 3 (NLRP3) inflammasome.[86]

4.2.2 Lead

Lead (Pb) is a ubiquitous heavy metal and environmental Pb levels have substantially increased since industrialization. Although occupational exposures are common, chronic Pb toxicity usually occurs via food, water, and soil and affects millions of children and adults.[87, 88] Despite stringent public health regulations and substantial lowering of blood Pb levels, Pb exposure still remains a significant problem. The developing nervous system is particularly susceptible to Pb exposure[89] and mechanisms including oxidative stress, mitochondrial dysfunction, and neuroinflammation have been identified as contributing to Pb-induced neurotoxicity.[90–92] Pb exposure in mice for 4 and 13 weeks induced perturbation in gut microbiota (*Ruminococcus*, *Blautia*, and *Oscillospira*) and altered fecal metabolic profile with a significant reduction in levels of three metabolites including vitamin E, bile acids, and cholesterol derivatives.[93] Another study revealed early lead exposure through maternal drinking water induced a sex-specific effect of reduced body weight in male offspring. However, regardless of sex, the microbiota revealed a greater abundance of *Desulfovibrionaceae*, *Barnesiella*, and *Clostridium* XIVb.[94] Further studies in mice evaluating chronic exposure to lead at 0.1 mg/L via drinking water for 15 weeks increased *Parabacteroides* and decreased *Dehalobacterium* in the cecum.[95]

4.2.3 Methylmercury

Mercurial compounds are major environmental pollutants arising from various sources such as volcanoes, oceanic sediments, forest fires, and combustion of fossil fuels. In aquatic resources, methylmercury is bioaccumulated at high levels in fish, shellfish, and sea mammals. Methylmercury is known to induce perturbation of intracellular Ca^{2+} levels, induce oxidative stress by generating reactive oxygen species (ROS), and forming protein complexes with thiol-containing compounds.[96] Rats provided with 10 mg/kg of methylmercury showed damaged intestinal walls and villus. Specifically, methylmercury induced a reduction in the relative abundance of *Lactobacillaceae*, *Bacteroideceae*, *Streptococcaceae*, and *Sutterellaceae*, while it enhanced *Desulfovibrionaceae*, *Helicobacteraceae*, *Peptococcaceae*, and *Rhodospirillaceae* with a decrease in metabolites and amino acids. Methylmercury also induced elevation of serum proinflammatory cytokines interleukin (IL)-1β and 6.[97] Human studies assessing the impact of methylmercury exposure in pregnant females revealed 17 microbial genera significantly correlated with mercury in maternal stool or hair.[98] The alpha diversity composition of the microbial community was significantly reduced during late gestation when compared with early gestation and 10 bacterial taxa correlated with methylmercury biomarkers from stool and blood.[99]

4.2.4 Arsenic

Groundwater contamination with arsenic is a major threat worldwide. Evidence from mice studies indicates that sodium arsenite supplementation (500 ppm in drinking water) for 2 weeks changed 2 phyla and 24 bacterial genera with more significant changes in metabolically important genera.[100] Arsenic was found to increase two bacterial phyla with increased plasma arginine metabolites including arginine, ornithine, asymmetric dimethylarginine (ADMA), and symmetric dimethylarginine (SDMA) levels.[101] Metagenomics and metabolomic profiling has revealed perturbed gut microbiota and microbial metabolites including amino acid derivatives, bile acids, lipids, fatty acids, isoflavones, and indole derivatives in mice exposed to 10 ppm arsenic for 4 weeks.[102] The role of the gut microbiome in arsenic detoxification is supported with findings indicating that microbiota perturbation or absence enhances arsenic toxicity and transfer of human gut commensal protects mice from arsenic-induced death attributed to transfer of *Faecalibacterium* bacteria.[103] Administration of arsenic induced a sex-specific effect on microbiota with females having decreased *Dorea* and increased *Akkermansia* modulating pathways of nitrogen, carbon, and sulfate metabolism.[104]

4.3 Organic compounds

4.3.1 Acrylamide

Acrylamide (ACR) is a water-soluble alkene with wide commercial application in cosmetics, paper, textiles, wastewater treatment plants, and molecular laboratories. It is present in certain foods that have been prepared by heating at very high temperatures like French fries, potato chips, and biscuits.[105–107] The cumulative neurotoxicity of ACR is linked to ataxia, muscle weakness, and cognitive impairments.Studies in experimental animals have demonstrated that ACR in its monomeric form induces neurotoxicity resembling neurological signs as observed with human intoxication (as reviewed by LoPachin and Lehning[108]). Further, human and animal ACR neurotoxicities are shown to be associated with the degeneration of distal axons and nerve terminals, Purkinje cell death, and oxidative stress, and forms Michael-type adducts with nucleophilic residues on proteins.[108–113] In addition, exposure to ACR has been recognized to induce developmental[114, 115] and reproductive toxicity.[116] It is now recognized that diet is the primary source of ACR exposure among the nonoccupationally exposed human population, and ACR intake during pregnancy is reported to have negative effects on fetal development.[117] We have adopted ACR in drinking water as a gestational rat model to recapitulate neurotoxic implications of ACR exposure on gestational parameters, extent of oxidative stress on maternal and fetal brain along with effects on cecal microbiota.[118] Our study showed the induction of oxidative stress, mitochondrial dysfunction, and a decrease in the number of *Bifidobacteria* in the cecum after exposure to ACR.

4.3.2 Polychlorinated biphenyls

Despite the cessation of large-scale use of polychlorinated biphenyls (PCBs), PCBs continue to pose a significant risk to human health because of improper disposal of electrical equipment. Several reviews of epidemiological findings show that PCB exposure during development promotes neuropsychological impairments with deficits in executive function and memory.[119, 120] It is also shown that developmental PCB exposure is associated with neurodevelopmental disorders such as ASD[121, 122] and attention-deficit disorder.[123] Animal model studies have demonstrated increased gut permeability and induction of proinflammatory cytokines interleukin (IL)-6 and tumor necrosis factor-α (TNF-α).[124] Fecal microarray analysis of mice receiving a PCB mixture revealed alterations in gut microbiota with decreased *Proteobacteria*.[125] Microbiome profiling of mouse models for exposure to PCBs has also suggested alterations in the gut microbiota.[126]

5 Microbiome implications for mitigating neurotoxicity

An increasing number of reports, which predominantly comprise studies in experimental animals, have employed manipulations of the gut microbiome for brain benefits. Microbiome-related interventional approaches have been applied for the amelioration of disease-related pathologies in lab animals, highlighting the potential for a novel approach for treating neurodegenerative diseases with neurotoxic components. Animal studies showing neurobiological changes following microbiota manipulation crucial for promoting healthy microbiome–gut–brain communication are starting to emerge and thus far have revealed significant benefits on the brain and behavior. Diet is recognized as one of the main factors to strongly and swiftly regulate the composition and diversity of the gut microbiota, making dietary manipulation a unique strategy to modulate the gut ecosystem with a lasting impact on the host. Certain dietary and bacterial origin food products are identified for use in specific age groups such as infants and early populations, improve host health and prevent GI problems by positively influencing the composition and/or activity of the GI microbiota. Such food components are now increasingly appreciated for their influence on brain development and biochemistry. Another front for the effects of these dietary adjuncts is their ability to influence several complex host behaviors.[127–130] To date, a few studies have attempted the use of dietary approaches to modulate microbiota and help abrogate chemical neurotoxicity.

5.1 Dietary interventions with prebiotics and probiotics

Prebiotics are selectively fermentable nondigestible food ingredients that specifically proliferate health-promoting gut bacteria. Their beneficial properties depend on the release of SCFAs.[131, 132] To date, specific prebiotics have been identified to influence host neurobiology and behavior. During pregnancy, intake of nondigestible oligosaccharides has been reported to have beneficial effects and may be used to mitigate certain pregnancy-related metabolic risks.[133] A few recent studies in gestational rat models reveal biochemical insights into the neuromodulatory role of nondigestible oligosaccharides against neurotoxic outcomes. Oral supplements of inulin (IN) to rats during gestation days (GD 6–19) abrogated the acrylamide-induced increase in malondialdehyde, ROS, and hydroperoxide levels in maternal and fetal brains of rats. Moreover, IN increased the maternal cecal *Bifidobacteria* numbers during GD 0–19 and restored the placental weights among acrylamide-exposed dams.[118] Further, IN supplements significantly attenuated acrylamide-induced oxidative stress markers in the fetal brain. Exemplifying an association between maternal microbiota and fetal brain, reports indicate that the maternal microbiome can influence neurodevelopment prenatal and postnatal neurodevelopment.[10, 134] Human studies have reported that IN intake enhances the relative abundances of three fecal genera, including *Anaerostipes*, *Bilophila*, and *Bifidobacterium*.[135] Another study with IN demonstrated enhanced fetal and placental weights, offered protection against mitochondrial dysfunction, and depletion of dopamine levels in maternal brain regions after rotenone exposure.[71] Pregnant dams provided a combination of prebiotics, fructo-oligosaccharides (FOS), and xylo-oligosaccharides (XOS) that attenuated acrylamide-associated protein oxidative damage and reduced glutathione levels.[136] Further, the combined prebiotic supplementation enhanced exploratory behavior of the dams upon indices of anxiety (open field exploration) and restored GABA levels.

Research interest in neurotransmitters has gained momentum in light of studies showing that certain gut microbes regulate neurochemicals that influence psychological behaviors.[137] For example, it has been determined that *Lactobacilli* and *Bifidobacteria* can produce GABA and acetylcholine.[138] The protective effects of combined oligosaccharide supplementation were discernible with the restoration of activity of cholinesterase activity and enhanced glutathione levels in the maternal brain.[139] These results suggest the influence of prebiotics on psychiatric symptoms and brain oxidative dysfunctions associated with neurotoxin exposure. However, further research must reveal the functional roles of gut microbiota in protecting against neurotoxicity.

Association between prebiotic intake and antiinflammatory response has been shown in a double-blind, placebo-controlled study with galactooligosaccharide (GOS)-mediated proliferation of beneficial *Bifidobacterium* spp. reducing the systemic levels of inflammatory markers IL-6, IL-1β, and TNF-α.[140] Neurodegenerative diseases involve persistent changes in neuronal function, possibly stemming from disturbances of neuroplasticity.[141] The neuroprotective actions of brain-derived neurotrophic factor (BDNF), a key proneuroplastic molecule, have received abundant experimental evidence.[142, 143] In two separate studies, it has been reported that rats fed with FOS and bimuno galacto-oligosaccharides (B-GOS) had increased levels of BDNF in hippocampus.[144, 145] The prebiotic-mediated effects on neuroplastic factors could conceivably contribute to mitigating consequences of insult by environmentally linked neurotoxicants.

The probiotic strain *Lactobacillus rhamnosus* reduced in vitro absorption of organophosphate pesticides parathion and chlorpyrifos.[146] In addition, dietary supplementation of the probiotic *Drosophila* reduced chlorpyrifos-induced mortality.[147] Administration of a probiotic bacterial mixture of *Lactobacillus rhamnosus* GG, *Bifidobacterium animalis* lactis, and *Lactobacillus acidophilus* to mice protected against dopaminergic cell loss and reduced accumulation of toxic metabolite 1-methyl-4-phenylpyridinium (MPP^+). Further, the study also showed that the probiotics had neuroprotective consequences by preventing rotenone-induced behavioral impairments, reactive gliosis, and trophic cytokines.[148] Another study using an MPTP model revealed that *Lactobacillus plantarum* PS128 mitigates motor deficits, an increase in Gram-negative *Enterobacteriaceae* and microbial modules related to lipopolysaccharides and peptidoglycan.[149] Further, the microbial treatment enhanced the antioxidant levels to reduce brain oxidative dysfunction and increased the levels of neurotransmitters.[149] Similarly, treatment of

MPTP-administered mice with lactic acid bacteria comprising three bacterial strains of *Lactobacillus plantarum* CRL 2130, *Streptococcus thermophilus* CRL 807, and *Streptococcus thermophilus* CRL 808 sufficiently restored motor behavior and reduced neuroinflammatory response.[150] It is shown that administration of probiotics (blend of *Streptococcus thermophilus*, *Lactobacillus acidophilus*, and *Bifidobacterium bifidum*) combined with herbal extracts of dandelion, coriander, and date palm seeds to rats reversed immunoreactivity of astrocytes in the cortex and cerebellum.[151] Collectively, these reports offer prospects for developing of new dietary strategies targeting the gut–brain axis for mitigating neurotoxicity.

6 Conclusions and future perspective

Accumulating evidence indicates that the microbiome is essential to human health and development, particularly the organization and function of the developing and adult brain. Perturbations of gut microbiota can profoundly affect brain development and has implications for adverse mental health later in life. The bidirectional gut–brain axis interaction involving the gut microbiome constitutes a key causal link between chemical exposure and neurotoxic outcomes throughout the lifespan. Studies in animal models have provided insight into chemical exposure, the microbiome, and brain function that influence the risk of neurodegenerative diseases and neurodevelopmental disorders. In particular, recent studies have uncovered the mechanisms of chemical agents-induced neurotoxicity with the possibility of direct and indirect effects by microbiota-dependent biotransformation, metabolic products, and immune responses. Moreover, manipulating the gut microbiota for brain benefits is an avenue that is gaining popularity. Studies applying genomic technologies have revealed the compelling influence of the microbiome on neurotoxic outcomes, inspiring further research to discern the complex interaction between gut bacteria and neurotoxic exposures. Further, such an understanding of mechanisms that trigger long-term sequelae of neurotoxicity informs about neurodegenerative and neurodevelopmental psychiatric disorders, identification of biomarkers, and even allows for manipulations to shape the gut microbiota. With a growing number of studies in animal models linking microbiome to the neurotoxicity, it is challenging when attempting to translate the relevance of these findings to humans showing more complex neurotoxic manifestations associated with several comorbidities. It is necessary to understand the extent of involvement of chemical-induced alterations in causing observable neurotoxicity and the specific effect of chemicals on microbial taxa implicated with such outcomes. To date, several of the published findings largely utilize sequencing studies based on phylogenetic markers assessing changes with bacterial taxa rather than the functions of microbial communities. Hence, additional work is needed to reveal the community phenotypes mediating neurotoxic outcomes.

Conflict of interest

None.

References

1. Julvez J, Grandjean P. Neurodevelopmental toxicity risks due to occupational exposure to industrial chemicals during pregnancy. *Ind Health*. 2009;47(5):459–468.
2. Grandjean P, Landrigan PJ. Developmental neurotoxicity of industrial chemicals. *Lancet*. 2006;368(9553):2167–2178.
3. Rice D, Barone Jr S. Critical periods of vulnerability for the developing nervous system: evidence from humans and animal models. *Environ Health Perspect*. 2000;108(Suppl 3):511–533.
4. Cannon JR, Greenamyre JT. The role of environmental exposures in neurodegeneration and neurodegenerative diseases. *Toxicol Sci*. 2011;124(2):225–250.
5. LoPachin RM, Barber DS, Gavin T. Molecular mechanisms of the conjugated α, β-unsaturated carbonyl derivatives: relevance to neurotoxicity and neurodegenerative diseases. *Toxicol Sci*. 2008;104(2):235–249.
6. Anger WK. Behavioral biomarkers to identify neurotoxic effects. In: *Use of Biomarkers in Assessing Health and Environmental Impacts of Chemical Pollutants*. Springer; 1993:159–168.
7. Food and Drug Administration. *Predicting neurotoxicity and behavioral dysfunction from preclinical toxicologic data*. Life Sciences Research Office, Federation of American Societies for Experimental Biology 9650 Rockville Pike; 1986.
8. Davison AN, Dobbing J. The developing brain. *Appl Neurochem*. 1968;4:253–286.
9. Dinan TG, Cryan JF. The impact of gut microbiota on brain and behaviour: implications for psychiatry. *Curr Opin Clin Nutr Metab Care*. 2015;18(6):552–558.
10. Borre YE, O'Keeffe GW, Clarke G, Stanton C, Dinan TG, Cryan JF. Microbiota and neurodevelopmental windows: implications for brain disorders. *Trends Mol Med*. 2014;20(9):509–518.
11. de la Fuente-Fernández R, Calne DB. Evidence for environmental causation of Parkinson's disease. *Parkinsonism Relat Disord*. 2002;8(4):235–241.
12. Foster HD. Why the preeminent risk factor in sporadic Alzheimer's disease cannot be genetic. *Med Hypotheses*. 2002;59(1):57–61.
13. Sampson TR, Debelius JW, Thron T, et al. Gut microbiota regulate motor deficits and neuroinflammation in a model of Parkinson's disease. *Cell*. 2016;167(6):1469–1480.
14. Yano JM, Yu K, Donaldson GP, et al. Indigenous bacteria from the gut microbiota regulate host serotonin biosynthesis. *Cell*. 2015;161(2):264–276.
15. Petrov VA, Saltykova IV, Zhukova IA, et al. Analysis of gut microbiota in patients with Parkinson's disease. *Bull Exp Biol Med*. 2017;162(6):734–737.
16. Lin A, Zheng W, He Y, et al. Gut microbiota in patients with Parkinson's disease in southern China. *Parkinsonism Relat Disord*. 2018;53:82–88.

17. Hill-Burns EM, Debelius JW, Morton JT, et al. Parkinson's disease and Parkinson's disease medications have distinct signatures of the gut microbiome. *Mov Disord.* 2017;32(5):739–749.
18. Vascellari S, Palmas V, Melis M, et al. Gut microbiota and metabolome alterations associated with Parkinson's disease. *mSystems.* 2020;5(5):e00561–20.
19. Vogt NM, Kerby RL, Dill-McFarland KA, et al. Gut microbiome alterations in Alzheimer's disease. *Sci Rep.* 2017;7(1):13537.
20. Zhuang ZQ, Shen LL, Li WW, et al. Gut microbiota is altered in patients with Alzheimer's disease. *J Alzheimers Dis.* 2018;63(4):1337–1346.
21. Kang D-W, Ilhan ZE, Isern NG, et al. Differences in fecal microbial metabolites and microbiota of children with autism spectrum disorders. *Anaerobe.* 2018;49:121–131.
22. Luna RA, Oezguen N, Balderas M, et al. Distinct microbiome-neuroimmune signatures correlate with functional abdominal pain in children with autism spectrum disorder. *Cell Mol Gastroenterol Hepatol.* 2017;3(2):218–230.
23. Fournier CN, Houser M, Tansey MG, Glass JD, Hertzberg VS. The gut microbiome and neuroinflammation in amyotrophic lateral sclerosis? Emerging clinical evidence. *Neurobiol Dis.* 2018;135, 104300.
24. Baizabal-Carvallo JF, Alonso-Juarez M. The link between gut dysbiosis and neuroinflammation in Parkinson's disease. *Neuroscience.* 2020;432:160–173.
25. Zhao Y, Zhang Y, Wang G, Han R, Xie X. Effects of chlorpyrifos on the gut microbiome and urine metabolome in mouse (Mus musculus). *Chemosphere.* 2016;153:287–293.
26. Muegge BD, Kuczynski J, Knights D, et al. Diet drives convergence in gut microbiome functions across mammalian phylogeny and within humans. *Science.* 2011;332(6032):970–974.
27. Scheperjans F, Aho V, Pereira PA, et al. Gut microbiota are related to Parkinson's disease and clinical phenotype. *Mov Disord.* 2015;30(3):350–358.
28. Liu S, Gao J, Zhu M, Liu K, Zhang H-L. Gut microbiota and dysbiosis in Alzheimer's disease: implications for pathogenesis and treatment. *Mol Neurobiol.* 2020;57(12):5026–5043.
29. Vuong HE, Hsiao EY. Emerging roles for the gut microbiome in autism spectrum disorder. *Biol Psychiatry.* 2017;81(5):411–423.
30. Round JL, Mazmanian SK. The gut microbiota shapes intestinal immune responses during health and disease. *Nat Rev Immunol.* 2009;9(5):313–323.
31. Forsythe P, Bienenstock J, Kunze WA. Vagal pathways for microbiome-brain-gut axis communication. In: *Microbial Endocrinology: The Microbiota-Gut-Brain Axis in Health and Disease.* Springer; 2014:115–133.
32. London L, Beseler C, Bouchard MF, et al. Neurobehavioral and neurodevelopmental effects of pesticide exposures. *Neurotoxicology.* 2012;33(4):887–896.
33. Ginsberg G, Hattis D, Sonawane B. Incorporating pharmacokinetic differences between children and adults in assessing children's risks to environmental toxicants. *Toxicol Appl Pharmacol.* 2004;198(2):164–183.
34. Claus SP, Guillou H, Ellero-Simatos S. The gut microbiota: a major player in the toxicity of environmental pollutants? *NPJ Biofilms Microbiomes.* 2016;2(1):1–11.
35. Wilson ID, Nicholson JK. Gut microbiome interactions with drug metabolism, efficacy, and toxicity. *Transl Res.* 2017;179:204–222.
36. Spanogiannopoulos P, Bess EN, Carmody RN, Turnbaugh PJ. The microbial pharmacists within us: a metagenomic view of xenobiotic metabolism. *Nat Rev Microbiol.* 2016;14(5):273.
37. Koppel N, Rekdal VM, Balskus EP. Chemical transformation of xenobiotics by the human gut microbiota. *Science.* 2017;356(6344):eaag2770.
38. Cuhra M, Bøhn T, Cuhra P. Glyphosate: too much of a good thing? *Front Environ Sci.* 2016;4:28.

39. Serrano-Niño JC, Cavazos-Garduño A, Hernandez-Mendoza A, et al. Assessment of probiotic strains ability to reduce the bioaccessibility of aflatoxin M1 in artificially contaminated milk using an in vitro digestive model. *Food Control.* 2013;31(1):202–207.
40. Dalile B, Van Oudenhove L, Vervliet B, Verbeke K. The role of short-chain fatty acids in microbiota–gut–brain communication. *Nat Rev Gastroenterol Hepatol.* 2019;16(8):461–478.
41. Haast RA, Kiliaan AJ. Impact of fatty acids on brain circulation, structure and function. *Prostaglandins Leukot Essent Fatty Acids.* 2015;92:3–14.
42. Tan J, McKenzie C, Potamitis M, Thorburn AN, Mackay CR, Macia L. The role of short-chain fatty acids in health and disease. In: *Advances in Immunology.* vol 121. Elsevier; 2014:91–119.
43. MacFabe DF. Enteric short-chain fatty acids: microbial messengers of metabolism, mitochondria, and mind: implications in autism spectrum disorders. *Microb Ecol Health Dis.* 2015;26(1):28177.
44. Unger MM, Spiegel J, Dillmann K-U, et al. Short chain fatty acids and gut microbiota differ between patients with Parkinson's disease and age-matched controls. *Parkinsonism Relat Disord.* 2016;32:66–72.
45. Erny D, de Angelis ALH, Jaitin D, et al. Host microbiota constantly control maturation and function of microglia in the CNS. *Nat Neurosci.* 2015;18(7):965–977.
46. Lach G, Schellekens H, Dinan TG, Cryan JF. Anxiety, depression, and the microbiome: a role for gut peptides. *Neurotherapeutics.* 2018;15(1):36–59.
47. Cryan JF, Dinan TG. More than a gut feeling: the microbiota regulates neurodevelopment and behavior. *Neuropsychopharmacology.* 2015;40(1):241.
48. Rea K, Dinan TG, Cryan JF. The microbiome: a key regulator of stress and neuroinflammation. *Neurobiol Stress.* 2016;4:23–33.
49. Fung TC, Olson CA, Hsiao EY. Interactions between the microbiota, immune and nervous systems in health and disease. *Nat Neurosci.* 2017;20(2):145.
50. Rea K, Dinan TG, Cryan JF. The brain-gut axis contributes to neuroprogression in stress-related disorders. In: *Neuroprogression in Psychiatric Disorders.* vol 31. Karger Publishers; 2017:152–161.
51. Defois C, Ratel J, Garrait G, et al. Food chemicals disrupt human gut microbiota activity and impact intestinal homeostasis as revealed by in vitro systems. *Sci Rep.* 2018;8(1):11006.
52. Glass CK, Saijo K, Winner B, Marchetto MC, Gage FH. Mechanisms underlying inflammation in neurodegeneration. *Cell.* 2010;140(6):918–934.
53. Tanner CM, Kamel F, Ross GW, et al. Rotenone, paraquat, and Parkinson's disease. *Environ Health Perspect.* 2011;119(6):866–872.
54. Freire C, Koifman S. Pesticide exposure and Parkinson's disease: epidemiological evidence of association. *Neurotoxicology.* 2012;33(5):947–971.
55. Shimohama S, Sawada H, Kitamura Y, Taniguchi T. Disease model: Parkinson's disease. *Trends Mol Med.* 2003;9(8):360–365.
56. Jackson-Lewis V, Lester D, Kozina E, Przedborski S, Smeyne RJ. From man to mouse: the MPTP model of Parkinson disease. In: *Movement Disorders.* Elsevier; 2015:287–306.
57. Uhl G, Javitch J, Snyder S. Normal MPTP binding in parkinsonian substantia nigra: evidence for extraneuronal toxin conversion in human brain. *Lancet.* 1985;325(8435):956–957.
58. Barcia C, Bahillo AS, Fernández-Villalba E, et al. Evidence of active microglia in substantia nigra pars compacta of parkinsonian monkeys 1 year after MPTP exposure. *Glia.* 2004;46(4):402–409.
59. Lai F, Jiang R, Xie W, et al. Intestinal pathology and gut microbiota alterations in a methyl-4-phenyl-1, 2, 3, 6-tetrahydropyridine (MPTP) mouse model of Parkinson's disease. *Neurochem Res.* 2018;43(10):1986–1999.

60. Xie W, Gao J, Jiang R, et al. Twice subacute MPTP administrations induced time-dependent dopaminergic neurodegeneration and inflammation in midbrain and ileum, as well as gut microbiota disorders in PD mice. *Neurotoxicology.* 2020;76:200–212.
61. Pu Y, Chang L, Qu Y, Wang S, Zhang K, Hashimoto K. Antibiotic-induced microbiome depletion protects against MPTP-induced dopaminergic neurotoxicity in the brain. *Aging (Albany NY).* 2019;11(17):6915.
62. Sun MF, Zhu YL, Zhou ZL, et al. Neuroprotective effects of fecal microbiota transplantation on MPTP-induced Parkinson's disease mice: gut microbiota, glial reaction and TLR4/TNF-α signaling pathway. *Brain Behav Immun.* 2018;70:48–60.
63. Betarbet R, Canet-Aviles RM, Sherer TB, et al. Intersecting pathways to neurodegeneration in Parkinson's disease: effects of the pesticide rotenone on DJ-1, α-synuclein, and the ubiquitin–proteasome system. *Neurobiol Dis.* 2006;22(2):404–420.
64. Chou AP, Li S, Fitzmaurice AG, Bronstein JM. Mechanisms of rotenone-induced proteasome inhibition. *Neurotoxicology.* 2010;31(4):367–372.
65. Greene JG, Noorian AR, Srinivasan S. Delayed gastric emptying and enteric nervous system dysfunction in the rotenone model of Parkinson's disease. *Exp Neurol.* 2009;218(1):154–161.
66. Drolet RE, Cannon JR, Montero L, Greenamyre JT. Chronic rotenone exposure reproduces Parkinson's disease gastrointestinal neuropathology. *Neurobiol Dis.* 2009;36(1):96–102.
67. Pan-Montojo F, Schwarz M, Winkler C, et al. Environmental toxins trigger PD-like progression via increased alpha-synuclein release from enteric neurons in mice. *Sci Rep.* 2012;2(1):1–12.
68. Yang X, Qian Y, Xu S, Song Y, Xiao Q. Longitudinal analysis of fecal microbiome and pathologic processes in a rotenone induced mice model of Parkinson's disease. *Front Aging Neurosci.* 2018;9:441.
69. Johnson ME, Stringer A, Bobrovskaya L. Rotenone induces gastrointestinal pathology and microbiota alterations in a rat model of Parkinson's disease. *Neurotoxicology.* 2018;65:174–185.
70. Perez-Pardo P, Dodiya HB, Engen PA, et al. Gut bacterial composition in a mouse model of Parkinson's disease. *Benefic Microbes.* 2018;9(5):799–814.
71. Krishna G, Muralidhara. Oral supplements of inulin during gestation offsets rotenone-induced oxidative impairments and neurotoxicity in maternal and prenatal rat brain. *Biomed Pharmacother.* 2018;104:751–762.
72. Peng J, Mao XO, Stevenson FF, Hsu M, Andersen JK. The herbicide paraquat induces dopaminergic nigral apoptosis through sustained activation of the JNK pathway. *J Biol Chem.* 2004;279(31):32626–32632.
73. Fei Q, McCormack AL, Di Monte DA, Ethell DW. Paraquat neurotoxicity is mediated by a Bak-dependent mechanism. *J Biol Chem.* 2008;283(6):3357–3364.
74. Lekova V. Role impact of the Gut microbiota in a Drosophila model for Parkinsonism. *Dissertations Available from ProQuest.* 2017;, AAI10683295. https://repository.upenn.edu/dissertations/AAI10683295.
75. Naudet N, Antier E, Gaillard D, et al. Oral exposure to paraquat triggers earlier expression of phosphorylated α-synuclein in the enteric nervous system of A53T mutant human α-synuclein transgenic mice. *J Neuropathol Exp Neurol.* 2017;76(12):1046–1057.
76. Slotkin TA. Developmental cholinotoxicants: nicotine and chlorpyrifos. *Environ Health Perspect.* 1999;107(Suppl 1):71–80.
77. Rauh VA, Perera FP, Horton MK, et al. Brain anomalies in children exposed prenatally to a common organophosphate pesticide. *Proc Natl Acad Sci U S A.* 2012;109(20):7871–7876.
78. Garcia SJ, Seidler FJ, Crumpton TL, Slotkin TA. Does the developmental neurotoxicity of chlorpyrifos involve glial targets? Macromolecule synthesis, adenylyl cyclase

signaling, nuclear transcription factors, and formation of reactive oxygen in C6 glioma cells. *Brain Res.* 2001;891(1–2):54–68.
79. Mileson BE, Chambers JE, Chen WL, et al. Common mechanism of toxicity: a case study of organophosphorus pesticides. *Toxicol Sci.* 1998;41(1):8–20.
80. Garcia SJ, Seidler FJ, Slotkin TA. Developmental neurotoxicity elicited by prenatal or postnatal chlorpyrifos exposure: effects on neurospecific proteins indicate changing vulnerabilities. *Environ Health Perspect.* 2003;111(3):297–303.
81. Li JW, Fang B, Pang GF, Zhang M, Ren FZ. Age-and diet-specific effects of chronic exposure to chlorpyrifos on hormones, inflammation and gut microbiota in rats. *Pestic Biochem Physiol.* 2019;159:68–79.
82. Fang B, Li JW, Zhang M, Ren FZ, Pang GF. Chronic chlorpyrifos exposure elicits diet-specific effects on metabolism and the gut microbiome in rats. *Food Chem Toxicol.* 2018;111:144–152.
83. Guilarte TR. Manganese neurotoxicity: new perspectives from behavioral, neuroimaging, and neuropathological studies in humans and non-human primates. *Front Aging Neurosci.* 2013;5:23.
84. Chi L, Gao B, Bian X, Tu P, Ru H, Lu K. Manganese-induced sex-specific gut microbiome perturbations in C57BL/6 mice. *Toxicol Appl Pharmacol.* 2017;331:142–153.
85. Wang Y, Telesford KM, Ochoa-Repáraz J, et al. An intestinal commensal symbiosis factor controls neuroinflammation via TLR2-mediated CD39 signalling. *Nat Commun.* 2014;5(1):1–10.
86. Wang H, Yang F, Xin R, et al. The gut microbiota attenuate neuroinflammation in manganese exposure by inhibiting cerebral NLRP3 inflammasome. *Biomed Pharmacother.* 2020;129:110449.
87. Roper WL, Houk VN, Falk H, Binder S. *Preventing Lead Poisoning in Young Children: A Statement by the Centers for Disease Control, October 1991.* Atlanta, GA (United States): Centers for Disease Control; 1991.
88. Hu H, Shih R, Rothenberg S, Schwartz BS. The epidemiology of lead toxicity in adults: measuring dose and consideration of other methodologic issues. *Environ Health Perspect.* 2007;115(3):455–462.
89. Gąssowska M, Baranowska-Bosiacka I, Moczyd\lowska J, et al. Perinatal exposure to lead (Pb) induces ultrastructural and molecular alterations in synapses of rat offspring. *Toxicology.* 2016;373:13–29.
90. Kumar VL, Muralidhara. Ameliorative effects of ferulic acid against lead acetate-induced oxidative stress, mitochondrial dysfunctions and toxicity in prepubertal rat brain. *Neurochem Res.* 2014;39(12):2501–2515.
91. Maiti AK, Saha NC, More SS, Panigrahi AK, Paul G. Neuroprotective efficacy of mitochondrial antioxidant MitoQ in suppressing peroxynitrite-mediated mitochondrial dysfunction inflicted by lead toxicity in the rat brain. *Neurotox Res.* 2017;31(3):358–372.
92. Chibowska K, Baranowska-Bosiacka I, Falkowska A, Gutowska I, Goschorska M, Chlubek D. Effect of lead (Pb) on inflammatory processes in the brain. *Int J Mol Sci.* 2016;17(12):2140.
93. Gao B, Chi L, Mahbub R, et al. Multi-omics reveals that lead exposure disturbs gut microbiome development, key metabolites, and metabolic pathways. *Chem Res Toxicol.* 2017;30(4):996–1005.
94. Wu J, Wen XW, Faulk C, et al. Perinatal lead exposure alters gut microbiota composition and results in sex-specific bodyweight increases in adult mice. *Toxicol Sci.* 2016;151(2):324–333.
95. Xia J, Jin C, Pan Z, Sun L, Fu Z, Jin Y. Chronic exposure to low concentrations of lead induces metabolic disorder and dysbiosis of the gut microbiota in mice. *Sci Total Environ.* 2018;631:439–448.

96. Atchison WD, Hare MF. Mechanisms of methylmercury-induced neurotoxicity. *FASEB J.* 1994;8(9):622–629.
97. Lin X, Zhao J, Zhang W, et al. Acute oral methylmercury exposure perturbs the gut microbiome and alters gut-brain axis related metabolites in rats. *Ecotoxicol Environ Saf.* 2020;190:110130.
98. Rothenberg SE, Keiser S, Ajami NJ, et al. The role of gut microbiota in fetal methylmercury exposure: insights from a pilot study. *Toxicol Lett.* 2016;242:60–67.
99. Rothenberg SE, Wagner CL, Alekseyenko AV, Azcarate-Peril MA. Longitudinal changes during pregnancy in gut microbiota and methylmercury biomarkers, and reversal of microbe-exposure correlations. *Environ Res.* 2019;172:700–712.
100. Li X, Brejnrod AD, Ernst M, et al. Heavy metal exposure causes changes in the metabolic health-associated gut microbiome and metabolites. *Environ Int.* 2019;126:454–467.
101. Dheer R, Patterson J, Dudash M, et al. Arsenic induces structural and compositional colonic microbiome change and promotes host nitrogen and amino acid metabolism. *Toxicol Appl Pharmacol.* 2015;289(3):397–408.
102. Lu K, Abo RP, Schlieper KA, et al. Arsenic exposure perturbs the gut microbiome and its metabolic profile in mice: an integrated metagenomics and metabolomics analysis. *Environ Health Perspect.* 2014;122(3):284–291.
103. Coryell M, McAlpine M, Pinkham NV, McDermott TR, Walk ST. The gut microbiome is required for full protection against acute arsenic toxicity in mouse models. *Nat Commun.* 2018;9(1):1–9.
104. Chi L, Bian X, Gao B, Ru H, Tu P, Lu K. Sex-specific effects of arsenic exposure on the trajectory and function of the gut microbiome. *Chem Res Toxicol.* 2016;29(6):949–951.
105. Smith EA, Oehme FW. Acrylamide and polyacrylamide: a review of production, use, environmental fate and neurotoxicity. *Rev Environ Health.* 1991;9(4):215–228.
106. Tareke E, Rydberg P, Karlsson P, Eriksson S, Törnqvist M. Analysis of acrylamide, a carcinogen formed in heated foodstuffs. *J Agric Food Chem.* 2002;50(17):4998–5006.
107. Dybing E, Sanner T. Risk assessment of acrylamide in foods. *Toxicol Sci.* 2003;75(1):7–15.
108. Lopachin Jr RM, Lehning EJ. Acrylamide-induced distal axon degeneration: a proposed mechanism of action. *Neurotoxicology.* 1994;15(2):247–259.
109. Spencer PS, Schaumburg HH. A review of acrylamide neurotoxicity part I. properties, uses and human exposure. *Can J Neurol Sci.* 1974;1(2):143–150.
110. LoPachin RM, Gavin T. Molecular mechanism of acrylamide neurotoxicity: lessons learned from organic chemistry. *Environ Health Perspect.* 2012;120(12):1650–1657.
111. LoPachin RM, Gavin T. Toxic neuropathies: mechanistic insights based on a chemical perspective. *Neurosci Lett.* 2015;596:78–83.
112. Prasad SN, Muralidhara. Neuroprotective efficacy of eugenol and isoeugenol in acrylamide-induced neuropathy in rats: behavioral and biochemical evidence. *Neurochem Res.* 2013;38(2):330–345.
113. Lopachin RM, Gavin T. Acrylamide-induced nerve terminal damage: relevance to neurotoxic and neurodegenerative mechanisms. *J Agric Food Chem.* 2008;56(15):5994–6003.
114. Wise LD, Gordon LR, Soper KA, Duchai DM, Morrissey RE. Developmental neurotoxicity evaluation of acrylamide in Sprague-Dawley rats. *Neurotoxicol Teratol.* 1995;17(2):189–198.
115. Ogawa B, Ohishi T, Wang L, et al. Disruptive neuronal development by acrylamide in the hippocampal dentate hilus after developmental exposure in rats. *Arch Toxicol.* 2011;85(8):987–994.
116. Tyl RW, Marr MC, Myers CB, Ross WP, Friedman MA. Relationship between acrylamide reproductive and neurotoxicity in male rats. *Reprod Toxicol.* 2000;14(2):147–157.
117. Duarte-Salles T, von Stedingk H, Granum B, et al. Dietary acrylamide intake during pregnancy and fetal growth-results from the Norwegian mother and child cohort study (MoBa). *Environ Health Perspect.* 2013;121(3):374–379.

118. Krishna G, Muralidhara. Inulin supplementation during gestation mitigates acrylamide-induced maternal and fetal brain oxidative dysfunctions and neurotoxicity in rats. *Neurotoxicol Teratol.* 2015;49:49–58.
119. Schantz SL, Widholm JJ, Rice DC. Effects of PCB exposure on neuropsychological function in children. *Environ Health Perspect.* 2003;111(3):357–576.
120. Pessah IN, Lein PJ, Seegal RF, Sagiv SK. Neurotoxicity of polychlorinated biphenyls and related organohalogens. *Acta Neuropathol.* 2019;138(3):363–387.
121. Granillo L, Sethi S, Keil KP, et al. Polychlorinated biphenyls influence on autism spectrum disorder risk in the MARBLES cohort. *Environ Res.* 2019;171:177–184.
122. Lyall K, Croen LA, Sjödin A, et al. Polychlorinated biphenyl and organochlorine pesticide concentrations in maternal mid-pregnancy serum samples: association with autism spectrum disorder and intellectual disability. *Environ Health Perspect.* 2017;125(3):474–480.
123. Eubig PA, Aguiar A, Schantz SL. Lead and PCBs as risk factors for attention deficit/hyperactivity disorder. *Environ Health Perspect.* 2010;118(12):1654–1667.
124. Phillips MC, Dheer R, Santaolalla R, et al. Intestinal exposure to PCB 153 induces inflammation via the ATM/NEMO pathway. *Toxicol Appl Pharmacol.* 2018;339:24–33.
125. Choi JJ, Eum SY, Rampersaud E, Daunert S, Abreu MT, Toborek M. Exercise attenuates PCB-induced changes in the mouse gut microbiome. *Environ Health Perspect.* 2013;121(6):725–730.
126. Rude KM, Pusceddu MM, Keogh CE, et al. Developmental exposure to polychlorinated biphenyls (PCBs) in the maternal diet causes host-microbe defects in weanling offspring mice. *Environ Pollut.* 2019;253:708–721.
127. Yen CH, Wang CH, Wu WT, Chen HL. Fructo-oligosaccharide improved brain β-amyloid, β-secretase, cognitive function, and plasma antioxidant levels in D-galactose-treated Balb/cJ mice. *Nutr Neurosci.* 2017;20(4):228–237.
128. Jia S, Lu Z, Gao Z, et al. Chitosan oligosaccharides alleviate cognitive deficits in an amyloid-β1–42-induced rat model of Alzheimer's disease. *Int J Biol Macromol.* 2016;83:416–425.
129. Savignac HM, Couch Y, Stratford M, et al. Prebiotic administration normalizes lipopolysaccharide (LPS)-induced anxiety and cortical 5-HT2A receptor and IL1-β levels in male mice. *Brain Behav Immun.* 2016;52:120–131.
130. McVey Neufeld K-A, O'Mahony SM, Hoban AE, et al. Neurobehavioural effects of Lactobacillus rhamnosus GG alone and in combination with prebiotics polydextrose and galactooligosaccharide in male rats exposed to early-life stress. *Nutr Neurosci.* 2019;22(6):425–434.
131. Gibson G, Scott K, Rastall R, et al. Dietary prebiotics: current status and new definition. *Food Sci Technol Bull Funct Food.* 2011;7:1–19.
132. Bindels LB, Delzenne NM, Cani PD, Walter J. Towards a more comprehensive concept for prebiotics. *Nat Rev Gastroenterol Hepatol.* 2015;12(5):303.
133. Qiu C, Coughlin KB, Frederick IO, Sorensen TK, Williams MA. Dietary fiber intake in early pregnancy and risk of subsequent preeclampsia. *Am J Hypertens.* 2008;21(8):903–909.
134. Pronovost GN, Hsiao EY. Perinatal interactions between the microbiome, immunity, and neurodevelopment. *Immunity.* 2019;50(1):18–36.
135. Vandeputte D, Falony G, Vieira-Silva S, et al. Prebiotic inulin-type fructans induce specific changes in the human gut microbiota. *Gut.* 2017;66(11):1968–1974.
136. Krishna G, Divyashri G, Prapulla SG, Muralidhara. A combination supplement of fructo-and xylo-oligosaccharides significantly abrogates oxidative impairments and neurotoxicity in maternal/fetal milieu following gestational exposure to acrylamide in rat. *Neurochem Res.* 2015;40(9):1904–1918.
137. Lyte M. Microbial endocrinology in the microbiome-gut-brain axis: how bacterial production and utilization of neurochemicals influence behavior. *PLoS Pathog.* 2013;9(11), e1003726.

138. Barrett E, Ross RP, O'toole PW, Fitzgerald GF, Stanton C. γ-Aminobutyric acid production by culturable bacteria from the human intestine. *J Appl Microbiol.* 2012;113(2):411–417.
139. Krishna G, Muralidhara. Oral supplements of combined fructo-and xylo-oligosaccharides during perinatal period significantly offsets acrylamide-induced oxidative impairments and neurotoxicity in rats. *J Physiol Pharmacol.* 2018;69(5):801–814.
140. Vulevic J, Juric A, Walton GE, et al. Influence of galacto-oligosaccharide mixture (B-GOS) on gut microbiota, immune parameters and metabonomics in elderly persons. *Br J Nutr.* 2015;114(4):586–595.
141. Teter B, Ashford JW. Neuroplasticity in Alzheimer's disease. *J Neurosci Res.* 2002;70(3):402–437.
142. Spina MB, Squinto SP, Miller J, Lindsay RM, Hyman C. Brain-derived neurotrophic factor protects dopamine neurons against 6-hydroxydopamine and N-methyl-4-phenylpyridinium ion toxicity: involvement of the glutathione system. *J Neurochem.* 1992;59(1):99–106.
143. Bousquet M, Gibrat C, Saint-Pierre M, Julien C, Calon F, Cicchetti F. Modulation of brain-derived neurotrophic factor as a potential neuroprotective mechanism of action of omega-3 fatty acids in a parkinsonian animal model. *Prog Neuropsychopharmacol Biol Psychiatry.* 2009;33(8):1401–1408.
144. Savignac HM, Corona G, Mills H, et al. Prebiotic feeding elevates central brain derived neurotrophic factor, N-methyl-D-aspartate receptor subunits and D-serine. *Neurochem Int.* 2013;63(8):756–764.
145. Williams S, Chen L, Savignac HM, Tzortzis G, Anthony DC, Burnet PW. Neonatal prebiotic (BGOS) supplementation increases the levels of synaptophysin, Glu N 2 A-subunits and BDNF proteins in the adult rat hippocampus. *Synapse.* 2016;70(3):121–124.
146. Trinder M, McDowell TW, Daisley BA, et al. Probiotic Lactobacillus rhamnosus reduces organophosphate pesticide absorption and toxicity to Drosophila melanogaster. *Appl Environ Microbiol.* 2016;82(20):6204–6213.
147. Daisley BA, Trinder M, McDowell TW, Collins SL, Sumarah MW, Reid G. Microbiota-mediated modulation of organophosphate insecticide toxicity by species-dependent interactions with lactobacilli in a Drosophila melanogaster insect model. *Appl Environ Microbiol.* 2018;84(9), e02820-17.
148. Srivastav S, Neupane S, Bhurtel S, et al. Probiotics mixture increases butyrate, and subsequently rescues the nigral dopaminergic neurons from MPTP and rotenone-induced neurotoxicity. *J Nutr Biochem.* 2019;69:73–86.
149. Liao JF, Cheng YF, You ST, et al. Lactobacillus plantarum PS128 alleviates neurodegenerative progression in 1-methyl-4-phenyl-1, 2, 3, 6-tetrahydropyridine-induced mouse models of Parkinson's disease. *Brain Behav Immun.* 2020;90:26–46.
150. Visñuk DP, de Giori GS, LeBlanc JG, de LeBlanc A de M. Neuroprotective effects associated with immune modulation by selected lactic acid bacteria in a Parkinson's disease model. *Nutrition.* 2020;79–80, 110995.
151. Abdel-Salam AM, Al Hemaid WA, Afifi AA, et al. Consolidating probiotic with dandelion, coriander and date palm seeds extracts against mercury neurotoxicity and for maintaining normal testosterone levels in male rats. *Toxicol Rep.* 2018;5:1069–1077.

CHAPTER 18

Summary and perspective for future research on the contribution of microbiota in visceral and neurological disorders

Tahira Farooqui[a] and Akhlaq A. Farooqui[b]
[a]Department of Entomology, The Ohio State University, Columbus, OH, United States
[b]Department of Molecular and Cellular Biochemistry, The Ohio State University, Columbus, OH, United States

1 Introduction

The gut microbiota is the microbial community found within the gut. The human gut microbiota is a complex system of mutualistic microorganisms, hosting an impressive 100 trillion bacteria from 500 to 1000 species representing 1%–3% of body mass and encoding > 4 million genes.[1,2] Colonization of microbes starts prenatally, through microbial transmission from mother to fetus.[3] Colonization of the human gut continues after birth and is modulated by factors including gestational age, mode of delivery (natural or by Caesarean section), diet (breastfeeding or infant formula), hygiene, and antibiotic exposure. The relationship between host and most of the gut microbiota has evolved as mutualistic or in a symbiotic manner.[4] The gut microbiota perform many important functions in the body, including enhancing energy extraction from the diet, producing vitamins, regulating the immune system, protecting against infectious agents, maintaining gut barrier integrity, and modulating metabolic and neurologic development, insulin resistance and insulin secretion, and behavioral and neurological functions of the host (Fig. 1).[5–10] The most common microbial phyla are Firmicutes and Bacteroidetes, which represent 70%–75% of total microbiome, which is defined as genetic material of all intestinal microbes (Fig. 2). Other bacteria occurring in lesser quantities are Proteobacteria, Actinobacteria, Fusobacteria, and Verrucomicrobia.[1] Since most mammalian cells have genes that encode for enzymes and structural proteins. The

Gut Microbiota in Neurologic and Visceral Diseases
https://doi.org/10.1016/B978-0-12-821039-0.00021-6

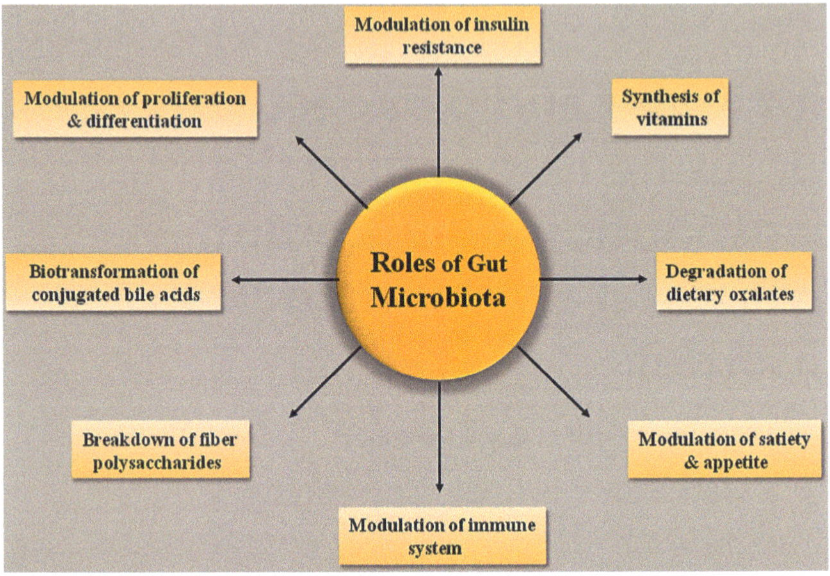

Fig. 1 Role of gut microbiota in metabolism.

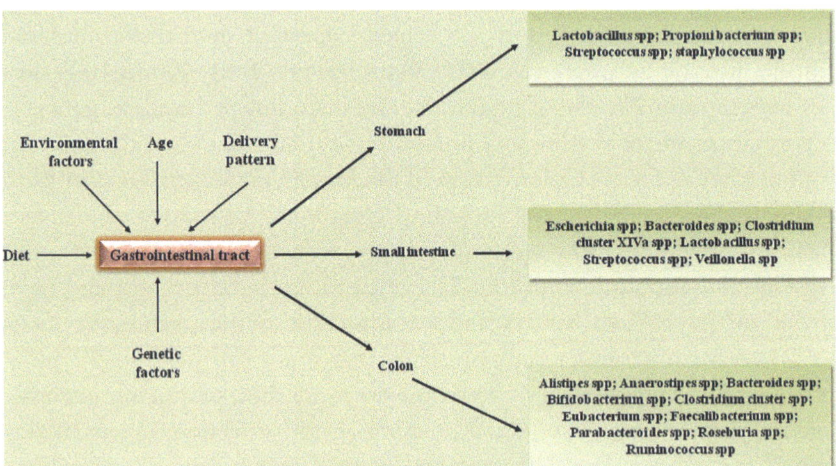

Fig. 2 Spatial distribution of bacteria in the gastrointestinal tract.

gut microbiome can be viewed as an anaerobic bioreactor programmed to synthesize molecules that not only direct the mammalian immune system, modify the mammalian epigenome, as well as regulate host metabolism by generating short-chain fatty acids (SCFAs).[11] The population of gut microbiota is affected by the environment, diet, exercise, medications, age, geographic factors, surgical interventions, and host genetics, particularly

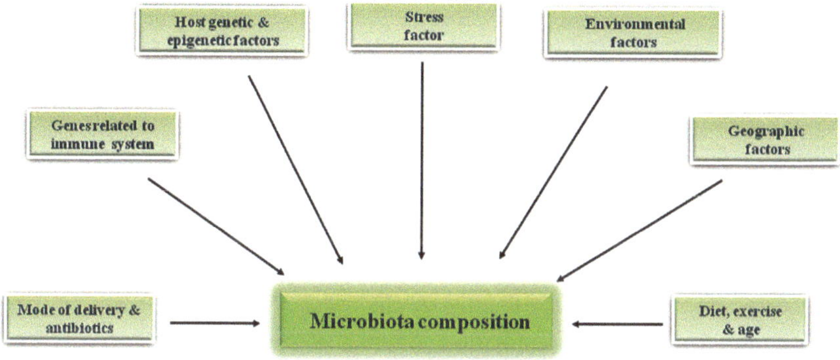

Fig. 3 Factors affecting composition of gut microbiota.

genes related to the immune system and metabolism (Fig. 3).[2, 12] Among these factors, diet, exercise, and age play a predominant role in modulating the composition of gut microbiota and promoting dysbiosis, a condition frequently associated with increased susceptibility to infections as well as to non-communicable diseases like obesity, metabolic syndromes, diabetes, allergy, and neurological disorders. Diet–gene interactions play a pivotal role in cognitive function and neurological disorders throughout life. Effects of diet on brain function in general and cognitive function in particular are multifactorial and very complex.

Long-term consumption of a Western diet alters the gut microbiota population leading to gut dysbiosis, obesity, hyperglycemia, insulin resistance, hypertension, mitochondrial dysfunction, oxidative stress, neuroinflammation, and cognitive dysfunction.[13, 14] Induction of these processes increases the risk of acute and chronic visceral and neurological diseases in both humans and animals.[15] In contrast, long-term consumption of a Mediterranean diet maintains normal homeostasis and retards not only the onset of neurotraumatic, neurodegenerative, and neuropsychiatric diseases, but also delays the onset of cardiovascular diseases, type 2 diabetes, obesity, and cancer.[15–17] Collective evidence suggests that diet has the ability to prevent and protect against acute and chronic neurological disorders. Furthermore, consumption of a healthy diet is necessary for maintaining the regenerative function of the brain. The ability of the brain to maintain homeostasis is challenged by aging, which is an important risk factor for neurotraumatic, neurodegenerative, and neuropsychiatric diseases.

Long-term consumption of a Western diet, which is high in protein, refined sugar, and fat, not only increases the level of a unique class of bile-tolerant microorganisms (*Alistipes*, *Bilophila*, and *Bacteroides*), but also increases levels of lipopolysaccharide (LPS), which elicits local inflammatory signals that increase gut mucosal permeability leading to deterioration of gut barrier and bacterial translocation. In addition, effects of a Western diet on brain function in general and cognitive function in particular are multifactorial and very complex and it is difficult to evaluate the precise benefits of specific nutrients on neurological and cognitive functions. In contrast, long-term consumption of a vegetarian diet, which is high in fiber and complex carbohydrates and low in fat, increases levels of *Roseburia*, *Eubacterium rectale*, and *Ruminococcus bromii* and decreases levels of LPS suggesting that microbial composition mirrors differences between herbivorous and carnivorous mammals. As stated, long-term consumption of a Mediterranean diet maintains normal homeostasis and retards not only the onset of neurological disorders, but also delays the onset of cardiovascular diseases, type 2 diabetes, obesity, and cancer.[15–17] Collective evidence suggests that diet has a profound and long-lasting effect on human health. The effects of specific dietary components such as dietary fibers not only goes beyond their nutritional value but also positively influences multiple aspects of human health such as microbiota composition. Industrial development in the modern era has been associated with a change in lifestyle leading to consumption of refined foods and decreased intake of whole grains, fruits, and vegetables, which are major sources of dietary fibers. This recent change in lifestyle has coincided with an increase in inflammatory diseases such as inflammatory bowel diseases, allergies, intestinal cancers, and neurological disorders.[15] Based on these observations, it is suggested that healthy diet and optimal gut microbiota have the ability to reduce oxidative stress and chronic inflammation as well as promote brain–immune–body interactions,[18, 19] which may slow down or even reverse the effect of aging on the brain. Collective evidence suggests that diet has the ability to prevent and protect against acute and chronic neurological disorders.

Maintaining a proper microbiota composition plays a critical role in protecting against pathogens and promoting the maintenance of host immune response.[20] Gut microbiota communicates with the brain by producing a number of neuroactive and immunocompetent substances (hormones, neurotransmitters, neuropeptides, and SCFAs (Table 1).[21-23] Among these bioactive compounds, SCFAs play an important role in harvesting extra energy from the undigested diet and maintaining glucose homeostasis.[24] SCFAs include acetate, propionate, and butyrate. Acetate and butyrate protect against diet-induced obesity without causing hypophagia, while propionate has shown to reduce food intake. The underlying mechanisms for these effects

Table 1 Bioactive substances released by gut microbiota.

Bioactive substances	Reference
Production of short-chain fatty acids	Kasubuchi et al.[21]
Production of tryptophan and indole metabolites	Wang and Wang[22]; Collins et al.[23]
Production of neurotoxins	Collins et al.[23]
Production of hormones	Wang and Wang[22]; Collins et al.[23]
Production of neurotransmitters	Wang and Wang[22]; Collins et al.,[23]
Production of LPS and peptidoglycans	Wang and Wang[22]; Collins et al.,[23]
Production of polysaccharide A	Wang and Wang[22]; Collins et al.[23]

are unclear. However, it is suggested that SCFAs produce their effects on signal transduction processes by interacting with G-protein coupled receptors (FFAR2, FFAR3, OLFR78, GPR109A) and acting as epigenetic regulators of gene expression. This process involves the inhibition of histone deacetylase (HDAC).[21] Butyrate and propionate not only inhibit HDAC but they also alter the expression of specific genes via conformational changes in the active site of HDAC leading to its inactivation. SCFAs have also been implicated in gastrointestinal functional, (neuro)immune regulation, and host metabolism. They also play an important role in stress-induced behavioral and physiological alterations. Administration of SCFAs to mice undergoing psychosocial stress alleviates enduring alterations in anhedonia and heightened stress-responsiveness, as well as stress-induced increases in intestinal permeability.[25]

In addition to SCFAs, the gut microbiota provides enzymes involved in the utilization of non-digestible carbohydrates, cholesterol reduction, and enzymes associated with the biosynthesis of vitamins (K and B group) and isoprenoids amino acids (e.g., lysine and threonine). Recent studies on the effect of fecal microbiota transplantation (FMT) from aged to young rats indicate that FMT impairs cognitive behavior in young recipient rats.[26] This is due to a decrease in the regional homogeneity of the medial prefrontal cortex and hippocampus; changes in synaptic structures and a decrease in dendritic spines; reduction in expression of brain-derived neurotrophic factor (BDNF), *N*-methyl-D-aspartate receptor NR1 subunit, and synaptophysin; and increase in expression of advanced glycation end products (AGEs) and receptor for AGEs (RAGE).[26] These studies suggest that FMT-mediated behavioral, brain structural, and functional alterations are related to cognitive decline. In addition, FMT also increases levels of proinflammatory cytokines and oxidative stress in young rats. These finding support the view that inflammation and oxidative stress may underlie gut-related cognitive decline during aging.[26] The major SCFAs are acetate, propionate, and butyrate, which not only contribute to the maintenance of the intestinal epithelium and permeability,[27] but also the regulation of glucose and lipid metabolism as well as immune and inflammatory responses.[28] This supports the view that gut microbiota plays several important roles in metabolism, immune systems, and inflammation.[29] In addition, microbiota also play a role in regulating host metabolism.[30–33] This is documented by the transfer of microbiota from lean and obese human twins into germ-free mice lacking a native gut microbiome, resulting in the conveyance of the metabolic phenotype of the host.[34, 35] As stated, microbiota depletion in mice confers significant protection against metabolic dysregulation induced by a high-fat diet such as obesity, glucose intolerance, and insulin resistance,[36] all of which are hallmarks of metabolic diseases including type 2 diabetes. Furthermore, the gut microbiome is also associated with host energy harvest by converting inaccessible nutrient sources such as plant polysaccharides and other complex carbohydrates into readily absorbable metabolites.[37] These studies suggest a key link between the gut microbiome and the release of several gut hormones such as CCK, PYY, GLP-1, GIP, and 5-HT. These hormones play an important role in insulin sensitivity, glucose tolerance, fat storage, and appetite.[15] Release of these hormones can be influenced by the presence of bacteria and their metabolites within the gut and, as such, microbial-mediated gut hormone release is an important component of microbial regulation of host metabolism.

2 Contribution of gut microbiota in the pathogenesis of visceral and neurological disorders

The recognition and monitoring of microbial antigens in the body is mainly performed by the innate immune system with pattern recognition receptors (PRRs) such as toll-like receptors (TLRs) and NOD-like receptors (NLRs).[38, 39] On one hand, interactions between TLR-4 and bacterial peptidoglycan or LPS promote mucosal low-grade inflammation and bacterial translocation[40] through the involvement of proinflammatory stimuli (LPS, lipids, fatty acids, and chemokines), which activate c-Jun N-terminal kinase (JNK) and IκB kinase (IKK)-β pathways intracellularly. IKKβ activation stimulates the family of nuclear factor (NF)-κB transcription factors and increases the expression of numerous mediators of chronic inflammation that can cause insulin resistance. On the other hand, the activation of the TLR4 pathways may also cause an increase in production of reactive oxygen and nitrogen species (ROS/RNS) and oxidative and nitrosative stress.

JNK activation also promotes the phosphorylation of insulin receptor substrate (IRS)-1 at serine sites, which inhibits normal signal transduction through the insulin receptor/IRS-1 axis, which also can result in insulin resistance.[41] Therefore, the metabolic endotoxemia induced by LPS derived from the gut microbiota is associated with inflammation and insulin resistance. By participating in carbohydrate, lipid, and amino acid metabolism, gut microbiota modulate the onset of several metabolic diseases, such as obesity, type 2 diabetes, dyslipidemia, non-alcoholic fatty liver disease (NAFLD), gout, vitamin deficiency, and atherosclerosis.[42–45]

Gut microbiota have been reported to communicate with the brain not only via the gut–brain axis but also through the development of gut dysbiosis, a process that contributes to the behavioral changes and development of neurological disorders. For example, interactions between gastrointestinal function and the brain modulate visceral events such as nausea, satiety, and pain, thus altering gastrointestinal secretions and motility.[23] The neuroendocrine, neuroimmune, sympathetic, and parasympathetic arms of the autonomic nervous system (ANS) and the enteric nervous system (ENS) are the key pathways through which the gut and the brain communicate with each other.[46] This may influence a broad spectrum of neuropsychiatric diseases.[23] Putative mechanisms by which gut microbiota communicate with the brain and influence behavior include gut-microbiota-derived products that enter the brain via cytokine release from the mucosal immune cells, via the release of gut hormones such as 5-HT from endocrine cells, or via afferent neural pathways, including the vagus nerve. Stress and emotions influence the composition of gut microbiota via the release of stress hormones or sympathetic neurotransmitters (GABA, 5-HT precursors etc.) as well as influence gut physiology by altering the habitat of the gut microbiota. Alternatively, host stress hormones such as noradrenalin may influence bacterial gene expression or signaling between bacteria, and this may change the microbial composition and activity of the microbiota.[23] Collective evidence suggests that there are three basic mechanisms associated with the communication between the gut and brain: direct neuronal communication mechanism, endocrine signaling mediators mechanism, and the immune system-mediated mechanism (Fig. 4).[47] Together, these systems create a highly integrated molecular communication network that links systemic imbalances with the development of neurodegeneration including insulin regulation, fat metabolism, oxidative stress markers, and immune signaling.[47] Age is a common

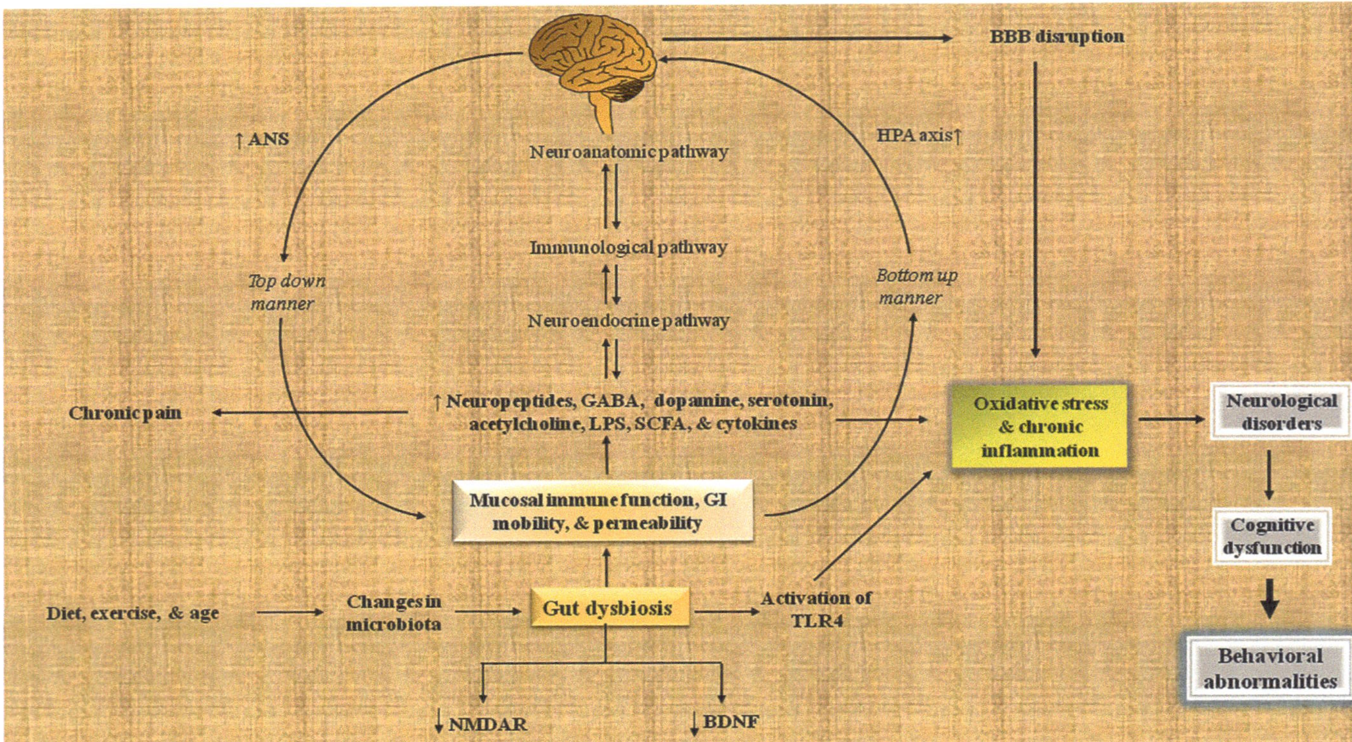

Fig. 4 Hypothetical diagram showing the effect of gut microbiota on the pathogenesis of neurological disorders. Autonomic nervous system (ANS); emotional motor system (EMS); hypothalamus–pituitary–adrenal axis (HPA axis); lipopolysaccharide (LPS); γ-aminobutyric acid (GABA); short-chain fatty acids (SCFAs); Toll-like receptor4 (TLR4); blood–brain barrier (BBB); brain-derived neurotrophic factor (BDNF); *N*-methyl-D-aspartic acid receptor (NMDA receptor).

factor in the development and onset of neurodegenerative and neuropsychiatric diseases and probiotics prevent many harmful effects of aging such as decreased neurotransmitter levels, chronic inflammation, oxidative stress, and apoptosis, all of which are factors that are proven aggravators of neurodegenerative and neuropsychiatric diseases. Indeed, patients with neurological disorders have a high rate of gastrointestinal comorbidities, and it has been proposed by some that management of the gut microbiota may prevent or alleviate the symptoms of these chronic diseases.

3 Future studies on gut dysbiosis and chronic visceral and neurological disorders

The mammalian gastrointestinal tract harbors trillions of commensal microorganisms (a wide array of bacteria, fungi, eukaryotic parasites, and viruses) collectively called the gut microbiota. The microbiota are a critical source of environmental stimuli and thus have a tremendous impact on the health of the host. Microbiota regulate homeostasis within the gut, and any alteration in their composition can lead to dysbiosis, a condition that contributes to obesity, insulin resistance, inflammatory bowel disease, allergy, autoimmune disease, type 2 diabetes, metabolic syndrome, and cancer. Our health, behaviors, and mood depend not only on our lifestyle (diet, exercise, and sleep) but also on the composition of our gut microbiota. Alterations in gut microbiota composition due to changes in lifestyle result in dysbiosis, a process that is frequently associated with increased susceptibility to infections as well as to non-communicable diseases like obesity, metabolic syndromes (e.g., diabetes and cardiovascular diseases), allergy, and other inflammatory diseases. Emerging evidence from more recent studies also demonstrate the existence of a bidirectional communication route linking gut and microbiota with the brain, thus suggesting that these microbes may play a role in neurological disorders as well as in host perception, behavior, and emotional response. Many of these neuropsychiatric disorders can be treated with transplantation of fecal microbiota from normal subjects to patients with neuropsychiatric diseases. To this end, many questions related to FMT remain unanswered.[48] For example, during FMT studies, is it possible to replace normal fecal matter with fecal matter produced in the culture? Can fresh feces be used or should feces be obtained from a stool bank? With what shall the stool be diluted? Which dilution material shall be used in different indications? What is the quantity of the stool to be used? Which method shall be used for application? What is the immunologic response emerging/changing after FMT? Does FMT have long-term risks?

4 Directions for future studies

FMT is a chosen application.[49] For this reason, its use may be restricted. In addition, as donor health screening takes time, a quick transplant may not be provided in case of emergency. To overcome this problem, different methods have been used. For the treatment of *Clostridium difficile* infection (CDI), it is proposed that the stool sample from the donor be placed into capsules that can be swallowed, which can be stored frozen at −80°C. Capsules can be warmed to −20°C 1–2h prior to use, and 15 capsules can be administered to patients orally for 2 days. Full remission can be achieved in 14 of 20 patients after 3 days.[50] Oral application of FMT in the form of capsules to regulate intestinal microbiota may become a routine procedure in the future.[34]

5 Conclusion

The mammalian gastrointestinal tract harbors trillions of commensal microorganisms, collectively known as the gut microbiota. The microbiota is a critical source of environmental stimuli and thus has a tremendous impact on the health of the host. The microbes within the microbiota regulate homeostasis within the gut, and any alteration in their composition can lead to disorders that include inflammatory bowel disease, allergy, autoimmune disease, diabetes, and cancer. The gut microbiota communicates with the brain via the microbiota–gut–brain axis. This axis plays an important role in brain development and function. The bidirectional communication between the gut microbiota and the brain involves three systems including the nervous, endocrine, and immune system pathways. A series of pre-clinical studies have indicated that gut dysbiosis and/or alterations of the gut microbiota can be implicated in the pathogeneses of neurological disorders and psychiatric conditions, including anxiety, depression, autism, and schizophrenia. The pathophysiology of neuropsychiatric disorders is multifactorial yet generally aggravated by changes in gut microbiota. To treat these diverse neuropsychiatric conditions effectively, a broad acting strategy is needed. It has been suggested that gut–brain axis signaling may play a pinnacle role in promoting resilience to several of these microbiota-mediated changes including HPA-axis activation, oxidative stress, neurotransmitter imbalances, and inflammation.

References

1. Eckburg PB, Bik EM, Bernstein CN, et al. Diversity of the human intestinal microbial flora. *Science.* 2005;308:1635–1638.
2. Foster JA, McVey Neufeld KA. Gut-brain axis: how the microbiome influences anxiety and depression. *Trends Neurosci.* 2013;36:305–312.

3. Rodriguez JM, Murphy K, Stanton C, et al. The composition of the gut microbiota throughout life, with an emphasis on early life. *Microb Ecol Health Dis.* 2015;26:26050.
4. Mazmanian SK, Round JL, Kasper DL. A microbial symbiosis factor prevents intestinal inflammatory disease. *Nature.* 2008;453:620–625.
5. Arpaia N, Campbell C, Fan X, et al. Metabolites produced by faecal microbiota transplantation commensal bacteria promote peripheral regulatory T-cell generation. *Nature.* 2013;504:451–455.
6. Chung H, Pamp SJ, Hill JA, et al. Gut immune maturation depends on colonization with a host-specific microbiota. *Cell.* 2012;149:1578–1593.
7. Kelly CJ, Zheng L, Campbell EL, et al. Crosstalk between microbiota-derived short-chain fatty acids and intestinal epithelial HIF augments tissue barrier function. *Cell Host Microbe.* 2015;17:662–671.
8. Rothschild D, Weissbrod O, Barkan E, et al. Environment dominates over host genetics in shaping human gut microbiota. *Nature.* 2018;555:210–215.
9. Sekirov I, Russell SL, Antunes LCM, Finlay BB. Gut microbiota in health and disease. *Physiol Rev.* 2010;90:859–904.
10. Zhang Q, Wu Y, Fei X. Effect of probiotics on glucose metabolism in patients with type 2 diabetes mellitus: a meta-analysis of randomized controlled trials. *Medicina.* 2016;52:28–34.
11. Kurilshikov A, Wijmenga C, Fu J, Zhernakova A. Host genetics and gut microbiome: challenges and perspectives. *Trends Immunol.* 2017;38:633–647.
12. Wu GD, Chen J, Hoffmann C, et al. Linking long-term dietary patterns with gut microbial enterotypes. *Science.* 2011;334:105–108.
13. Farooqui AA. *Metabolic Syndrome-An Important Risk Factor for Stroke, Alzheimer Disease, and Depression.* New York: Springer; 2013.
14. Farooqui AA. *High Calorie Diet and the Human Brain.* Springer, International Publishing Switzerland; 2015.
15. Farooqui AA. *Insulin Resistance as a Risk Factor in Visceral and Neurological Disorders.* San Diego, CA: Academic Press-An Imprint of Elsevier; 2020.
16. Farooqui AA. *Phytochemical, Signal Transduction, and Neurological Disorders.* New York: Springer; 2012.
17. Sofi F, Macchi C, Abbate R, Gensini GF, Casini A. Mediterranean diet and health. *Biofactors.* 2013;39:335–342.
18. Chesnokova V, Pechnick RN, Wawrowsky K. Chronic peripheral inflammation, hippocampal neurogenesis, and behavior. *Brain Behav Immun.* 2016;58:1–8.
19. Kizil C, Kyritsis N, Brand M. Effects of inflammation on stem cells: together they strive? *EMBO Rep.* 2015;16:416–426.
20. Dieterich W, Schink M, Zopf Y. Microbiota in the gastrointestinal tract. *Med Sci (Basel).* 2018;6:116.
21. Kasubuchi M, Hasegawa S, Hiramatsu T, Ichimura A, Kimura I. Dietary gut microbial metabolites, short-chain fatty acids, and host metabolic regulation. *Nutrients.* 2015;7:2839–2849.
22. Wang HX, Wang YP. Gut microbiota-brain axis. *Chin Med J (Engl).* 2016;129:2373–2380.
23. Collins SM, Surette M, Bercik P. The interplay between the intestinal microbiota and the brain. *Nat Rev Microbiol.* 2012;10:735–742.
24. Canfora EE, Jocken JW, Blaak EE. Short-chain fatty acids in control of body weight and insulin sensitivity. *Nat Rev Endocrinol.* 2015;11:577.
25. van de Wouw M, Boehme M, Lyte JM, et al. Short-chain fatty acids: microbial metabolites that alleviate stress-induced brain-gut axis alterations. *J Physiol.* 2018;596:4923–4944.
26. Li Y, Yin Y, Wang R, et al. Age-related shifts in gut microbiota contribute to cognitive decline in aged rats. *Aging (Albany NY).* 2020;12:7801–7817.

27. Morrison DJ, Preston T. Formation of short chain fatty acids by the gut microbiota and their impact on human metabolism. *Gut Microbes.* 2016;7:189–200.
28. Gudi R, Suber J, Brown R, Johnson BM, Vasu C. Pretreatment with yeast-derived complex dietary polysaccharides suppresses gut inflammation, alters the microbiota composition, and increases immune regulatory short-chain fatty acid production in C57BL/6 mice. *J Nutr.* 2020;150(5):1291–1302.
29. Sivaprakasam S, Prasad PD, Singh N. Benefits of short-chain fatty acids and their receptors in inflammation and carcinogenesis. *Pharmacol Ther.* 2016;164:144–151.
30. Aguirre M, Eck A, Koenen ME, et al. Diet drives quick changes in the metabolic activity and composition of human gut microbiota in a validated in vitro gut model. *Res Microbiol.* 2016;167:114–125.
31. Brubaker PL. Linking the gut microbiome to metabolism through endocrine hormones. *Endocrinology.* 2018;159:2978–2979.
32. Fabbiano S, Suarez-Zamorano N, Chevalier C, et al. Functional gut microbiota remodeling contributes to the caloric restriction-induced metabolic improvements. *Cell Metab.* 2018;28, 907.e7–921.e7.
33. Molinaro A, Caesar R, Holm LM, et al. Host-microbiota interaction induces bi-phasic inflammation and glucose intolerance in mice. *Mol Metab.* 2017;6:1371–1380.
34. Borody TJ, Brandt LJ, Paramsothy S. Therapeutic faecal microbiota transplantation: current status and future developments. *Curr Opin Gastroenterol.* 2014;30:97–105.
35. Ridaura VK, Faith JJ, Rey FE, et al. Gut microbiota from twins discordant for obesity modulate metabolism in mice. *Science.* 2013;341:1241214.
36. Suarez-Zamorano N, Fabbiano S, Chevalier C, et al. Microbiota depletion promotes browning of white adipose tissue and reduces obesity. *Nat Med.* 2015;21:1497–1501.
37. Tremaroli V, Backhed F. Functional interactions between the gut microbiota and host metabolism. *Nature.* 2012;489:242–249.
38. Agier J, Pastwinska J, Brzezinska-Blaszczyk E. An overview of mast cell pattern recognition receptors. *Inflamm Res.* 2018;67:737–746.
39. O'Neill LAJ, Golenbock D, Bowie AG. The history of toll-like receptors—redefining innate immunity. *Nat Rev Immunol.* 2013;13:453–460.
40. Amar J, Chabo C, Waget A, et al. Intestinal mucosal adherence and translocation of commensal bacteria at the early onset of type 2 diabetes: molecular mechanisms and probiotic treatment. *EMBO Mol Med.* 2011;3:559–572.
41. Shoelson SE, Lee J, Goldfine AB. Inflammation and insulin resistance. *J Clin Invest.* 2006;116:1793–1801.
42. Boulange CL, Neves AL, Chilloux J, Nicholson JK, Dumas ME. Impact of the gut microbiota on inflammation, obesity, and metabolic disease. *Genome Med.* 2016;8:42.
43. Guo Z, Zhang J, Wang Z, et al. Intestinal microbiota distinguish gout patients from healthy humans. *Sci Rep.* 2016;6:20602.
44. Leung C, Rivera L, Furness JB, Angus PW. The role of the gut microbiota in NAFLD. *Nat Rev Gastroenterol Hepatol.* 2016;13:412–425.
45. Tang WH, Kitai T, Hazen SL. Gut microbiota in cardiovascular health and disease. *Circ Res.* 2017;120:1183–1196.
46. Cryan JF, Dinan TG. Mind-altering microorganisms: the impact of the gut microbiota on brain and behaviour. *Nat Rev Neurosci.* 2012;13:701–712.
47. Carabotti M, Scirocco A, Maselli MA, Severi C. The gut-brain axis: interactions between enteric microbiota, central and enteric nervous systems. *Ann Gastroenterol.* 2015;28:203–209.
48. Vandenplas Y, Pierard D, De Greef E. Fecal microbiota transplantation: just a fancy trend? *J Pediatr Gastroenterol Nutr.* 2015;61:4–7.
49. Evrensel A, Ceylan ME. The role of fecal microbiota transplantation in psychiatric treatment. *Anatolian J Psychiatry.* 2015;16:380.
50. Youngster I, Russell GH, Pindar C, Ziv-Baran T, Sauk J, Hohmann EL. Oral, capsulized, frozen fecal microbiota transplantation for relapsing Clostridium difficile infection. *JAMA.* 2014;312:1772–1778.

Index

Note: Page numbers followed by *f* indicate figures and *t* indicate tables.

A

B

C

D

E

F

H

I

K

L

M

N

O

FSC
www.fsc.org
MIX
Papier | Fördert
gute Waldnutzung
FSC® C083411

Kolibri 360 GmbH
Ferdinand-Jühlke-Straße 7
99095 Erfurt, Deutschland
produktsicherheit@kolibri360.de